Introduction to Dynamical Systems

This book provides a broad introduction to the subject of dynamical systems, suitable for a one- or two-semester graduate course. In the first chapter, the authors introduce over a dozen examples, and then use these examples throughout the book to motivate and clarify the development of the theory. Topics include topological dynamics, symbolic dynamics, ergodic theory, hyperbolic dynamics, one-dimensional dynamics, complex dynamics, and measure-theoretic entropy. The authors top off the presentation with some beautiful and remarkable applications of dynamical systems to such areas as number theory, data storage, and Internet search engines.

This book grew out of lecture notes from the graduate dynamical systems course at the University of Maryland, College Park, and reflects not only the tastes of the authors, but also to some extent the collective opinion of the Dynamics Group at the University of Maryland, which includes experts in virtually every major area of dynamical systems.

Michael Brin is Professor Emeritus of Mathematics at the University of Maryland. He is the author of over 30 papers, three of which appeared in the Annals of Mathematics, and he has lectured at conferences and universities around the world. His main research areas are dynamical systems and Riemannian geometry. In 2008, he established the Michael Brin Prize in Dynamical Systems.

Garrett Stuck is a former Professor of Mathematics at the University of Maryland and has held visiting positions at the Institut des Hautes Études Scientifiques in Paris and the Mathematical Sciences Research Institute in Berkeley. He has coauthored several textbooks, including *The Mathematica Primer*. Dr. Stuck is also a founder of Chalkfree, Inc. He currently works in the finance industry.

Introduction to Dynamical Systems

MICHAEL BRIN

GARRETT STUCK

Shaftesbury Road, Cambridge CB2 8EA, United Kingdom

One Liberty Plaza, 20th Floor, New York, NY 10006, USA

477 Williamstown Road, Port Melbourne, VIC 3207, Australia

314–321, 3rd Floor, Plot 3, Splendor Forum, Jasola District Centre, New Delhi – 110025, India

103 Penang Road, #05–06/07, Visioncrest Commercial, Singapore 238467

Cambridge University Press is part of Cambridge University Press & Assessment, a department of the University of Cambridge.

We share the University's mission to contribute to society through the pursuit of education, learning and research at the highest international levels of excellence.

www.cambridge.org
Information on this title: www.cambridge.org/9781107538948

First published 2002
Paperback edition 2015 with corrections

A catalogue record for this publication is available from the British Library

Library of Congress Cataloging-in-Publication data
Brin,Michael.
Introduction to dynamical systems / Michael Brin,Garrett Stuck.
p. cm.
Includes bibliographical references and index.
ISBN 0-521-80841-5
1. Differentiable dynamical systems. I. Stuck,Garrett, 1961– II. Title.
QA614.8.B75 2002
514′.74 – dc21 2002022281

ISBN 978-0-521-80841-5 Hardback
ISBN 978-1-107-53894-8 Paperback

To Eugenia, Pamela, Sergey, Sam, Jonathan, and Catherine for their patience and support. And in memory of our dear colleague Dan Rudolph.

Contents

Introduction

The purpose of this book is to provide a broad and general introduction to the subject of dynamical systems, suitable for a one- or two-semester graduate course. We introduce the principal themes of dynamical systems both through examples and by explaining and proving fundamental and accessible results. We make no attempt to be exhaustive in our treatment of any particular area.

This book grew out of lecture notes from the graduate dynamical systems course at the University of Maryland, College Park. The choice of topics reflects not only the tastes of the authors, but also to a large extent the collective opinion of the Dynamics Group at the University of Maryland, which includes experts in virtually every major area of dynamical systems.

Early versions of this book have been used by several instructors at Maryland, the University of Bonn, and Pennsylvania State University. Experience has shown that, with minor omissions, the first five chapters of the book can be covered in a one-semester course. Instructors who wish to cover a different set of topics may safely omit some of the sections at the end of Chapter 1, §§2.7–2.8, §§3.5–3.8, and §§4.8–4.12, and then choose from topics in later chapters. Examples from Chapter 1 are used throughout the book. Chapter 6 depends on Chapter 5, but the other chapters are essentially independent. Every section ends with exercises (starred exercises are the most difficult).

The exposition of most of the concepts and results in this book have been refined over the years by various authors. Since most of these ideas have appeared so often and in so many variants in the literature, we have not attempted to identify the original sources. In many cases, we followed the written exposition from specific sources listed in the bibliography. These sources cover particular topics in greater depth than we do here, and we recommend them for further reading. We also benefited from the advice and guidance of a number of specialists, including Joe Auslander, Werner Ballmann, Ken Berg, Mike Boyle, Boris Hasselblatt, Michael Jakobson, Anatole Katok, Michal Misiurewicz, and Dan Rudolph. We thank them for their contributions. We are especially grateful to Vitaly Bergelson for his contributions to the treatment of applications of topological dynamics and ergodic theory to combinatorial number theory. We thank the students who used early versions of this book in our classes, and who found many typos, errors, and omissions. Thanks, also, to the colleagues who used the hardcover edition of this text, and whose comments and corrections have been included in this new paperback edition.

CHAPTER ONE

Examples and basic concepts

Dynamical systems is the study of the long-term behavior of evolving systems. The modern theory of dynamical systems originated at the end of the nineteenth century with fundamental questions concerning the stability and evolution of the solar system. Attempts to answer those questions led to the development of a rich and powerful field with applications to physics, biology, meteorology, astronomy, economics, and other areas.

By analogy with celestial mechanics, the evolution of a particular state of a dynamical system is referred to as an *orbit*. A number of themes appear repeatedly in the study of dynamical systems, including properties of individual orbits; periodic orbits; typical behavior of orbits; statistical properties of orbits; randomness vs. determinism; entropy; chaotic behavior; and stability under perturbation of individual orbits and patterns. We introduce some of these themes through the examples in this chapter.

We use the following notation throughout the book: $\mathbb{N}$ is the set of positive integers; $\mathbb{N}_0 = \mathbb{N} \cup \{0\}$; $\mathbb{Z}$ is the set of integers; $\mathbb{Q}$ is the set of rational numbers; $\mathbb{R}$ is the set of real numbers; $\mathbb{C}$ is the set of complex numbers; $\mathbb{R}^+$ is the set of positive real numbers; $\mathbb{R}_0^+ = \mathbb{R}^+ \cup \{0\}$.

1.1 The notion of a dynamical system

A *discrete-time dynamical system* consists of a non-empty set X and a map $f: X \to X$. For $n \in \mathbb{N}$, the nth iterate of f is the n-fold composition $f^n = f \circ \cdots \circ f$; we define f^0 to be the identity map, denoted Id. If f is invertible, then $f^{-n} = f^{-1} \circ \cdots \circ f^{-1}$ (n times). Since $f^{n+m} = f^n \circ f^m$, these iterates form a group if f is invertible, and a semigroup otherwise.

Although we have defined dynamical systems in a completely abstract setting, where X is simply a set, in practice X usually has additional structure that is preserved by the map f. For example, (X, f) could be a measure space and a measure-preserving map; a topological space and a continuous map; a metric space and an isometry; or a smooth manifold and a differentiable map.

A *continuous-time dynamical system* consists of a space X and a one-parameter family of maps of $\{f^t : X \to X\}$, $t \in \mathbb{R}$ or $t \in \mathbb{R}_0^+$, that forms a one-parameter group or semigroup, i.e. $f^{t+s} = f^t \circ f^s$ and $f^0 = \mathrm{Id}$. The dynamical system is called a *flow* if the time t ranges over $\mathbb{R}$, and a *semiflow* if t ranges over $\mathbb{R}_0^+$. For a flow, the *time-t map* f^t is invertible, since $f^{-t} = (f^t)^{-1}$. Note that for a fixed t_0, the iterates $(f^{t_0})^n = f^{t_0 n}$ form a discrete-time dynamical system.

We will use the term *dynamical system* to refer to either discrete-time or continuous-time dynamical systems. Most concepts and results in dynamical systems have both discrete-time and continuous-time versions. The continuous-time version can often be deduced from the discrete-time version. In this book, we focus mainly on discrete-time dynamical systems, where the results are usually easier to formulate and prove.

To avoid having to define basic terminology in four different cases, we write the elements of a dynamical system as f^t, where t ranges over $\mathbb{Z}, \mathbb{N}_0, \mathbb{R}$, or $\mathbb{R}_0^+$, as appropriate. For $x \in X$, we define the *positive semiorbit* $\mathcal{O}_f^+(x) = \bigcup_{t \geqslant 0} f^t(x)$. In the invertible case, we define the *negative semiorbit* $\mathcal{O}_f^-(x) = \bigcup_{t \leqslant 0} f^t(x)$, and the *orbit* $\mathcal{O}_f(x) = \mathcal{O}_f^+(x) \cup \mathcal{O}_f^-(x) = \bigcup_t f^t(x)$ (we omit the subscript "f" if the context is clear). A point $x \in X$ is a *periodic point* of *period* $T > 0$ if $f^T(x) = x$. The orbit of a periodic point is called a *periodic orbit*. If $f^t(x) = x$ for all t, then x is a *fixed point*. If x is periodic, but not fixed, then the smallest positive T (if it exists), such that $f^T(x) = x$, is called the *minimal period* of x. If $f^s(x)$ is periodic for some $s > 0$, we say that x is *eventually periodic*. In invertible dynamical systems, eventually periodic points are periodic.

For a subset $A \subset X$ and $t > 0$, let $f^t(A)$ be the image of A under f^t, and let $f^{-t}(A)$ be the preimage under f^t, i.e. $f^{-t}(A) = (f^t)^{-1}(A) = \{x \in X : f^t(x) \in A\}$. Note that $f^{-t}(f^t(A))$ contains A but, for a non-invertible dynamical system, is generally not equal to A. A subset $A \subset X$ is *f-invariant* if $f^t(A) \subset A$ for all t; *forward f-invariant* if $f^t(A) \subset A$ for all $t \geqslant 0$; and *backward f-invariant* if $f^{-t}(A) \subset A$ for all $t \geqslant 0$.

In order to classify dynamical systems, we need a notion of equivalence. Let $f^t : X \to X$ and $g^t : Y \to Y$ be dynamical systems. A *semiconjugacy* from (Y, g) to (X, f) (or, briefly, from g to f) is a surjective map $\pi : Y \to X$ satisfying $f^t \circ \pi = \pi \circ g^t$, for all t. We express this formula schematically by saying that the following diagram commutes:

$$\begin{array}{ccc} Y & \xrightarrow{g} & Y \\ \pi\downarrow & & \downarrow\pi \\ X & \xrightarrow{f} & X \end{array}$$

An invertible semiconjugacy is called a *conjugacy*. If there is a conjugacy from one dynamical system to another, the two systems are said to be *conjugate*; conjugacy is an equivalence relation. To study a particular dynamical system, we often look for a conjugacy or semiconjugacy with a better understood model. To classify dynamical systems, we study equivalence classes determined by conjugacies preserving some specified structure. Note that for some classes of dynamical systems (e.g. measure-preserving transformations) the word *isomorphism* is used instead of "conjugacy".

If there is a semiconjugacy π from g to f, then (X, f) is a *factor* of (Y, g), and (Y, g) is an *extension* of (X, f). The map $\pi : Y \to X$ is also called a *factor map* or *projection*. The simplest example of an extension is the *direct product*

$$(f_1 \times f_2)^t : X_1 \times X_2 \to X_1 \times X_2$$

of two dynamical systems $f_i^t : X_i \to X_i$, $i = 1, 2$, where $(f_1 \times f_2)^t(x_1, x_2) = (f_1^t(x_1), f_2^t(x_2))$. Projection of $X_1 \times X_2$ onto X_1 or X_2 is a semiconjugacy, so (X_1, f_1) and (X_2, f_2) are factors of $(X_1 \times X_2, f_1 \times f_2)$.

An extension (Y, g) of (X, f) with factor map $\pi : Y \to X$ is called a *skew product* over (X, f) if $Y = X \times F$ and π is the projection onto the first factor or, more generally, if Y is a fiber bundle over X with projection π.

Exercise 1.1.1. Show that the complement of a forward invariant set is backward invariant, and vice versa. Show that if f is bijective, then an invariant set A satisfies $f^t(A) = A$ for all t. Show that this is false, in general, if f is not bijective.

Exercise 1.1.2. Suppose (X, f) is a factor of (Y, g) by a semiconjugacy $\pi : Y \to X$. Show that if $y \in Y$ is a periodic point, then $\pi(y) \in X$ is periodic. Give an example to show that the preimage of a periodic point does not necessarily contain a periodic point.

1.2 Circle rotations

Consider the unit circle $S^1 = [0, 1]/\sim$, where $\sim$ indicates that 0 and 1 are identified. Addition mod 1 makes S^1 an abelian group. The natural distance on $[0, 1]$ induces a distance on S^1; specifically,

$$d(x, y) = \min(|x - y|, 1 - |x - y|).$$

Lebesgue measure on $[0, 1]$ gives a natural measure λ on S^1, also called Lebesgue measure.

We can also describe the circle as the set $S^1 = \{z \in \mathbb{C}\colon |z| = 1\}$, with complex multiplication as the group operation. The two notations are related by $z = e^{2\pi ix}$, which is an isometry if we divide arc length on the multiplicative circle by 2π. We will generally use the additive notation for the circle.

For $\alpha \in \mathbb{R}$, let R_α be the rotation of S^1 by angle $2\pi\alpha$, i.e.

$$R_\alpha x = x + \alpha \mod 1.$$

The collection $\{R_\alpha\colon \alpha \in [0, 1)\}$ is a commutative group with composition as group operation, $R_\alpha \circ R_\beta = R_\gamma$, where $\gamma = \alpha + \beta \mod 1$. Note that R_α is an isometry: it preserves the distance d. It also preserves Lebesgue measure λ, i.e. the Lebesgue measure of a set is the same as the Lebesgue measure of its preimage.

If $\alpha = p/q$ is rational, then $R_\alpha^q = \mathrm{Id}$, so every orbit is periodic. On the other hand, if α is irrational, then every positive semiorbit is dense in S^1. Indeed, the pigeon-hole principle implies that, for any $\epsilon > 0$, there are $m, n < 1/\epsilon$ such that $m < n$ and $d(R_\alpha^m, R_\alpha^n) < \epsilon$. Thus R^{n-m} is rotation by an angle less than ϵ, so every positive semiorbit is ϵ-*dense* in S^1 (i.e. comes within distance ϵ of every point in S^1). Since ϵ is arbitrary, every positive semiorbit is dense.

For α irrational, density of every orbit of R_α implies that S^1 is the only R_α-invariant closed non-empty subset. A dynamical system with no proper closed non-empty invariant subsets is called *minimal*. In Chapter 4, we show that any measurable R_α-invariant subset of S^1 has either measure zero or full measure. A measurable dynamical system with this property is called *ergodic*.

Circle rotations are examples of an important class of dynamical systems arising as group translations. Given a group G and an element $h \in G$, define maps $L_h\colon G \to G$ and $R_h\colon G \to G$ by

$$L_h g = hg \quad \text{and} \quad R_h g = gh.$$

These maps are called *left* and *right translation* by h. If G is commutative, $L_h = R_h$.

A *topological group* is a topological space G with a group structure such that group multiplication $(g, h) \mapsto gh$ and the inverse $g \mapsto g^{-1}$ are continuous maps. A continuous homomorphism of a topological group to itself is called an *endomorphism*; an invertible endomorphism is an *automorphism*. Many important examples of dynamical systems arise as translations or endomorphisms of topological groups.

Exercise 1.2.1. Show that for any $k \in \mathbb{Z}, k \neq 0$, there is a continuous semiconjugacy from R_α to $R_{k\alpha}$.

Exercise 1.2.2. Prove that for any finite sequence of decimal digits there is an integer $n > 0$ such that the decimal representation of 2^n starts with that sequence of digits.

Exercise 1.2.3. Let G be a topological group. Prove that for each $g \in G$, the closure $H(g)$ of the set $\{g^n\}_{n=-\infty}^{\infty}$ is a commutative subgroup of G. Thus, if G has a minimal left translation, then G is abelian.

***Exercise 1.2.4.** Show that R_α and R_β are conjugate by a homeomorphism if and only if $\alpha = \pm\beta \bmod 1$.

1.3 Expanding endomorphisms of the circle

For $m \in \mathbb{Z}$, $|m| > 1$, define the *times-m* map $E_m \colon S^1 \to S^1$ by

$$E_m x = mx \bmod 1.$$

This map is a non-invertible group endomorphism of S^1. Every point has $|m|$ preimages. In contrast to a circle rotation, E_m expands arc length and distances between nearby points by a factor of $|m|$: if $d(x, y) \leqslant 1/(2|m|)$, then $d(E_m x, E_m y) = |m| d(x, y)$. A map (of a metric space) that expands distances between nearby points by a factor of at least $\mu > 1$ is called *expanding*.

The map E_m preserves Lebesgue measure λ on S^1 in the following sense: if $A \subset S^1$ is measurable, then $\lambda(E_m^{-1}(A)) = \lambda(A)$ (Exercise 1.3.1). Note, however, that for a sufficiently small interval I, $\lambda(E_m(I)) = |m|\lambda(I)$. We will show later that E_m is ergodic (Proposition 4.4.2).

Fix a positive integer $m > 1$. We will now construct a semiconjugacy from another natural dynamical system to E_m. Let $\Sigma = \{0, \dots, m-1\}^{\mathbb{N}}$ be the set of sequences of elements in $\{0, \dots, m-1\}$. The *shift* $\sigma : \Sigma \to \Sigma$ discards the first element of a sequence and shifts the remaining elements one place to the left:

$$\sigma((x_1, x_2, x_3, \dots)) = (x_2, x_3, x_4, \dots).$$

A base-m expansion of $x \in [0, 1]$ is a sequence $(x_i)_{i \in \mathbb{N}} \in \Sigma$ such that $x = \Sigma_{i=1}^{\infty} x_i / m^i$. In analogy with decimal notation, we write $x = 0.x_1 x_2 x_3 \cdots$. Base-m expansions are not always unique: a fraction whose denominator is a power of m is represented both by a sequence with trailing $(m-1)$'s and a sequence with trailing 0's. For example, in base 5, we have $0.144\cdots = 0.200\cdots = 2/5$.

Define a map

$$\phi\colon \Sigma \to [0,1], \qquad \phi((x_i)_{i\in\mathbb{N}}) = \sum_{i=1}^{\infty} x_i/m^i.$$

We can consider ϕ as a map into S^1 by identifying 0 and 1. This map is surjective, and one-to-one except on the countable set of sequences with trailing 0's or $(m-1)$'s. If $x = 0.x_1x_2x_3\cdots \in [0,1)$, then $E_m x = 0.x_2x_3\cdots$. Thus $\phi\circ\sigma = E_m\circ\phi$, so ϕ is a semiconjugacy from σ to E_m.

We can use the semiconjugacy of E_m with the shift σ to deduce properties of E_m. For example, a sequence $(x_i)\in\Sigma$ is a periodic point for σ with period k if and only if it is a periodic sequence with period k, i.e. $x_{k+i} = x_i$ for all i. It follows that the number of periodic points of σ of period k is m^k. More generally, (x_i) is eventually periodic for σ if and only if the sequence (x_i) is eventually periodic. A point $x\in S^1 = [0,1]/\sim$ is periodic for E_m with period k if and only x has a base-m expansion $x = 0.x_1x_2\cdots$ that is periodic with period k. Therefore the number of periodic points of E_m of period k is $m^k - 1$ (since 0 and 1 are identified).

Let $\mathcal{F}_m = \bigcup_{k=1}^{\infty}\{0,\dots,m-1\}^k$ be the set of all finite sequences of elements of the set $\{0,\dots,m-1\}$. A subset $A\subset[0,1]$ is dense if and only if every finite sequence $w\in\mathcal{F}_m$ occurs at the beginning of the base m expansion of some element of A. It follows that the set of periodic points is dense in S^1. The orbit of a point $x = 0.x_1x_2\cdots$ is dense in S^1 if and only if every finite sequence from $\mathcal{F}_m$ appears in the sequence (x_i). Since $\mathcal{F}_m$ is countable, we can construct such a point by concatenating all elements of $\mathcal{F}_m$.

Although ϕ is not one-to-one, we can construct a right inverse to ϕ. Consider the partition of $S^1 = [0,1]/\sim$ into intervals:

$$P_k = [k/m, (k+1)/m), \qquad 0\leqslant k\leqslant m-1.$$

For $x\in[0,1]$, define $\psi_i(x) = k$ if $E_m^i x\in P_k$. The map $\psi\colon S^1\to\Sigma$, given by $x\mapsto(\psi_i(x))_{i=0}^{\infty}$, is a right inverse for ϕ, i.e. $\phi\circ\psi = \mathrm{Id}\colon S^1\to S^1$. In particular, $x\in S^1$ is uniquely determined by the sequence $(\psi_i(x))$.

The use of partitions to code points by sequences is the principal motivation for *symbolic dynamics*, the study of shifts on sequence spaces, which is the subject of the next section and Chapter 3.

Exercise 1.3.1. Prove that $\lambda(E_m^{-1}([a,b])) = \lambda([a,b])$ for any interval $[a,b]\subset[0,1]$.

Exercise 1.3.2. Prove that $E_k \circ E_l = E_l \circ E_k = E_{kl}$. When is $E_k \circ R_\alpha = R_\alpha \circ E_k$?

Exercise 1.3.3. Show that the set of points with dense orbits is uncountable.

Exercise 1.3.4. Prove that the set

$$C = \{x \in [0,1] \colon E_3^k x \notin (1/3, 2/3)\ \forall\, k \in \mathbb{N}_0\}$$

is the standard middle-thirds Cantor set.

***Exercise 1.3.5.** Show that the set of points with dense orbits under E_m has Lebesgue measure 1.

1.4 Shifts and subshifts

In this section we generalize the notion of shift space introduced in the previous section. For an integer $m > 1$ set $\mathcal{A}_m = \{1, \ldots, m\}$. We refer to $\mathcal{A}_m$ as an *alphabet*, and its elements as *symbols*. A finite sequence of symbols is called a *word*. Let $\Sigma_m = \mathcal{A}_m^{\mathbb{Z}}$ be the set of infinite two-sided sequences of symbols in $\mathcal{A}_m$ and $\Sigma_m^+ = \mathcal{A}_m^{\mathbb{N}}$ the set of infinite one-sided sequences. We say that a sequence $x = (x_i)$ contains the word $w = w_1 w_2 \cdots w_k$ (or that w *occurs* in x) if there is some j such that $w_i = x_{j+i}$ for $i = 1, \ldots, k$.

Given a one- or two-sided sequence $x = (x_i)$, let $\sigma(x) = (\sigma(x)_i)$ be the sequence obtained by shifting x one step to the left, i.e. $\sigma(x)_i = x_{i+1}$. This defines a self-map of both Σ_m and Σ_m^+ called the *shift*. The pair (Σ_m, σ) is called the *full two-sided shift*; (Σ_m^+, σ) is the *full one-sided shift*. The two-sided shift is invertible. For a one-sided sequence, the leftmost symbol disappears, so the one-sided shift is non-invertible and every point has m preimages. Both shifts have m^n periodic points of period n.

The shift spaces Σ_m and Σ_m^+ are compact topological spaces in the product topology. This topology has a basis consisting of *cylinders*

$$C^{n_1,\ldots,n_k}_{j_1,\ldots,j_k} = \{x = (x_l) \colon x_{n_i} = j_i,\ i = 1, \ldots, k\},$$

where $n_1 < n_2 < \cdots < n_k$ are indices in $\mathbb{Z}$ or $\mathbb{N}$, and $j_i \in \mathcal{A}_m$. Since the preimage of a cylinder is a cylinder, σ is continuous on Σ_m^+ and is a homeomorphism of Σ_m. The metric

$$d(x, x') = 2^{-l}, \text{ where } l = \min\{|i| \colon x_i \neq x_i'\},$$

generates the product topology on Σ_m and Σ_m^+ (Exercise 1.4.3). In Σ_m, the open ball $B(x, 2^{-l})$ is the symmetric cylinder $C^{-l,-l+1,\ldots,l}_{x_{-l},x_{-l+1},\ldots,x_l}$ and in Σ_m^+,

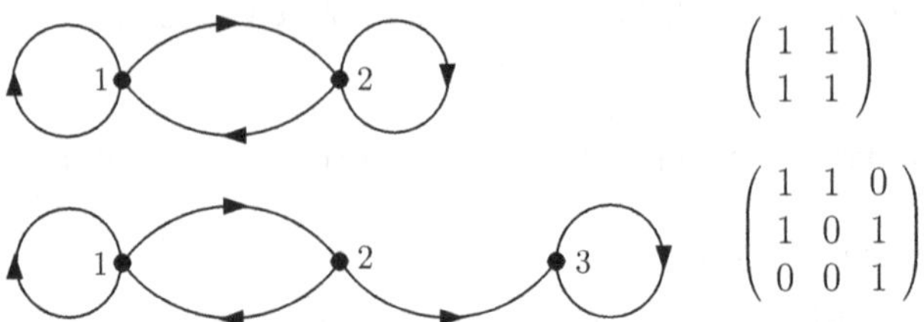

Figure 1.1. Examples of directed graphs with labeled vertices and the corresponding adjacency matrices.

$B(x, 2^{-l}) = C^{1,\ldots,l}_{x_1,\ldots,x_l}$. The shift is expanding on Σ_m^+; if $d(x, x') < 1/2$, then $d(\sigma(x), \sigma(x')) = 2d(x, x')$.

In the product topology, periodic points are dense and there are dense orbits (Exercise 1.4.5).

Now we describe a natural class of closed shift-invariant subsets of the full shift spaces. These *subshifts* can be described in terms of *adjacency matrices* or their associated *directed graphs*. An adjacency matrix $A = (a_{ij})$ is an $m \times m$ matrix whose entries are 0's and 1's. Associated to A is a directed graph Γ_A with m vertices such that a_{ij} is the number of edges from the ith vertex to the jth vertex. Conversely, if Γ is a finite directed graph with vertices $v_1, \ldots, v_m$, then Γ determines an adjacency matrix B, and $\Gamma = \Gamma_B$. Figure 1.1 shows two adjacency matrices and the associated graphs.

Given an $m \times m$ adjacency matrix $A = (a_{ij})$, we say that a word or infinite sequence x (in the alphabet $\mathcal{A}_m$) is *allowed* if $a_{x_i x_{i+1}} > 0$ for every i; or, equivalently, if there is a directed edge from x_i to x_{i+1} for every i. A word or sequence that is not allowed is said to be *forbidden*. Let $\Sigma_A \subset \Sigma_m$ be the set of allowed two-sided sequences (x_i), and $\Sigma_A^+ \subset \Sigma_m^+$ be the set of allowed one-sided sequences. We can view a sequence $(x_i) \in \Sigma_A$ (or Σ_A^+) as an infinite walk along directed edges in the graph Γ_A, where x_i is the index of the vertex visited at time i. The sets Σ_A and Σ_A^+ are closed shift-invariant subsets of Σ_m and Σ_m^+, and inherit the subspace topology. The pairs (Σ_A, σ) and (Σ_A^+, σ) are called the two-sided and one-sided *vertex shifts* determined by A.

A point $(x_i) \in \Sigma_A$ (or Σ_A^+) is periodic of period n if and only if $x_{i+n} = x_i$ for every i. The number of periodic points of period n (in Σ_A or Σ_A^+) is equal to the trace of A^n (Exercise 1.4.2).

Exercise 1.4.1. Let A be a matrix of zeros and ones. A vertex v_i can be *reached* (in n steps) from a vertex v_j if there is a path (consisting of n edges) from v_i to v_j along directed edges of Γ_A. What properties of A correspond to the following properties of Γ_A?

(a) Any vertex can be reached from some other vertex.
(b) There are no terminal vertices, i.e. there is at least one directed edge starting at each vertex.
(c) Any vertex can be reached in one step from any other vertex.
(d) Any vertex can be reached from any other vertex in exactly n steps.

Exercise 1.4.2. Let A be an $m \times m$ matrix of zeros and ones. Prove that:

(a) the number of fixed points in Σ_A (or Σ_A^+) is the trace of A;
(b) the number of allowed words of length $n+1$ beginning with the symbol i and ending with j is the i, jth entry of A^n; and
(c) the number of periodic points of period n in Σ_A (or Σ_A^+) is the trace of A^n.

Exercise 1.4.3. Verify that the metrics on Σ_m and Σ_m^+ generate the product topology.

Exercise 1.4.4. Show that the semiconjugacy $\phi \colon \Sigma \to [0,1]$ of §1.3 is continuous with respect to the product topology on Σ.

Exercise 1.4.5. Assume that all entries of some power of A are positive. Show that in the product topology on Σ_A and Σ_A^+, periodic points are dense and there are dense orbits.

1.5 Quadratic maps

The expanding maps of the circle introduced in §1.3 are *linear maps* in the sense that they come from linear maps of the real line. The simplest nonlinear dynamical systems in dimension one are the quadratic maps:

$$q_\mu(x) = \mu x(1-x), \quad \mu > 0.$$

Figure 1.2 shows the graph of q_3 and successive images $x_i = q_3^i(x_0)$ of a point x_0.

If $\mu > 1$ and $x \notin [0,1]$, then $q_\mu^n(x) \to -\infty$ as $n \to \infty$. For this reason we focus our attention on the interval $[0,1]$. For $\mu \in [0,4]$, the interval $[0,1]$ is forward invariant under q_μ. For $\mu > 4$, the interval $(1/2 - \sqrt{1/4 - 1/\mu}, 1/2 + \sqrt{1/4 - 1/\mu})$ maps outside $[0,1]$; we show in Chapter 7 that the set of points Λ_μ whose forward orbits stay in $[0,1]$ is a Cantor set, and (Λ_μ, q_μ) is equivalent to the full one-sided shift on two symbols.

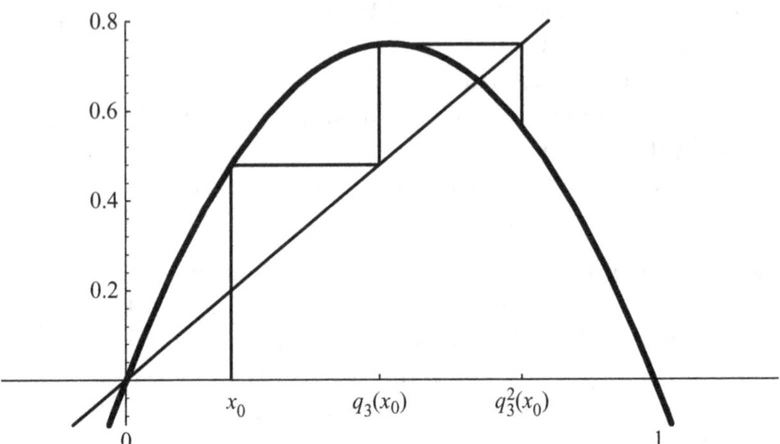

Figure 1.2. Quadratic map q_3.

Let X be a locally compact metric space and $f\colon X \to X$ a continuous map. A fixed point p of f is *attracting* if it has a neighborhood U such that $\overline{U}$ is compact, $f(\overline{U}) \subset U$, and $\bigcap_{n\geqslant 0} f^n(U) = \{p\}$. A fixed point p is *repelling* if it has a neighborhood U such that $\overline{U} \subset f(U)$ and $\bigcap_{n\geqslant 0} f^{-n}(U) = \{p\}$. Note that if f is invertible, then p is attracting for f if and only if it is repelling for f^{-1}, and vice versa. A fixed point p is called *isolated* if there is a neighborhood of p that contains no other fixed points.

If x is a periodic point of f of period n, then we say that x is an *attracting* (*repelling*) *periodic point* if x is an attracting (repelling) fixed point of f^n. We also say that the periodic orbit $\mathcal{O}(x)$ is attracting or repelling, respectively.

The fixed points of q_μ are 0 and $1 - 1/\mu$. Note that $q'_\mu(0) = \mu$ and $q'_\mu(1 - 1/\mu) = 2 - \mu$. Thus 0 is attracting for $\mu < 1$ and repelling for $\mu > 1$, and $1 - 1/\mu$ is attracting for $\mu \in (1, 3)$ and repelling for $\mu \notin [1, 3]$ (Exercise 1.5.4).

The maps q_μ, $\mu > 4$, have interesting and complicated dynamical behavior. In particular, periodic points abound. For example,

$$q_\mu([1/\mu, 1/2]) \supset [1 - 1/\mu, 1],$$

$$q_\mu([1 - 1/\mu, 1]) \supset [0, 1 - 1/\mu] \supset [1/\mu, 1/2].$$

Hence, $q_\mu^2([1/\mu, 1/2]) \supset [1/\mu, 1/2]$, so the Intermediate Value Theorem implies that q_μ^2 has a fixed point $p_2 \in [1/\mu, 1/2]$. Thus p_2 and $q_\mu(p_2)$ are non-fixed periodic points of period 2. This approach to showing existence of periodic points applies to many one-dimensional maps. We exploit this technique in Chapter 7 to prove the Sharkovsky Theorem (Theorem 7.3.1) which asserts, for example, that for continuous self-maps of the interval, the existence of an orbit of period three implies the existence of periodic orbits of all orders.

Exercise 1.5.1. Show that for any $x \notin [0, 1]$, $q_\mu^n(x) \to -\infty$ as $n \to \infty$.

Exercise 1.5.2. Show that a repelling fixed point is an isolated fixed point.

Exercise 1.5.3. Suppose p is an attracting fixed point for f. Show that there is a neighborhood U of p such that the forward orbit of every point in U converges to p.

Exercise 1.5.4. Let $f: \mathbb{R} \to \mathbb{R}$ be a C^1 map and p a fixed point. Show that if $|f'(p)| < 1$, then p is attracting, and if $|f'(p)| > 1$, then p is repelling.

Exercise 1.5.5. Are 0 and $1 - 1/\mu$ attracting or repelling for $\mu = 1$? for $\mu = 3$?

Exercise 1.5.6. Show the existence of a non-fixed periodic point of q_μ of period 3, for $\mu > 4$.

Exercise 1.5.7. Is the period 2 orbit $\{p_2, q_\mu(p_2)\}$ attracting or repelling for $\mu > 4$?

1.6 The Gauss transformation

Let $[x]$ denote the greatest integer less than or equal to x, for $x \in \mathbb{R}$. The map $\varphi: [0, 1] \to [0, 1]$ defined by

$$\varphi(x) = \begin{cases} 1/x - [1/x] & \text{if } x \in (0, 1], \\ 0 & \text{if } x = 0 \end{cases}$$

was studied by Carl Gauss, and is now called the *Gauss transformation*. Note that φ maps each interval $(1/(n+1), 1/n]$ continuously and monotonically onto $[0, 1)$; it is discontinuous at $1/n$ for all $n \in \mathbb{N}$. Figure 1.3 shows the graph of φ.

Gauss discovered a natural invariant measure μ for φ. The Gauss measure of an interval $A = (a, b)$ is

$$\mu(A) = \frac{1}{\log 2} \int_a^b \frac{dx}{1+x} = (\log 2)^{-1} \log \frac{1+b}{1+a}.$$

This measure is φ-invariant in the sense that $\mu(\varphi^{-1}(A)) = \mu(A)$ for any interval $A = (a, b)$. To prove invariance, note that the preimage of (a, b)

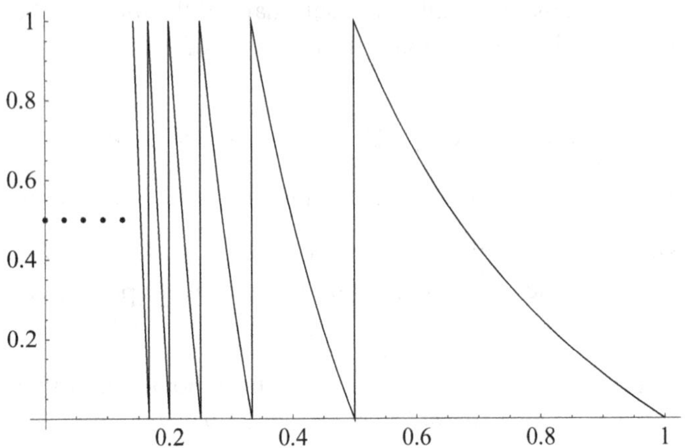

Figure 1.3. Gauss transformation.

consists of infinitely many intervals; in the interval $(1/(n+1), 1/n)$, the preimage is $(1/(n+b), 1/(n+a))$. Thus

$$\mu(\varphi^{-1}((a,b))) = \mu\left(\bigcup_{n=1}^{\infty}\left(\frac{1}{n+b}, \frac{1}{n+a}\right)\right)$$
$$= \frac{1}{\log 2}\sum_{n=1}^{\infty}\log\left(\frac{n+a+1}{n+a}\cdot\frac{n+b}{n+b+1}\right) = \mu((a,b)).$$

Note that in general $\mu(\varphi(A)) \neq \mu(A)$.

The Gauss transformation is closely related to continued fractions. The expression

$$[a_1, a_2, \ldots, a_n] = \cfrac{1}{a_1 + \cfrac{1}{a_2 + \cdots \cfrac{1}{a_n}}}, \qquad a_1, \ldots, a_n \in \mathbb{N},$$

is called a *finite continued fraction*. For $x \in (0,1]$ we have $x = 1/([\frac{1}{x}] + \varphi(x))$. More generally, if $\varphi^{n-1}(x) \neq 0$, set $a_i = [1/\varphi^{i-1}(x)] \geqslant 1$ for $i \leqslant n$. Then

$$x = \cfrac{1}{a_1 + \cfrac{1}{a_2 + \cfrac{1}{\cdots + \cfrac{1}{a_n + \varphi^n(x)}}}}.$$

Note that x is rational if and only if $\varphi^m(x) = 0$ for some $m \in \mathbb{N}$ (Exercise 1.6.2). Thus any rational number is uniquely represented by a finite continued fraction.

For an irrational number $x \in (0,1)$, the sequence of finite continued fractions

$$[a_1, a_2, \dots, a_n] = \cfrac{1}{a_1 + \cfrac{1}{a_2 + \cfrac{1}{\dots + \cfrac{1}{a_n}}}}$$

converges to x (where $a_i = [1/\varphi^{i-1}(x)]$) (Exercise 1.6.4). This is expressed concisely with the infinite continued fraction notation

$$x = [a_1, a_2, \dots] = \cfrac{1}{a_1 + \cfrac{1}{a_2 + \cdots}}.$$

Conversely, given a sequence $(b_i)_{i\in\mathbb{N}}$, $b_i \in \mathbb{N}$, the sequence $[b_1, b_2, \dots, b_n]$ converges, as $n \to \infty$, to a number $y \in [0,1]$, and the representation $y = [b_1, b_2, \dots]$ is unique (Exercise 1.6.4). Hence $\varphi(y) = [b_2, b_3, \dots]$, since $b_n = [1/\varphi^{n-1}(y)]$.

We summarize this discussion by saying that the continued fraction representation conjugates the Gauss transformation and the shift on the space of finite or infinite integer-valued sequences $(b_i)_{i=1}^{\omega}$, $\omega \in \mathbb{N} \cup \{\infty\}$, $b_i \in \mathbb{N}$. (By convention, the shift of a finite sequence is obtained by deleting the first term; the empty sequence represents 0.) As an immediate consequence, we obtain a description of the eventually periodic points of φ (see Exercise 1.6.3).

Exercise 1.6.1. What are the fixed points of the Gauss transformation?

Exercise 1.6.2. Show that $x \in [0,1]$ is rational if and only if $\varphi^m(x) = 0$ for some $m \in \mathbb{N}$.

Exercise 1.6.3. Show that:

(a) a number with periodic continued fraction expansion satisfies a quadratic equation with integer coefficients; and

(b) a number with eventually periodic continued fraction expansion satisfies a quadratic equation with integer coefficients.

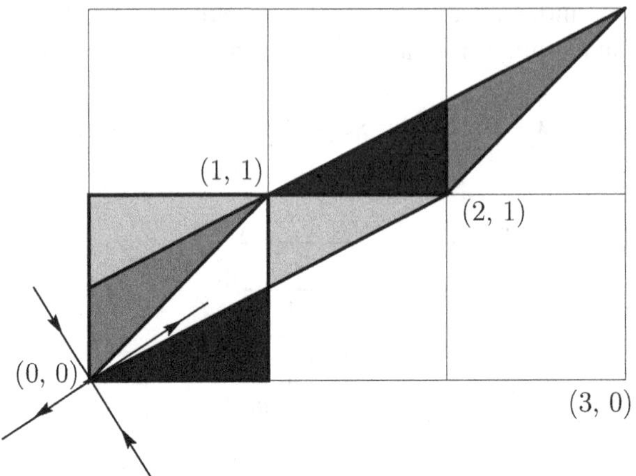

Figure 1.4. The image of the torus under A.

Thes converse of the second statement is also true, but is more difficult to prove (Archibald, 1970; Hardy and Wright, 1979).

***Exercise 1.6.4.** Show that given any infinite sequence $b_k \in \mathbb{N}, k = 1, 2, \ldots,$ the sequence $[b_1, \ldots, b_n]$ of finite continued fractions converges. Show that for any irrational $x \in \mathbb{R}$, the continued fraction $[a_1, a_2, \ldots], a_i = [1/\phi^{i-1}(x)]$, converges to x, and that this continued fraction representation is unique.

1.7 Hyperbolic toral automorphisms

Consider the linear map of $\mathbb{R}^2$ given by the matrix $A = \begin{pmatrix} 2 & 1 \\ 1 & 1 \end{pmatrix}$. The eigenvalues are $\lambda = (3 + \sqrt{5})/2 > 1$ and $1/\lambda$. The map expands by a factor of λ in the direction of the eigenvector $v_\lambda = ((1 + \sqrt{5})/2, 1)$, and contracts by $1/\lambda$ in the direction of $v_{1/\lambda} = ((1 - \sqrt{5})/2, 1)$. The eigenvectors are perpendicular because A is symmetric.

Since A has integer entries, it preserves the integer lattice $\mathbb{Z}^2 \subset \mathbb{R}^2$ and induces a map (which we also call A) of the *torus* $\mathbb{T}^2 = \mathbb{R}^2/\mathbb{Z}^2$. The torus can be viewed as the unit square $[0, 1] \times [0, 1]$ with opposite sides identified: $(x_1, 0) \sim (x_1, 1)$ and $(0, x_2) \sim (1, x_2)$, $x_1, x_2 \in [0, 1]$. The map A is given in coordinates by

$$A \begin{pmatrix} x_1 \\ x_2 \end{pmatrix} = \begin{pmatrix} (2x_1 + x_2) \mod 1 \\ (x_1 + x_2) \mod 1 \end{pmatrix}$$

(see Figure 1.4). Note that $\mathbb{T}^2$ is a commutative group and A is an automorphism, since A^{-1} is also an integer matrix.

The periodic points of $A\colon \mathbb{T}^2 \to \mathbb{T}^2$ are the points with rational coordinates (Exercise 1.7.1).

The lines in $\mathbb{R}^2$ parallel to the eigenvector v_λ project to a family W^u of parallel lines on $\mathbb{T}^2$. For $x \in \mathbb{T}^2$, the line $W^u(x)$ through x is called the *unstable manifold* of x. The family W^u partitions $\mathbb{T}^2$ and is called the *unstable foliation* of A. This foliation is invariant in the sense that $A(W^u(x)) = W^u(Ax)$. Moreover, A expands each line in W^u by a factor of λ. Similarly, the *stable foliation* W^s is obtained by projecting the family of lines in $\mathbb{R}^2$ parallel to $v_{1/\lambda}$. This foliation is also invariant under A, and A contracts each *stable manifold* $W^s(x)$ by $1/\lambda$. Since the slopes of v_λ and $v_{1/\lambda}$ are irrational, each of the stable and unstable manifolds is dense in $\mathbb{T}^2$ (Exercise 1.11.1).

In a similar way, any $n \times n$ integer matrix B induces a group endomorphism of the n-torus $\mathbb{T}^n = \mathbb{R}^n/\mathbb{Z}^n = [0,1]^n/\sim$. The map is invertible (an automorphism) if and only if B^{-1} is an integer matrix, which happens if and only if $|\det B| = 1$ (Exercise 1.7.2). If B is invertible and the eigenvalues do not lie on the unit circle, then $B\colon \mathbb{T}^n \to \mathbb{T}^n$ has expanding and contracting subspaces of complementary dimensions, and is called a *hyperbolic toral automorphism*. The stable and unstable manifolds of a hyperbolic toral automorphism are dense in $\mathbb{T}^n$ (§5.10). This is easy to show in dimension two (Exercises 1.7.3 and 1.11.1).

Hyperbolic toral automorphisms are prototypes of the more general class of *hyperbolic dynamical systems*. These systems have uniform expansion and contraction in complementary directions at every point. We discuss them in detail in Chapter 5.

Exercise 1.7.1. Consider the automorphism of $\mathbb{T}^2$ corresponding to a nonsingular 2×2 integer matrix whose eigenvalues are not roots of 1.

(a) Prove that every point with rational coordinates is eventually periodic.

(b) Prove that every eventually periodic point has rational coordinates.

Exercise 1.7.2. Prove that the inverse of an $n \times n$ integer matrix B is also an integer matrix if and only if $|\det B| = 1$.

Exercise 1.7.3. Show that the eigenvalues of a two-dimensional hyperbolic toral automorphism are irrational (so the stable and unstable manifolds are dense by Exercise 1.11.1).

Exercise 1.7.4. Show that the number of fixed points of a hyperbolic toral automorphism A is $|\det(A - I)|$ (hence the number of periodic points of period n is $|\det(A^n - I)|$).

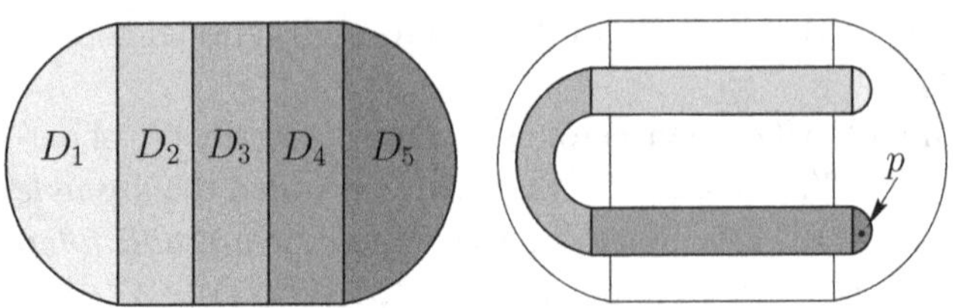

Figure 1.5. The horseshoe map.

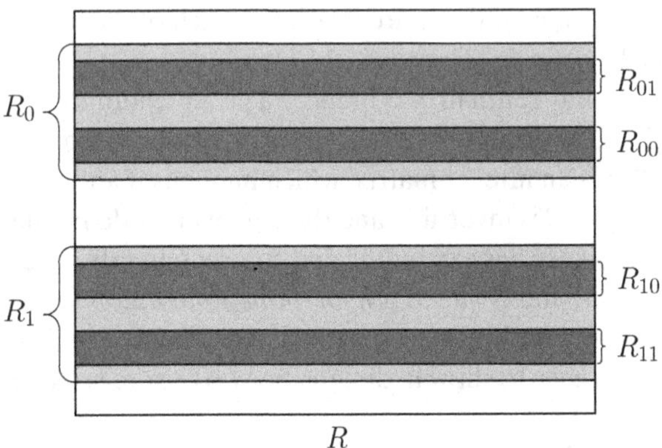

Figure 1.6. Horizontal rectangles.

1.8 The horseshoe

Consider a region $D \subset \mathbb{R}^2$ consisting of two semicircular regions D_1 and D_5 together with a unit square $R = D_2 \cup D_3 \cup D_4$ (see Figure 1.5).

Let $f \colon D \to D$ be a differentiable map that stretches and bends D into a horseshoe as shown in Figure 1.5. Assume also that f stretches $D_2 \cup D_4$ uniformly in the horizontal direction by a factor of $\mu > 2$ and contracts uniformly in the vertical direction by $\lambda < 1/2$. Since $f(D_5) \subset D_5$, the Brouwer Fixed Point Theorem implies the existence of a fixed point $p \in D_5$.

Set $R_0 = f(D_2) \cap R$ and $R_1 = f(D_4) \cap R$. Note that $f(R) \cap R = R_0 \cup R_1$. The set $f^2(R) \cap f(R) \cap R = f^2(R) \cap R$ consists of four horizontal rectangles R_{ij}, $i, j \in \{0, 1\}$, of height λ^2 (see Figure 1.6). More generally, for any finite sequence $\omega_0, \ldots, \omega_n$ of zeros and ones,

$$R_{\omega_0\omega_1\ldots\omega_n} = R_{\omega_0} \cap f(R_{\omega_1}) \cap \cdots \cap f^n(R_{\omega_n})$$

is a horizontal rectangle of height λ^n, and $f^n(R) \cap R$ is the union of 2^n such rectangles. For an infinite sequence $\omega = (\omega_i) \in \{0, 1\}^{\mathbb{N}_0}$, let

$R_\omega = \bigcap_{i=0}^{\infty} f^i(R_{\omega_i})$. The set $H^+ = \bigcap_{n=0}^{\infty} f^n(R) = \bigcup_\omega R_\omega$ is the product of a horizontal interval of length 1 and a vertical Cantor set C^+ (a Cantor set is a compact, perfect, totally disconnected set). Note that $f(H^+) = H^+$.

We now construct, in a similar way, a set H^- using preimages. Observe that $f^{-1}(R_0) = f^{-1}(R) \cap D_2$ and $f^{-1}(R_1) = f^{-1}(R) \cap D_4$ are vertical rectangles of width μ^{-1}. For any sequence $\omega_{-m}, \omega_{-m+1s}, \dots, \omega_{-1}$ of zeros and ones, $\bigcap_{i=1}^{m} f^{-i}(R_{\omega_i})$ is a vertical rectangle of width μ^{-m}, and $H^- = \bigcap_{i=1}^{\infty} f^{-i}(R)$ is the product of a vertical interval (of length 1) and a horizontal Cantor set C^-.

The *horseshoe set* $H = H^+ \cap H^- = \bigcap_{i=-\infty}^{\infty} f^i(R)$ is the product of the Cantor sets C^- and C^+ and is closed and f-invariant. It is *locally maximal*, i.e. there is an open set U containing H such that any f-invariant subset of U containing H coincides with H (Exercise 1.8.2). The map $\phi\colon \Sigma_2 = \{0,1\}^{\mathbb{Z}} \to H$ that assigns to each infinite sequence $\omega = (\omega_i) \in \Sigma_2$ the unique point $\phi(\omega) = \bigcap_{-\infty}^{\infty} f^i(R_{\omega_i})$ is a bijection (Exercise 1.8.3). Note that

$$f(\phi(\omega)) = \bigcap_{-\infty}^{\infty} f^{i+1}(R_{\omega_i}) = \phi(\sigma_r(\omega)),$$

where σ_r is the right shift in Σ_2, $\sigma_r(\omega)_{i+1} = \omega_i$. Thus, ϕ conjugates $f|H$ and the full two-sided 2-shift.

The *horseshoe* was introduced by S. Smale in the 1960s as an example of a *hyperbolic set* that "survives" small perturbations. We discuss hyperbolic sets in Chapter 5.

Exercise 1.8.1. Draw a picture of $f^{-1}(R) \cap f(R)$ and $f^{-2}(R) \cap f^2(R)$.

Exercise 1.8.2. Prove that H is a locally maximal f-invariant set.

Exercise 1.8.3. Prove that ϕ is a bijection, and that both ϕ and ϕ^{-1} are continuous.

1.9 The solenoid

Consider the solid torus $\mathcal{T} = S^1 \times D^2$, where $S^1 = [0,1] \bmod 1$ and $D^2 = \{(x,y) \in \mathbb{R}^2 : x^2 + y^2 \leqslant 1\}$. Fix $\lambda \in (0, 1/2)$ and define $F\colon \mathcal{T} \to \mathcal{T}$ by

$$F(\phi, x, y) = \left(2\phi, \lambda x + \frac{1}{2}\cos 2\pi\phi, \lambda y + \frac{1}{2}\sin 2\pi\phi\right).$$

The map F stretches by a factor of 2 in the S^1-direction, contracts by a factor of λ in the D^2-direction, and wraps the image twice inside $\mathcal{T}$ (see Figure 1.7).

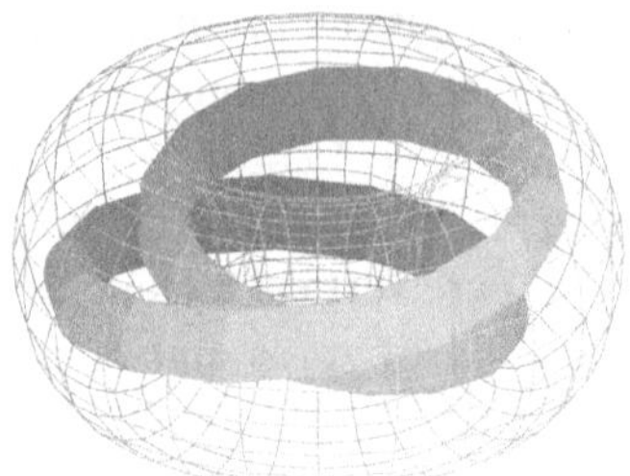

Figure 1.7. The solid torus and its image $F(\mathcal{T})$.

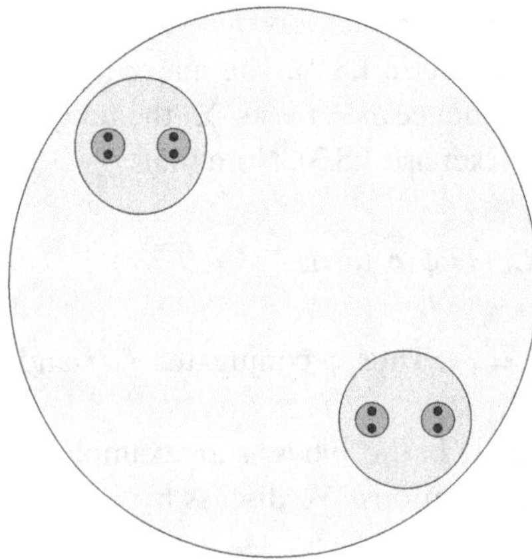

Figure 1.8. A cross-section of the solenoid.

The image $F(\mathcal{T})$ is contained in the interior $\mathrm{int}(\mathcal{T})$ of $\mathcal{T}$, and $F^{n+1}(\mathcal{T}) \subset \mathrm{int}(F^n(\mathcal{T}))$. Note that F is one-to-one (Exercise 1.9.1). A slice $F(\mathcal{T}) \cap \{\phi = \text{const}\}$ consists of two disks of radius λ centered at diametrically opposite points at distance $1/2$ from the center of the slice. A slice $F^n(\mathcal{T}) \cap \{\phi = \text{const}\}$ consists of 2^n disks of radius λ^n: two disks inside each of the 2^{n-1} disks of $F^{n-1}(\mathcal{T}) \cap \{\phi = \text{const}\}$. Slices of $F(\mathcal{T}), F^2(\mathcal{T})$ and $F^3(\mathcal{T})$ for $\phi = 1/8$ are shown in Figure 1.8.

The set $S = \bigcap_{n=0}^{\infty} F^n(\mathcal{T})$ is called a *solenoid*. It is a closed F-invariant subset of $\mathcal{T}$ on which F is bijective (Exercise 1.9.1). It can be shown that S is locally the product of an interval with a Cantor set in the two-dimensional disk.

The solenoid is an *attractor* for F. In fact, any neighborhood of S contains $F^n(\mathcal{T})$ for n sufficiently large, so the forward orbit of every point in

$\mathcal{T}$ converges to S. Moreover, S is a hyperbolic set, and is therefore called a *hyperbolic attractor*. We give a precise definition of attractors in §1.13.

Let Φ denote the set of sequences $(\phi_i)_{i=0}^{\infty}$, where $\phi_i \in S^1$ and $\phi_i = 2\phi_{i+1} \bmod 1$ for all i. The product topology on $(S^1)^{\mathbb{N}_0}$ induces the subspace topology on Φ. The space Φ is a commutative group under componentwise addition (mod 1). The map $(\phi, \psi) \mapsto \phi - \psi$ is continuous, so Φ is a topological group. The map $\alpha\colon \Phi \to \Phi$, $(\phi_0, \phi_1, \dots) \mapsto (2\phi_0, \phi_0, \phi_1, \dots)$ is a group automorphism and a homeomorphism (Exercise 1.9.3).

For $s \in S$, the first (angular) coordinates of the preimages $F^{-n}(s) = (\phi_n, x_n, y_n)$ form a sequence $h(s) = (\phi_0, \phi_1, \dots) \in \Phi$. This defines a map $h\colon S \to \Phi$. The inverse of h is the map $(\phi_0, \phi_1, \dots) \mapsto \bigcap_{n=0}^{\infty} F^n(\{\phi_n\} \times D^2)$, and h is a homeomorphism (Exercise 1.9.2). Note that $h\colon S \to \Phi$ conjugates F and α, i.e. $h \circ F = \alpha \circ h$. This conjugation allows one to study properties of (S, F) by studying properties of the algebraic system (Φ, α).

Exercise 1.9.1. Prove that (a) $F\colon \mathcal{T} \to \mathcal{T}$ is injective; and (b) $F\colon S \to S$ is bijective.

Exercise 1.9.2. Prove that for every $(\phi_0, \phi_1, \dots) \in \Phi$ the intersection $\bigcap_{n=0}^{\infty} F^n(\{\phi_n\} \times D^2)$ consists of a single point s, and $h(s) = (\phi_0, \phi_1, \dots)$. Show that h is a homeomorphism.

Exercise 1.9.3. Show that Φ is a topological group and α is an automorphism and homeomorphism.

Exercise 1.9.4. Find the fixed point of F and all periodic points of period 2.

1.10 Flows and differential equations

Flows arise naturally from first-order autonomous differential equations. Suppose $\dot{x} = F(x)$ is a differential equation in $\mathbb{R}^n$, where $F\colon \mathbb{R}^n \to \mathbb{R}^n$ is continuously differentiable. For each point $x \in \mathbb{R}^n$, there is a unique solution $f^t(x)$ starting at x at time 0 and defined for all t in some neighborhood of 0. To simplify matters, we will assume that the solution is defined for all $t \in \mathbb{R}$; this will be the case, for example, if F is bounded, or is dominated in norm by a linear function. For fixed $t \in \mathbb{R}$, the time-t map $x \mapsto f^t(x)$ is a C^1 diffeomorphism of $\mathbb{R}^n$. Because the equation is autonomous, $f^{t+s}(x) = f^t(f^s(x))$, i.e. f^t is a flow.

Conversely, given a flow $f^t : \mathbb{R}^n \to \mathbb{R}^n$, if the map $(t, x) \mapsto f^t(x)$ is differentiable, then f^t is the time-t map of the differential equation

$$\dot{x} = \frac{d}{dt}\bigg|_{t=0} f^t(x).$$

Here are some examples. Consider the linear autonomous differential equation $\dot{x} = Ax$ in $\mathbb{R}^n$, where A is a real $n \times n$ matrix. The flow of this differential equation is $f^t(x) = e^{At}x$, where e^{At} is the matrix exponential. If A is non-singular, the flow has exactly one fixed point at the origin. If all the eigenvalues of A have negative real part, then every orbit approaches the origin and the origin is asymptotically stable. If some eigenvalue has positive real part, then the origin is unstable.

Most differential equations that arise in applications are non-linear. The differential equation governing an ideal frictionless pendulum is one of the most familiar:

$$\ddot{\theta} + \sin\theta = 0.$$

This equation cannot be solved in closed form, but it can be studied by qualitative methods. It is equivalent to the system

$$\begin{aligned} \dot{x} &= y \\ \dot{y} &= -\sin x. \end{aligned}$$

The energy E of the system is the sum of the kinetic and potential energies, $E(x, y) = 1 - \cos x + y^2/2$. One can show (Exercise 1.10.2) by differentiating $E(x, y)$ with respect to t that E is constant along solutions of the differential equation. Equivalently, if f^t is the flow in $\mathbb{R}^2$ of this differential equation, then E is *invariant* by the flow, i.e. $E(f^t(x, y)) = E(x, y)$, for all $t \in \mathbb{R}$, $(x, y) \in \mathbb{R}^2$. A function that is constant on the orbits of a dynamical system is called a *first integral* of the system.

The fixed points in the phase plane for the undamped pendulum are $(k\pi, 0), k \in \mathbb{Z}$. The points $(2k\pi, 0)$ are local minima of the energy. The points $((2k+1)\pi, 0)$ are saddle points.

Now consider the damped pendulum $\ddot{\theta} + \gamma\dot{\theta} + \sin\theta = 0$, or the equivalent system

$$\begin{aligned} \dot{x} &= y \\ \dot{y} &= -\sin x - \gamma y. \end{aligned}$$

A simple calculation shows that $\dot{E} < 0$ except at the fixed points $(k\pi, 0)$, $k \in \mathbb{Z}$, which are the local extrema of the energy. Thus the energy is strictly decreasing along every non-constant solution. In particular, every trajectory approaches a critical point of the energy, and almost every trajectory approaches a local minimum.

The energy of the pendulum is an example of a *Lyapunov function*, i.e. a continuous function that is non-increasing along the orbits of the flow. Any strict local minimum of a Lyapunov function is an asymptotically stable equilibrium point of the differential equation. Moreover, any bounded orbit must converge to the maximal invariant subset M of the set of points satisfying $\dot{E} = 0$. In the case of the damped pendulum, $M = \{(k\pi, 0) \colon k \in \mathbb{Z}\}$.

Here is another class of examples that appears frequently in applications, particularly optimization problems. Given a smooth function $f \colon \mathbb{R}^n \to \mathbb{R}$, the flow of the differential equation

$$\dot{x} = \operatorname{grad} f(x)$$

is called the *gradient flow* of f. The function $-f$ is a Lyapunov function for the gradient flow. The trajectories are the projections to $\mathbb{R}^n$ of paths of steepest ascent along the graph of f, and are orthogonal to the level sets of f (Exercise 1.10.3).

A *Hamiltonian system* is a flow in $\mathbb{R}^{2n}$ given by a system of differential equations of the form

$$\dot{q}_i = \frac{\partial H}{\partial p_i}, \qquad \dot{p}_i = -\frac{\partial H}{\partial q_i}, \qquad i = 1, \ldots, n,$$

where the *Hamiltonian function* $H(p, q)$ is assumed to be smooth. Since the divergence of the right-hand side is 0, the flow preserves volume. The Hamiltonian function is a first integral, so that the level surfaces of H are invariant under the flow. If for some $C \in \mathbb{R}$ the level surface $H(p, q) = C$ is compact, the restriction of the flow to the level surface preserves a finite measure with smooth density. Hamiltonian flows have many applications in physics and mathematics. For example, the flow associated with the undamped pendulum is a Hamiltonian flow, where the Hamiltonian function is the total energy of the pendulum (Exercise 1.10.5).

Exercise 1.10.1. Show that the scalar differential equation $\dot{x} = x \log x$ induces the flow $f^t(x) = x^{\exp(t)}$ on the line.

Exercise 1.10.2. Show that the energy is constant along solutions of the undamped pendulum equation, and strictly decreasing along non-constant solutions of the damped pendulum equation.

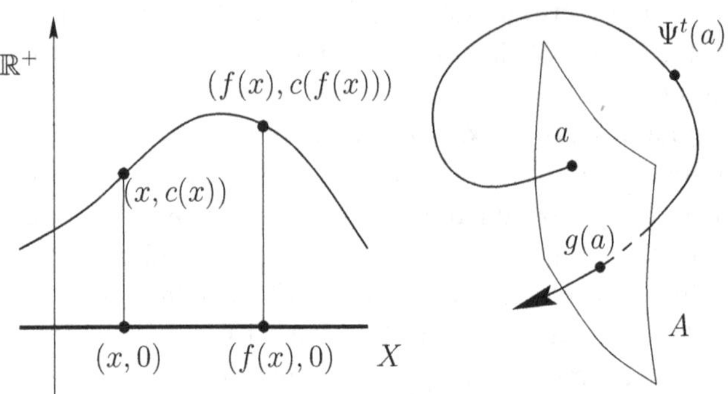

Figure 1.9. Suspension and cross-section.

Exercise 1.10.3. Show that $-f$ is a Lyapunov function for the gradient flow of f, and that the trajectories are orthogonal to the level sets of f.

Exercise 1.10.4. Prove that any differentiable one-parameter group of linear maps of $\mathbb{R}$ is the flow of a differential equation $\dot{x} = kx$.

Exercise 1.10.5. Show that the flow of the undamped pendulum is a Hamiltonian flow.

1.11 Suspension and cross-section

There are natural constructions for passing from a map to a flow, and vice versa. Given a map $f: X \to X$, and a function $c: X \to \mathbb{R}^+$ bounded away from 0, consider the quotient space

$$X_c = \{(x,t) \in X \times \mathbb{R}^+ : 0 \leqslant t \leqslant c(x)\}/\sim,$$

where $\sim$ is the equivalence relation $(x, c(x)) \sim (f(x), 0)$. The *suspension* of f with *ceiling function* c is the semiflow $\phi^t: X_c \to X_c$ given by $\phi^t(x,s) = (f^n(x), s')$, where n and s' satisfy

$$\sum_{i=0}^{n-1} c(f^i(x)) + s' = t + s, \quad 0 \leqslant s' \leqslant c(f^n(x)).$$

In other words, flow along $\{x\} \times \mathbb{R}^+$ to $(x, c(x))$, then jump to $(f(x), 0)$ and continue along $\{f(x)\} \times \mathbb{R}^+$, and so on (see Figure 1.9). A suspension flow is also called a *flow under a function*.

Conversely, a *cross-section* of a flow or semiflow $\psi^t: Y \to Y$ is a subset $A \subset Y$ with the following property: for every $y \in Y$, the set $T_y = \{t \in \mathbb{R}^+ : \psi^t(y) \in A\}$ is a non-empty discrete subset of $\mathbb{R}^+$. For $a \in A$, let $\tau(a) = \min T_a$ be the *return time* to A. Define the *first return map* $g\colon A \to A$ by $g(a) = \psi^{\tau(a)}(a)$, i.e. $g(a)$ is the first point after a in $\mathcal{O}^+_\psi(x) \cap A$ (see Figure 1.9). The first return map is often called the *Poincaré map*. Since the dimension of the cross-section is less by 1, in many cases maps in dimension n present the same level of difficulty as flows in dimension $n+1$.

Suspension and cross-section are inverse constructions: the suspension of g with ceiling function τ is ψ^t, and $X \times \{0\}$ is a cross-section of ϕ with first return map f. If ϕ is a suspension of f, then the dynamical properties of f and ϕ are closely related, e.g. the periodic orbits of f correspond to the periodic orbits of ϕ. Both of these constructions can be tailored to specific settings (topological, measurable, smooth, etc.).

As an example, consider the 2-torus $\mathbb{T}^2 = \mathbb{R}^2/\mathbb{Z}^2 = S^1 \times S^1$, with topology and metric induced from the topology and metric on $\mathbb{R}^2$. Fix $\alpha \in \mathbb{R}$, and define the linear flow $\phi^t_\alpha\colon \mathbb{T}^2 \to \mathbb{T}^2$ by

$$\phi^t_\alpha(x, y) = (x + \alpha t, y + t) \bmod 1.$$

Note that ϕ^t_α is the suspension of the circle rotation R_α with ceiling function 1, and $S^1 \times \{0\}$ is a cross-section for ϕ_α with constant return time $\tau(y) = 1$ and first return map R_α. The flow ϕ^t_α consists of left translations by the elements $g^t = (\alpha t, t) \bmod 1$, which form a one-parameter subgroup of $\mathbb{T}^2$.

Exercise 1.11.1. Show that if α is irrational, then every orbit of ϕ_α is dense in T^2, and if α is rational, then every orbit of ϕ_α is periodic.

Exercise 1.11.2. Let ϕ^t be a suspension of f. Show that a periodic orbit of ϕ^t corresponds to a periodic orbit of f, and that a dense orbit of ϕ^t corresponds to a dense orbit of f.

***Exercise 1.11.3.** Suppose 1, s, and αs are real numbers that are linearly independent over $\mathbb{Q}$. Show that every orbit of the time s map ϕ^s_α is dense in T^2.

1.12 Chaos and Lyapunov exponents

A dynamical system is *deterministic* in the sense that the evolution of the system is described by a specific map, so that the present (the initial state) completely determines the future (the forward orbit of the state). At the same time, dynamical systems often appear to be *chaotic* in that they have *sensitive dependence on initial conditions*, i.e. minor changes in the initial state lead to dramatically different long-term behavior. Specifically, a dynamical system (X, f) has sensitive dependence on initial conditions on a subset $X' \subset X$ if there is $\epsilon > 0$ such that for every $x \in X'$ and $\delta > 0$ there are $y \in X$ and $n \in \mathbb{N}$ for which $d(x, y) < \delta$ and $d(f^n(x), f^n(y)) > \epsilon$. Although there is no universal agreement on a definition of *chaos*, it is generally agreed that a chaotic dynamical system should exhibit sensitive dependence on initial conditions. Chaotic systems are usually assumed to have some additional properties, e.g. existence of a dense orbit.

The study of chaotic behavior has become one of the central issues in dynamical systems during the last two decades. In practice, the term *chaos* has been applied to a variety of systems that exhibit some type of random behavior. This random behavior is observed experimentally in some situations, and in others follows from specific properties of the system. Often a system is declared to be chaotic based on the observation that a typical orbit appears to be randomly distributed, and different orbits appear to be uncorrelated. The variety of views and approaches in this area precludes a universal definition of chaos.

The simplest example of a chaotic system is the circle endomorphism $(S^1, E_m), m > 1$ (§1.3). Distances between points x and y are expanded by a factor of m if $d(x, y) \leqslant 1/(2m)$, so any two points are moved at least $1/(2m)$ apart by some iterate of E_m, so E_m has sensitive dependence on initial conditions. A typical orbit is dense (§1.3), and is *uniformly distributed* on the circle (Proposition 4.4.2).

The simplest non-linear chaotic dynamical systems in dimension one are the quadratic maps $q_\mu(x) = \mu x(1 - x), \mu \geqslant 4$, restricted to the forward invariant set $\Lambda_\mu \subset [0, 1]$ (see §1.5 and Chapter 7).

Sensitive dependence on initial conditions is usually associated with positive *Lyapunov exponents*. Let f be a differentiable map of an open subset $U \subset \mathbb{R}^m$ into itself, and let $df(x)$ denote the derivative of f at x. For $x \in U$ and a non-zero vector $v \in \mathbb{R}^m$ define the *Lyapunov exponent* $\chi(x, v)$ by

$$\chi(x, v) = \overline{\lim_{n\to\infty}} \frac{1}{n} \log \|df^n(x)v\|.$$

If f has uniformly bounded first derivatives, then χ is well-defined for every $x \in U$ and every non-zero vector v.

The Lyapunov exponent measures the exponential growth rate of tangent vectors along orbits, and has the following properties:

$$\chi(x, \lambda v) = \chi(x, v) \text{ for all real } \lambda \neq 0,$$

$$\chi(x, v + w) \leqslant \max(\chi(x, v), \chi(x, w)),$$

$$\chi(f(x), df(x)v) = \chi(x, v). \tag{1.1}$$

See Exercise 1.12.1.

If $\chi(x, v) = \chi > 0$ for some vector v, then there is a sequence $n_j \to \infty$ such that for every $\eta > 0$

$$\|df^{n_j}(x)v\| \geqslant e^{(\chi-\eta)n_j}\|v\|.$$

This implies that, for a fixed j, there is a point $y \in U$ such that

$$d(f^{n_j}(x), f^{n_j}(y)) \geqslant \frac{1}{2}e^{(\chi-\eta)n_j}d(x, y).$$

In general, this does not imply sensitive dependence on initial conditions, since the distance between x and y cannot be controlled. However, most dynamical systems with positive Lyapunov exponents have sensitive dependence on initial conditions.

Conversely, if two close points are moved far apart by f^n, by the Intermediate Value Theorem, there must exist points x and directions v for which $\|df^n(x)v\| > \|v\|$. Therefore we expect f to have positive Lyapunov exponents if it has sensitive dependence on initial conditions, though this is not always the case.

The circle endomorphisms E_m, $m > 1$ have a positive exponent at all points. A quadratic map $q_\mu, \mu > 2 + \sqrt{5}$, has a positive exponent at any point whose forward orbit does not contain 0.

Exercise 1.12.1. Prove (1.1).

Exercise 1.12.2. Compute the Lyapunov exponent for E_m.

Exercise 1.12.3. Compute the Lyapunov exponents for the solenoid, §1.9.

Exercise 1.12.4. Using a computer, calculate the first 100 points in the orbit of $\sqrt{2} - 1$ under the map E_2. What proportion of these points is contained in each of the intervals $[0, 1/4)$, $[1/4, 1/2)$, $[1/2, 3/4)$, and $[3/4, 1)$?

1.13 Attractors

Let X be a compact topological space, and $f\colon X \to X$ a continuous map. Generalizing the notion of an attracting fixed point, we say that a compact set $C \subset X$ is an *attractor* if there is an open set U containing C such that $f(\overline{U}) \subset U$ and $C = \bigcap_{n\geqslant 0} f^n(U)$. It follows that $f(C) = C$, since $f(C) = \bigcap_{n\geqslant 1} f^n(U) \subset C$; on the other hand, $C = \bigcap_{n\geqslant 1} f^n(U) = f(C)$, since $f(U) \subset U$. Moreover, the forward orbit of any point $x \in U$ converges to C, i.e. for any open set V containing C, there is some $N > 0$ such that $f^n(x) \in V$ for all $n \geqslant N$. To see this, observe that X is covered by V together with the open sets $X\backslash f^n(\overline{U})$, $n \geqslant 0$. By compactness, there is a finite subcover, and since $f^n(U) \subset f^{n-1}(U)$, we conclude that there is some $N > 0$ such that $X = V \cup (X \setminus f^n(\overline{U}))$ for all $n \geqslant N$. Thus $f^n(x) \in f^n(U) \subset V$ for $n \geqslant N$.

The *basin of attraction* of C is the set $BA(C) = \bigcup_{n \geqslant 0} f^{-n}(U)$. The basin $BA(C)$ is precisely the set of points whose forward orbits converge to C (Exercise 1.13.1).

An open set $U \subset X$ such that $\overline{U}$ is compact and $f(\overline{U}) \subset U$ is called a *trapping region* for f. If U is a trapping region, then $\bigcap_{n \geqslant 0} f^n(U)$ is an attractor. For flows generated by differential equations, any region with the property that along the boundary the vector field points into the region, is a trapping region for the flow. In practice, the existence of an attractor is proved by constructing a trapping region. An attractor can be studied experimentally by numerically approximating orbits that start in the trapping region.

The simplest examples of attractors are: the intersection of the images of the whole space (if the space is compact); attracting fixed points; and attracting periodic orbits. For flows, the examples include asymptotically stable fixed points and asymptotically stable periodic orbits.

Many dynamical systems have attractors of a more complicated nature. For example, recall that the solenoid S (§1.9) is a (hyperbolic) attractor for $(\mathcal{T}, F)$. Locally, S is the product of an interval with a Cantor set. The structure of hyperbolic attractors is relatively well understood. However, some non-linear systems have attractors that are chaotic (with sensitive dependence on initial conditions) but not hyperbolic. These attractors are called *strange attractors*. The best-known examples of strange attractors are the Hénon attractor and the Lorenz attractor.

The study of strange attractors began with the publication by E. N. Lorenz in 1963 of the paper "Deterministic non-periodic flow" (Lorenz, 1963). In the process of investigating meteorological models, Lorenz studied the non-linear system of differential equations

$$\begin{aligned} \dot{x} &= \sigma(y - x) \\ \dot{y} &= Rx - y - xz \\ \dot{z} &= -bz + xy, \end{aligned} \tag{1.2}$$

now called the *Lorenz system*. He observed that at parameter values $\sigma = 10$, $b = 8/3$, and $R = 28$, the solutions of (1.2) eventually start revolving alternately about two repelling equilibrium points at $(\pm\sqrt{72}, \pm\sqrt{72}, 27)$. The number of times the solution revolves about one equilibrium before switching to the other has no discernible pattern. There is a trapping region U that contains 0 but not the two repelling equilibrium points. The attractor contained in U is called the Lorenz attractor. It is an extremely complicated

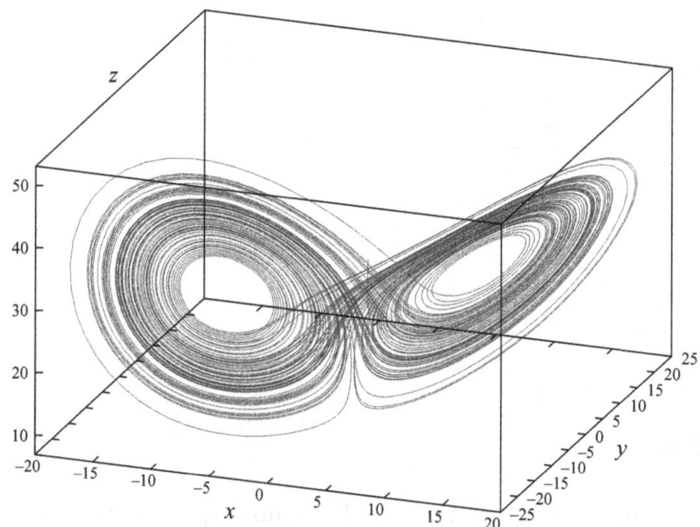

Figure 1.10. Lorenz attractor.

set consisting of uncountably many orbits (including a saddle fixed point at 0), and non-fixed periodic orbits that are known to be knotted (Williams, 1984). The attractor is not hyperbolic in the usual sense, though it has strong expansion and contraction, and sensitive dependence on initial conditions. The attractor persists for small changes in the parameter values (see Figure 1.10).

The Hénon map $H = (f, g)\colon \mathbb{R}^2 \to \mathbb{R}^2$ is defined by

$$f(x, y) = a - by - x^2$$
$$g(x, y) = x$$

where a and b are constants (Hénon, 1976). The Jacobian of the derivative dH equals b. If $b \neq 0$, the Hénon map is invertible; the inverse is

$$(x, y) \mapsto (y, (a - x - y^2)/b).$$

The map changes area by a factor of $|b|$, and reverses orientation if $b < 0$.

For the specific parameter values $a = 1.4$ and $b = -0.3$, Hénon showed that there is a trapping region U homeomorphic to a disk. His numerical experiments suggested that the resulting attractor has a dense orbit and sensitive dependence on initial conditions, though these properties have not been rigorously proved. Figure 1.11 shows a long segment of an orbit starting in the trapping region, which is believed to approximate the attractor. It is known that for a large set of parameter values $a \in [1, 2]$, $b \in [-1, 0]$,

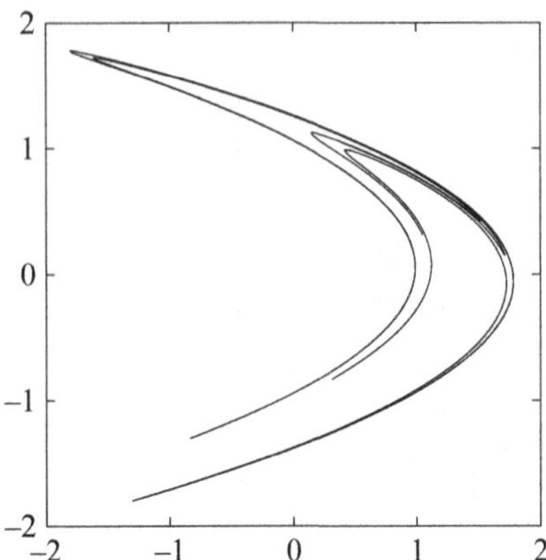

Figure 1.11. Hénon attractor.

the attractor has a dense orbit and a positive Lyapunov exponent, but is not hyperbolic (Benedicks and Carleson, 1991).

Exercise 1.13.1. Let A be an attractor. Show that $x \in BA(A)$ if and only if the forward orbit of x converges to A.

Exercise 1.13.2. Find a trapping region for the flow generated by the Lorenz equations with parameter values $\sigma = 10, b = 8/3$, and $R = 28$.

Exercise 1.13.3. Find a trapping region for the Hénon map with parameter values $a = 1.4, b = -0.3$.

Exercise 1.13.4. Using a computer, plot the first 1000 points in an orbit of the Hénon map starting in a trapping region.

CHAPTER TWO

Topological dynamics

A *topological dynamical system* is a topological space X and either a continuous map $f: X \to X$ or a continuous (semi)flow f^t on X, i.e. a (semi)flow f^t for which the map $(t, x) \mapsto f^t(x)$ is continuous. To simplify the exposition, we assume that X is locally compact, metrizable, and second countable, though many of the results in this chapter are true under weaker assumptions on X. As we noted earlier, we will focus our attention on discrete-time systems, though all general results in this chapter are valid for continuous-time systems as well.

Let X and Y be topological spaces. Recall that a continuous map $f: X \to Y$ is a *homeomorphism* if it is one-to-one and the inverse is continuous.

Let $f: X \to X$ and $g: Y \to Y$ be topological dynamical systems. A *topological semiconjugacy* from g to f is a surjective continuous map $h: Y \to X$ such that $f \circ h = h \circ g$. If h is a homeomorphism, it is called a *topological conjugacy* and f and g are said to be *topologically conjugate* or *isomorphic*. Topologically conjugate dynamical systems have identical topological properties. Consequently, all the properties and invariants we introduce in this chapter, including minimality, topological transitivity, topological mixing and topological entropy, are preserved by topological conjugacy.

Throughout this chapter, a metric space X with metric d is denoted (X, d). If $x \in X$ and $r > 0$, then $B(x, r)$ denotes the open ball of radius r centered at x. If (X, d) and (Y, d') are metric spaces, then $f: X \to Y$ is an isometry if $d'(f(x_1), f(x_2)) = d(x_1, x_2)$ for all $x_1, x_2 \in X$.

2.1 Limit sets and recurrence

Let $f: X \to X$ be a topological dynamical system. Let x be a point in X. A point $y \in X$ is an *ω-limit point* of x if there is a sequence of natural numbers $n_k \to \infty$ (as $k \to \infty$) such that $f^{n_k}(x) \to y$. The *ω-limit set* of x is the set

$\omega(x) = \omega_f(x)$ of all ω-limit points of x. Equivalently,

$$\omega(x) = \bigcap_{n\in\mathbb{N}} \overline{\bigcup_{i\geqslant n} f^i(x)}.$$

If f is invertible, the *α-limit set* of x is $\alpha(x) = \alpha_f(x) = \bigcap_{n\in\mathbb{N}} \overline{\bigcup_{i\geqslant n} f^{-i}(x)}$. A point in $\alpha(x)$ is an *α-limit point* of x. Both the α- and ω-limit sets are closed and f-invariant (Exercise 2.1.1).

A point x is called *(positively) recurrent* if $x \in \omega(x)$; the set $\mathcal{R}(f)$ of recurrent points is f-invariant. Periodic points are recurrent.

A point x is *non-wandering* if for any neighborhood U of x there exists $n \in \mathbb{N}$ such that $f^n(U) \cap U \neq \emptyset$. The set $NW(f)$ of non-wandering points is closed, f-invariant, and contains $\omega(x)$ and $\alpha(x)$ for all $x \in X$ (Exercise 2.1.2). Every recurrent point is non-wandering, and in fact $\overline{\mathcal{R}(f)} \subset NW(f)$ (Exercise 2.1.3); in general, however, $NW(f) \not\subset \overline{\mathcal{R}(f)}$ (Exercise 2.1.10).

Recall the notation $\mathcal{O}(x) = \bigcup_{n\in\mathbb{Z}} f^n(x)$ for an invertible f, and $\mathcal{O}^+(x) = \bigcup_{n\in\mathbb{N}_0} f^n(x)$.

PROPOSITION 2.1.1

1. *Let f be a homeomorphism, $y \in \overline{\mathcal{O}(x)}$, and $z \in \overline{\mathcal{O}(y)}$. Then $z \in \overline{\mathcal{O}(x)}$.*
2. *Let f be a continuous map, $y \in \overline{\mathcal{O}^+(x)}$, and $z \in \overline{\mathcal{O}^+(y)}$. Then $z \in \overline{\mathcal{O}^+(x)}$.*

Proof. Exercise 2.1.7. □

Let X be compact. A closed, non-empty, forward f-invariant subset $Y \subset X$ is a *minimal set* for f if it contains no proper, closed, non-empty, forward f-invariant subset. A compact invariant set Y is minimal if and only if the forward orbit of every point in Y is dense in Y (Exercise 2.1.4). Note that a periodic orbit is a minimal set. If X itself is a minimal set, we say that f is minimal.

PROPOSITION 2.1.2. *Let $f: X \to X$ be a topological dynamical system. If X is compact, then X contains a minimal set for f.*

Proof. The proof is a straightforward application of Zorn's Lemma. Let $\mathcal{C}$ be the collection of non-empty, closed, f-invariant subsets of X, with the partial ordering given by inclusion. Then $\mathcal{C}$ is not empty since $X \in \mathcal{C}$. Suppose $\mathcal{K} \subset \mathcal{C}$ is a totally ordered subset. Then any finite intersection of elements of $\mathcal{K}$ is non-empty, so by the finite intersection property for compact sets, $\bigcap_{K\in\mathcal{K}} K \neq \emptyset$. Thus, by Zorn's Lemma, $\mathcal{C}$ contains a minimal element, which is a minimal set for f. □

In a compact topological space, every point in a minimal set is recurrent (Exercise 2.1.4), so the existence of minimal sets implies the existence of recurrent points.

A subset $A \subset \mathbb{N}$ (or $\mathbb{Z}$) is *relatively dense* (or *syndetic*) if there is $k > 0$ such that $\{n, n+1, \dots, n+k\} \cap A \neq \emptyset$ for any n. A point $x \in X$ is *almost periodic* if for any neighborhood U of x, the set $\{i \in \mathbb{N}\colon f^i(x) \in U\}$ is relatively dense in $\mathbb{N}$.

PROPOSITION 2.1.3. *If X is a compact Hausdorff space and $f\colon X \to X$ is continuous, then $\overline{\mathcal{O}^+(x)}$ is minimal for f if and only if x is almost periodic.*

Proof. Suppose x is almost periodic and $y \in \overline{\mathcal{O}^+(x)}$. We need to show that $x \in \overline{\mathcal{O}^+(y)}$. Let U be a neighborhood of x. There is an open set $U' \subset X, x \in U' \subset U$, and an open set $V \subset X \times X$ containing the diagonal, such that if $x_1 \in U'$ and $(x_1, x_2) \in V$, then $x_2 \in U$. Since x is almost periodic, there is $K \in \mathbb{N}$ such that for every $j \in \mathbb{N}$ we have that $f^{j+k}(x) \in U'$ for some $0 \leqslant k \leqslant K$. Let $V' = \bigcap_{i=0}^{K} f^{-i}(V)$. Note that V' is open and contains the diagonal of $X \times X$. There is a neighborhood $W \ni y$ such that $W \times W \subset V'$. Choose n such that $f^n(x) \in W$, and choose k such that $f^{n+k}(x) \in U'$ with $0 \leqslant k \leqslant K$. Then $(f^{n+k}(x), f^k(y)) \in V$ and hence $f^k(y) \in U$.

Conversely, suppose x is not almost periodic. Then there is a neighborhood U of x such that $A = \{i\colon f^i(x) \in U\}$ is not relatively dense. Thus there are sequences $a_i \in \mathbb{N}$ and $k_i \in \mathbb{N}$, $k_i \to \infty$, such that $f^{a_i+j}(x) \notin U$ for $j = 0, \dots, k_i$. Let y be a limit point of $\{f^{a_i}(x)\}$. By passing to a subsequence, we may assume that $f^{a_i}(x) \to y$. Fix $j \in \mathbb{N}$. Note that $f^{a_i+j}(x) \to f^j(y)$, and $f^{a_i+j}(x) \notin U$ for i sufficiently large. Thus $f^j(y) \notin U$ for all $j \in \mathbb{N}$, so $x \notin \overline{\mathcal{O}^+(y)}$, which implies that $\overline{\mathcal{O}^+(x)}$ is not minimal. □

Recall that an irrational circle rotation R_α is minimal (§1.2). Therefore every point is non-wandering, recurrent, and almost periodic. An expanding endomorphism E_m of the circle has dense orbits (§1.3), but is not minimal because it has periodic points. Every point is non-wandering, but not all points are recurrent (Exercise 2.1.5).

Exercise 2.1.1. Show that the α- and ω-limit sets of a point are closed invariant sets.

Exercise 2.1.2. Show that the set of non-wandering points is closed, is f-invariant, and contains $\omega(x)$ and $\alpha(x)$ for all $x \in X$.

Exercise 2.1.3. Show that $\overline{\mathcal{R}(f)} \subset NW(f)$.

Exercise 2.1.4. Let X be compact, $f: X \to X$ continuous.

(a) Show that $Y \subset X$ is minimal if and only if $\omega(y) = Y$ for every $y \in Y$.

(b) Show that Y is minimal if and only if the forward orbit of every point in Y is dense in Y.

Exercise 2.1.5. Show that there are points that are non-recurrent and not eventually periodic for an expanding circle endomorphism E_m.

Exercise 2.1.6. For a hyperbolic toral automorphism $A: \mathbb{T}^2 \to \mathbb{T}^2$, show that:

(a) $\mathcal{R}(A)$ is dense, and hence $NW(A) = \mathbb{T}^2$, but

(b) $\mathcal{R}(A) \neq \mathbb{T}^2$.

Exercise 2.1.7. Prove Proposition 2.1.1.

Exercise 2.1.8. Prove that a homeomorphism $f: X \to X$ is minimal if and only if for each non-empty open set $U \subset X$ there is $n \in \mathbb{N}$ such that $\bigcup_{k=-n}^{n} f^k(U) = X$.

Exercise 2.1.9. Prove that a homeomorphism f of a compact metric space X is minimal if and only if for every $\epsilon > 0$ there is $N = N(\epsilon) \in \mathbb{N}$ such that for every $x \in X$ the set $\{x, f(x), \ldots, f^N(x)\}$ is ϵ-dense in X.

***Exercise 2.1.10.** Give an example of a dynamical system where $NW(f) \not\subset \overline{\mathcal{R}(f)}$.

2.2 Topological transitivity

A topological dynamical system $f: X \to X$ is *topologically transitive* if there is a point $x \in X$ whose forward orbit is dense in X. If X has no isolated points, this condition is equivalent to the existence of a point whose ω-limit set is dense in X (Exercise 2.2.1).

PROPOSITION 2.2.1. *Let $f: X \to X$ be a continuous map of a locally compact Hausdorff space X. Suppose that for any two non-empty open sets U and V there is $n \in \mathbb{N}$ such that $f^n(U) \cap V \neq \emptyset$. Then f is topologically transitive.*

Proof. The hypothesis implies that given any open set $V \subset X$, the set $\bigcup_{n\in\mathbb{N}} f^{-n}(V)$ is dense in X, since it intersects every open set. Let $\{V_i\}$ be a countable basis for the topology of X. Then $Y = \bigcap_i \bigcup_{n\in\mathbb{N}} f^{-n}(V_i)$ is a countable intersection of open, dense sets, and is therefore non-empty by the Baire Category Theorem. The forward orbit of any point $y \in Y$ enters each V_i, hence is dense in X. □

In most topological spaces, existence of a dense full orbit for a homeomorphism implies existence of a dense forward orbit, as we show in the next proposition. Note, however, that density of a particular full orbit $\mathcal{O}(x)$ does not imply density of the corresponding forward orbit $\mathcal{O}^+(x)$ (see Exercise 2.2.2).

PROPOSITION 2.2.2. *Let $f\colon X \to X$ be a homeomorphism of a compact metric space, and suppose that X has no isolated points. If there is a dense full orbit $\mathcal{O}(x)$, then there is a dense forward orbit $\mathcal{O}^+(y)$.*

Proof. Since $\overline{\mathcal{O}(x)} = X$, the orbit $\mathcal{O}(x)$ visits every non-empty open set U at least once, and therefore infinitely many times because X has no isolated points. Hence there is a sequence n_k, with $|n_k| \to \infty$, such that $f^{n_k}(x) \in B(x, 1/k)$ for $k \in \mathbb{N}$, i.e. $f^{n_k}(x) \to x$ as $k \to \infty$. Thus $f^{n_k+l}(x) \to f^l(x)$ for any $l \in \mathbb{Z}$. There are either infinitely many positive or infinitely many negative indices n_k, and it follows that either $\mathcal{O}(x) \subset \overline{\mathcal{O}^+(x)}$ or $\mathcal{O}(x) \subset \overline{\mathcal{O}^-(x)}$. In the former case, $\overline{\mathcal{O}^+(x)} = X$, and we are done. In the latter case, let U, V be non-empty open sets. Since $\overline{\mathcal{O}^-(x)} = X$, there are integers $i < j < 0$ such that $f^i(x) \in U$ and $f^j(x) \in V$, so $f^{j-i}(U) \cap V \neq \emptyset$. Hence, by Proposition 2.2.1, f is topologically transitive. □

Exercise 2.2.1. Show that if X has no isolated points and $\mathcal{O}^+(x)$ is dense, then $\omega(x)$ is dense. Give an example to show that this is not true if X has isolated points.

Exercise 2.2.2. Give an example of a dynamical system with a dense full orbit but no dense forward orbit.

Exercise 2.2.3. Is the product of two topologically transitive systems topologically transitive? Is a factor of a topologically transitive system topologically transitive?

Exercise 2.2.4. Let $f\colon X \to X$ be a homeomorphism. Show that if f has a non-constant first integral or Lyapunov function (§1.10), then it is not topologically transitive.

Exercise 2.2.5. Let $f\colon X \to X$ be a topological dynamical system with at least two orbits. Show that if f has an attracting periodic point, then it is not topologically transitive.

Exercise 2.2.6. Let α be irrational and $f\colon T^2 \to T^2$ be the homeomorphism of the 2-torus given by $f(x, y) = (x + \alpha, x + y)$.

(a) Prove that every non-empty, open, f-invariant set is dense, i.e. f is topologically transitive.

(b) Suppose the forward orbit of (x_0, y_0) is dense. Prove that for every $y \in S^1$ the forward orbit of (x_0, y) is dense. Moreover, if the set $\bigcup_{k=0}^{n} f^k(x_0, y_0)$ is ϵ-dense, then $\bigcup_{k=0}^{n} f^k(x_0, y)$ is ϵ-dense.
(c) Prove that every forward orbit is dense, i.e. f is minimal.

2.3 Topological mixing

A topological dynamical system $f: X \to X$ is *topologically mixing* if for any two non-empty open sets $U, V \subset X$, there is $N > 0$ such that $f^n(U) \cap V \neq \emptyset$ for $n \geqslant N$. Topological mixing implies topological transitivity by Proposition 2.2.1, but not vice versa. For example, an irrational circle rotation is minimal and therefore topologically transitive, but not topologically mixing (Exercise 2.3.1).

The following propositions establish topological mixing for some of the examples from Chapter 1.

PROPOSITION 2.3.1. *Any hyperbolic toral automorphism $A: \mathbb{T}^2 \to \mathbb{T}^2$ is topologically mixing.*

Proof. By Exercise 1.7.3, for each $x \in \mathbb{T}^2$, the unstable manifold $W^u(x)$ of A is dense in $\mathbb{T}^2$. Thus for every $\epsilon > 0$, the collection of balls of radius ϵ centered at points of $W^u(x)$ covers $\mathbb{T}^2$. By compactness, a finite subcollection of these balls also covers $\mathbb{T}^2$. Hence there is a bounded segment $S_0 \subset W^u(x)$ whose ϵ-neighborhood covers $\mathbb{T}^2$. Since group translations of $\mathbb{T}^2$ are isometries, the ϵ-neighborhood of any translate $L_g S_0 = g + S_0 \subset W^u(g + x)$ covers $\mathbb{T}^2$. To summarize: for every $\epsilon > 0$, there is $L(\epsilon) > 0$ such that every segment S of length $L(\epsilon)$ in an unstable manifold is ϵ-dense in $\mathbb{T}^2$, i.e. $d(y, S) \leqslant \epsilon$ for every $y \in \mathbb{T}^2$.

Let U and V be non-empty open sets in $\mathbb{T}^2$. Choose $y \in V$ and $\epsilon > 0$ such that $\overline{B(y, \epsilon)} \subset V$. The open set U contains a segment of length $\delta > 0$ in some unstable manifold $W^u(x)$. Let λ, $|\lambda| > 1$, be the expanding eigenvalue of A, and choose $N > 0$ such that $|\lambda|^N \delta \geqslant L(\epsilon)$. Then for any $n \geqslant N$, the image $A^n U$ contains a segment of length at least $L(\epsilon)$ in some unstable manifold, so $A^n U$ is ϵ-dense in $\mathbb{T}^2$ and therefore intersects V. □

PROPOSITION 2.3.2. *The full two-sided shift (Σ_m, σ) and the full one-sided shift (Σ_m^+, σ) are topologically mixing.*

Proof. Recall from §1.4 that the topology on Σ_m has a basis consisting of open metric balls $B(\omega, 2^{-l}) = \{\omega' : \omega'_i = \omega_i, |i| \leqslant l\}$. Thus it suffices to show that for any two balls $B(\omega, 2^{-l_1})$ and $B(\omega', 2^{-l_2})$, there is $N > 0$ such that $\sigma^n B(\omega, 2^{-l_1}) \cap B(\omega', 2^{-l_2}) \neq \emptyset$ for $n \geqslant N$. Elements of $\sigma^n B(\omega, 2^{-l_1})$ are sequences with specified values in the places $-n - l_1, \dots, -n + l_1$. Therefore the intersection is non-empty when $-n + l_1 < -l_2$, i.e. $n \geqslant N = l_1 + l_2 + 1$. This proves that (Σ_m, σ) is topologically mixing; the proof for (Σ_m^+, σ) is an exercise (Exercise 2.3.4). □

COROLLARY 2.3.3. *The horseshoe (H, f) (§1.8) is topologically mixing.*

Proof. The horseshoe (H, f) is topologically conjugate to the full two-shift (Σ_2, σ) (see Exercise 1.8.3). □

PROPOSITION 2.3.4. *The solenoid (S, F) is topologically mixing.*

Proof. Recall (Exercise 1.9.2) that (S, F) is topologically conjugate to (Φ, α), where

$$\Phi = \left\{(\phi_i) \colon \phi_i \in S^1,\ \phi_i = 2\phi_{i+1},\ \forall i\right\} \subset \prod_{i=0}^{\infty} S^1 = \mathbb{T}^\infty$$

and $(\phi_0, \phi_1, \phi_2, \dots) \overset{\alpha}{\mapsto} (2\phi_0, \phi_0, \phi_1, \dots)$. Thus it suffices to show that (Φ, α) is topologically mixing. The topology in $\mathbb{T}^\infty$ has a basis consisting of open sets $\prod_{i=0}^{\infty} I_k$, where the I_j's are open in S^1 and all but finitely many are equal to S^1. Let $U = (I_0 \times I_1 \times \cdots \times I_k \times S^1 \times S^1 \cdots) \cap \Phi$ and $V = (J_0 \times J_1 \times \cdots \times J_l \times S^1 \times S^1 \cdots) \cap \Phi$ be non-empty open sets from this basis. Choose $m > 0$ so that $2^m I_0 = S^1$. Then for $n > m + l$, the first $n - m$ components of

$$\alpha^n(U) = \left(2^n I_0 \times 2^{n-1} I_0 \times \cdots \times I_0 \times I_1 \times \cdots \times I_k \times S^1 \times S^1 \cdots\right) \cap \Phi$$

are S^1, so $\alpha^n(U) \cap V \neq \emptyset$. □

Exercise 2.3.1. Show that a circle rotation is not topologically mixing. Show that an isometry is not topologically mixing if there is more than one point in the space.

Exercise 2.3.2. Show that expanding endomorphisms of S^1 are topologically mixing (see §1.3).

Exercise 2.3.3. Show that a factor of a topologically mixing system is also topologically mixing.

Exercise 2.3.4. Prove that (Σ_m^+, σ) is topologically mixing.

2.4 Expansiveness

A homeomorphism $f: X \to X$ is *expansive* if there is $\delta > 0$ such that for any two distinct points $x, y \in X$, there is some $n \in \mathbb{Z}$ such that $d(f^n(x), f^n(y)) \geqslant \delta$. A non-invertible continuous map $f: X \to X$ is *positively expansive* if there is $\delta > 0$ such that for any two distinct points $x, y \in X$, there is some $n \geqslant 0$ such that $d(f^n(x), f^n(y)) \geqslant \delta$. Any number $\delta > 0$ with this property is called an *expansiveness constant* for f.

Among the examples from Chapter 1, the following are expansive (or positively expansive): the circle endomorphisms E_m, $|m| \geqslant 2$; the full and one-sided shifts; the hyperbolic toral automorphisms; the horseshoe; and the solenoid (Exercise 2.4.2). For sufficiently large values of the parameter μ, the quadratic map q_μ is expansive on the invariant set Λ_μ. Circle rotations, group translations, and other equicontinuous homeomorphisms (see §2.7) are not expansive.

PROPOSITION 2.4.1. *Let f be a homeomorphism of an infinite compact metric space X. Then for every $\epsilon > 0$ there are distinct points $x_0, y_0 \in X$ such that $d(f^n(x_0), f^n(y_0)) \leqslant \epsilon$ for all $n \in \mathbb{N}_0$.*

Proof (King, 1990). Fix $\epsilon > 0$. Let E be the set of natural numbers m for which there is a pair $x, y \in X$ such that

$$d(x, y) \geqslant \epsilon \quad \text{and} \quad d(f^n(x), f^n(y)) \leqslant \epsilon \quad \text{for} \quad n = 1, \ldots, m. \tag{2.1}$$

Let $M = \sup E$ if $E \neq \emptyset$ and $M = 0$ if $E = \emptyset$.

If $M = \infty$, then for every $m \in \mathbb{N}$ there is a pair x_m, y_m satisfying (2.1). By compactness, there is a sequence $m_k \to \infty$ such that the limits

$$\lim_{k\to\infty} x_{m_k} = x', \qquad \lim_{k\to\infty} y_{m_k} = y'$$

exist. By (2.1), $d(x', y') \geqslant \epsilon$ and, since f^j is continuous,

$$d(f^j(x'), f^j(y')) = \lim_{k\to\infty} d\big(f^j(x_{m_k}), f^j(y_{m_k})\big) \leqslant \epsilon$$

for every $j \in \mathbb{N}$. Thus $x_0 = f(x'), y_0 = f(y')$ are the desired points.

Suppose now that M is finite. Since any finite collection of iterates of f is equicontinuous, there is $\delta > 0$ such that if $d(x, y) < \delta$, then $d(f^n(x), f^n(y)) < \epsilon$ for $0 \leqslant n \leqslant M$; the definition of M then implies that $d(f^{-1}(x), f^{-1}(y)) < \epsilon$. By induction, we conclude that $d(f^{-j}(x), f^{-j}(y)) < \epsilon$ for $j \in \mathbb{N}$ whenever $d(x, y) < \delta$. By compactness, there is a finite collection $\mathcal{B}$ of open $\delta/2$ balls that covers X. Let K be the cardinality of $\mathcal{B}$. Since X is infinite, we can choose a set $W \subset X$ consisting of $K+1$ distinct points. By the pigeon-hole principle, for each $j \in \mathbb{Z}$, there are distinct points

$a_j, b_j \in W$ such that $f^j(a_j)$ and $f^j(b_j)$ belong to the same ball $B_j \in \mathcal{B}$, so $d(f^j(a_j), f^j(b_j)) < \delta$. Thus $d(f^n(a_j), f^n(b_j)) < \epsilon$ for $-\infty < n \leqslant j$. Since W is finite, there are distinct $x_0, y_0 \in W$ such that

$$a_j = x_0 \quad \text{and} \quad b_j = y_0$$

for infinitely many positive j and hence $d(f^n(x_0), f^n(y_0)) < \epsilon$ for all $n \geqslant 0$. □

COROLLARY 2.4.2. *Let f be an expansive homeomorphism of an infinite compact metric space X. Then there are $x_0, y_0 \in X$ such that $d(f^n(x_0), f^n(y_0)) \to 0$ as $n \to \infty$.*

Proof. Let $\delta > 0$ be an expansiveness constant for f. By Proposition 2.4.1, there are $x_0, y_0 \in X$ such that $d(f^n(x_0), f^n(y_0)) < \delta$ for all $n \in \mathbb{N}$. Suppose $d(f^n(x_0), f^n(y_0)) \nrightarrow 0$. Then by compactness, there is a sequence $n_k \to \infty$ such that $f^{n_k}(x_0) \to x'$ and $f^{n_k}(y_0) \to y'$ with $x' \neq y'$. Then $f^{n_k+m}(x_0) \to f^m(x')$ and $f^{n_k+m}(y_0) \to f^m(y')$ for any $m \in \mathbb{Z}$. For k large, $n_k + m > 0$ and hence $d(f^m(x'), f^m(y')) \leqslant \delta$ for all $m \in \mathbb{Z}$, which contradicts expansiveness. □

Exercise 2.4.1. Prove that every isometry of a compact metric space to itself is surjective and therefore a homeomorphism.

Exercise 2.4.2. Show that the expanding circle endomorphisms E_m, $|m| \geqslant 2$, the full one- and two-sided shifts, the hyperbolic toral automorphisms, the horseshoe, and the solenoid are expansive, and compute expansiveness constants for each.

2.5 Topological entropy

Topological entropy is the exponential growth rate of the number of essentially different orbit segments of length n. It is a topological invariant that measures the complexity of the orbit structure of a dynamical system. Topological entropy is analogous to measure-theoretic entropy, which we introduce in Chapter 9.

Let (X, d) be a compact metric space, and $f: X \to X$ a continuous map. For each $n \in \mathbb{N}$, the function

$$d_n(x, y) = \max_{0 \leqslant k \leqslant n-1} d(f^k(x), f^k(y))$$

measures the maximum distance between the first n iterates of x and y. Each d_n is a metric on X, $d_n \geqslant d_{n-1}$, and $d_1 = d$. Moreover, the d_i are all equivalent metrics in the sense that they induce the same topology on X (Exercise 2.5.1).

Fix $\epsilon > 0$. A subset $A \subset X$ is (n, ϵ)-*spanning* if for every $x \in X$ there is $y \in A$ such that $d_n(x, y) < \epsilon$. By compactness, there are finite (n, ϵ)-spanning sets. Let $\operatorname{span}(n, \epsilon, f)$ be the minimum cardinality of an (n, ϵ)-spanning set.

A subset $A \subset X$ is (n, ϵ)-*separated* if any two distinct points in A are at least ϵ apart in the metric d_n. Any (n, ϵ)-separated set is finite. Let $\operatorname{sep}(n, \epsilon, f)$ be the maximum cardinality of an (n, ϵ)-separated set.

Let $\operatorname{cov}(n, \epsilon, f)$ be the minimum cardinality of a covering of X by sets of d_n-diameter less than ϵ (the diameter of a set is the supremum of distances between pairs of points in the set). Again, by compactness, $\operatorname{cov}(n, \epsilon, f)$ is finite.

The quantities $\operatorname{span}(n, \epsilon, f)$, $\operatorname{sep}(n, \epsilon, f)$, and $\operatorname{cov}(n, \epsilon, f)$ count the number of orbit segments of length n that are distinguishable at scale ϵ. These quantities are related by the following lemma.

LEMMA 2.5.1. $\operatorname{cov}(n, 2\epsilon, f) \leqslant \operatorname{span}(n, \epsilon, f) \leqslant \operatorname{sep}(n, \epsilon, f) \leqslant \operatorname{cov}(n, \epsilon, f)$.

Proof. Suppose A is an (n, ϵ)-spanning set of minimum cardinality. Then the open balls of radius ϵ centered at the points of A cover X. By compactness, there exists $\epsilon_1 < \epsilon$ such that the balls of radius ϵ_1 centered at the points of A also cover X. Their diameter is $2\epsilon_1 < 2\epsilon$, so $\operatorname{cov}(n, 2\epsilon, f) \leqslant \operatorname{span}(n, \epsilon, f)$. The other inequalities are left as an exercise (Exercise 2.5.2). □

Let

$$h_\epsilon(f) = \overline{\lim_{n\to\infty}} \frac{1}{n} \log(\operatorname{cov}(n, \epsilon, f)). \tag{2.2}$$

The quantity $\operatorname{cov}(n, \epsilon, f)$ increases monotonically as ϵ decreases, so $h_\epsilon(f)$ does as well. Thus the limit

$$h_{\text{top}} = h(f) = \lim_{\epsilon\to 0^+} h_\epsilon(f)$$

exists; it is called the *topological entropy* of f. The inequalities in Lemma 2.5.1 imply that equivalent definitions of $h(f)$ can be given using $\operatorname{span}(n, \epsilon, f)$ or $\operatorname{sep}(n, \epsilon, f)$, i.e.

$$h(f) = \lim_{\epsilon\to 0^+} \overline{\lim_{n\to\infty}} \frac{1}{n} \log(\operatorname{span}(n, \epsilon, f)) \tag{2.3}$$

$$= \lim_{\epsilon\to 0^+} \overline{\lim_{n\to\infty}} \frac{1}{n} \log(\operatorname{sep}(n, \epsilon, f)). \tag{2.4}$$

LEMMA 2.5.2. *The limit* $\lim_{n\to\infty} \frac{1}{n} \log(\operatorname{cov}(n, \epsilon, f)) = h_\epsilon(f)$ *exists and is finite.*

Proof. Let U have d_m-diameter less than ϵ and V have d_n-diameter less than ϵ. Then $U \cap f^{-m}(V)$ has d_{m+n}-diameter less than ϵ. Hence

$$\operatorname{cov}(m+n, \epsilon, f) \leqslant \operatorname{cov}(m, \epsilon, f) \cdot \operatorname{cov}(n, \epsilon, f),$$

so the sequence $a_n = \log(\operatorname{cov}(n, \epsilon, f)) \geqslant 0$ is subadditive. A standard lemma from calculus implies that a_n/n converges to a finite limit as $n \to \infty$ (Exercise 2.5.3). □

It follows from Lemmas 2.5.1 and 2.5.2 that the lim sups in Eqs. (2.2), (2.3), and (2.4) are finite. Moreover, the corresponding lim infs are finite, and

$$h(f) = \lim_{\epsilon \to 0^+} \varliminf_{n \to \infty} \frac{1}{n} \log(\operatorname{cov}(n, \epsilon, f)) \tag{2.5}$$

$$= \lim_{\epsilon \to 0^+} \varliminf_{n \to \infty} \frac{1}{n} \log(\operatorname{span}(n, \epsilon, f)) \tag{2.6}$$

$$= \lim_{\epsilon \to 0^+} \varliminf_{n \to \infty} \frac{1}{n} \log(\operatorname{sep}(n, \epsilon, f)). \tag{2.7}$$

The topological entropy is either $+\infty$ or a finite non-negative number. There are dramatic differences between dynamical systems with positive entropy and dynamical systems with zero entropy. Any isometry has zero topological entropy (Exercise 2.5.4). In the next section we show that topological entropy is positive for several of the examples from Chapter 1.

PROPOSITION 2.5.3. *The topological entropy of a continuous map $f: X \to X$ does not depend on the choice of a particular metric generating the topology of X.*

Proof. Suppose d and d' are metrics generating the topology of X. For $\epsilon > 0$, let $\delta(\epsilon) = \sup\{d'(x, y) : d(x, y) \leqslant \epsilon\}$. By compactness, $\delta(\epsilon) \to 0$ as $\epsilon \to 0$. If U is a set of d_n-diameter less than ϵ, then U has d'_n-diameter at most $\delta(\epsilon)$. Thus $\operatorname{cov}'(n, \delta(\epsilon), f) \leqslant \operatorname{cov}(n, \epsilon, f)$, where cov and cov′ correspond to the metrics d and d', respectively. Hence

$$\lim_{\delta \to 0^+} \varliminf_{n \to \infty} \frac{1}{n} \log(\operatorname{cov}'(n, \delta, f)) \leqslant \lim_{\epsilon \to 0^+} \varliminf_{n \to \infty} \frac{1}{n} \log(\operatorname{cov}(n, \epsilon, f)).$$

Interchanging d and d' gives the opposite inequality. □

COROLLARY 2.5.4. *Topological entropy is an invariant of topological conjugacy.*

Proof. Suppose $f: X \to X$ and $g: Y \to Y$ are topologically conjugate dynamical systems, with conjugacy $\phi: Y \to X$. Let d be a metric on X. Then

$d'(y_1, y_2) = d(\phi(y_1), \phi(y_2))$ is a metric on Y generating the topology of Y. Since ϕ is an isometry of (X, d) and (Y, d'), and the entropy is independent of the metric by Proposition 2.5.3, it follows that $h(f) = h(g)$. □

PROPOSITION 2.5.5. *Let $f: X \to X$ be a continuous map of a compact metric space X.*

1. *$h(f^m) = m \cdot h(f)$ for $m \in \mathbb{N}$.*
2. *If f is invertible, then $h(f^{-1}) = h(f)$. Thus $h(f^m) = |m| \cdot h(f)$ for all $m \in \mathbb{Z}$.*
3. *If A_i, $i = 1, \ldots, k$ are closed (not necessarily disjoint) forward f-invariant subsets of X, whose union is X, then*
$$h(f) = \max_{1 \leqslant i \leqslant k} h(f|A_i).$$
In particular, if A is a closed forward invariant subset of X, then $h(f|A) \leqslant h(f)$.

Proof

1. Note that
$$\max_{0 \leqslant i < n} d(f^{mi}(x), f^{mi}(y)) \leqslant \max_{0 \leqslant j < mn} d(f^j(x), f^j(y)).$$
Thus $\operatorname{span}(n, \epsilon, f^m) \leqslant \operatorname{span}(mn, \epsilon, f)$, so $h(f^m) \leqslant m \cdot h(f)$. Conversely, for $\epsilon > 0$, there is $\delta(\epsilon) > 0$ such that $d(x, y) < \delta(\epsilon)$ implies that $d(f^i(x), f^i(y)) < \epsilon$ for $i = 0, \ldots, m$. Then $\operatorname{span}(n, \delta(\epsilon), f^m) \geqslant \operatorname{span}(mn, \epsilon, f)$, so $h(f^m) \geqslant m \cdot h(f)$.
2. The nth image of an (n, ϵ)-separated set for f is (n, ϵ)-separated for f^{-1}, and vice versa.
3. Any (n, ϵ)-separated set in A_i is (n, ϵ)-separated in X, so $h(f|A_i) \leqslant h(f)$. Conversely, the union of (n, ϵ)-spanning sets for the A_i's is an (n, ϵ)-spanning set for X. Thus if $\operatorname{span}_i(n, \epsilon, f)$ is the minimum cardinality of an (n, ϵ)-spanning subset of A_i, then
$$\operatorname{span}(n, \epsilon, f) \leqslant \sum_{i=1}^{k} \operatorname{span}_i(n, \epsilon, f) \leqslant k \cdot \max_{1 \leqslant i \leqslant k} \operatorname{span}_i(n, \epsilon, f).$$
Therefore
$$\overline{\lim_{n\to\infty}} \frac{1}{n} \log(\operatorname{span}(n, \epsilon, f)) \leqslant \overline{\lim_{n\to\infty}} \frac{1}{n} \log k + \overline{\lim_{n\to\infty}} \frac{1}{n} \log(\max_{1 \leqslant i \leqslant k} \operatorname{span}_i(n, \epsilon, f))$$
$$= 0 + \max_{1 \leqslant i \leqslant k} \overline{\lim_{n\to\infty}} \frac{1}{n} \log(\operatorname{span}_i(n, \epsilon, f)).$$
The result follows by taking the limit as $\epsilon \to 0$. □

PROPOSITION 2.5.6. *Let (X, d^X) and (Y, d^Y) be compact metric spaces and $f\colon X \to X$, $g\colon Y \to Y$ continuous maps. Then:*

1. *$h(f \times g) = h(f) + h(g)$; and*
2. *if g is a factor of f (or equivalently, f is an extension of g), then $h(f) \geqslant h(g)$.*

Proof. To prove (1), note that the metric $d((x, y), (x', y')) = \max\{d^X(x, x'), d^Y(y, y')\}$ generates the product topology on $X \times Y$, and

$$d_n((x, y), (x', y')) = \max\left\{d_n^X(x, x'), d_n^Y(y, y')\right\}.$$

If $U \subset X$ and $V \subset Y$ have diameter less than ϵ, then $U \times V$ has d-diameter less than ϵ. Hence

$$\operatorname{cov}(n, \epsilon, f \times g) \leqslant \operatorname{cov}(n, \epsilon, f) \cdot \operatorname{cov}(n, \epsilon, g),$$

so $h(f \times g) \leqslant h(f) + h(g)$. On the other hand, if $A \subset X$ and $B \subset Y$ are (n, ϵ)-separated, then $A \times B$ is (n, ϵ)-separated for d. Hence

$$\operatorname{sep}(n, \epsilon, f \times g) \geqslant \operatorname{sep}(n, \epsilon, f) \cdot \operatorname{sep}(n, \epsilon, g),$$

so, by (2.7), $h(f \times g) \geqslant h(f) + h(g)$.

The proof of (2) is left as an exercise (Exercise 2.5.5). □

PROPOSITION 2.5.7. *Let (X, d) be a compact metric space, and $f\colon X \to X$ an expansive homeomorphism with expansiveness constant δ. Then $h(f) = h_\epsilon(f)$ for any $\epsilon < \delta$.*

Proof. Fix γ and ϵ with $0 < \gamma < \epsilon < \delta$. We will show that $h_{2\gamma}(f) = h_\epsilon(f)$. By monotonicity, it suffices to show that $h_{2\gamma}(f) \leqslant h_\epsilon(f)$.

By expansiveness, for distinct points x and y, there is some $i \in \mathbb{Z}$ such that $d(f^i(x), f^i(y)) \geqslant \delta > \epsilon$. Since the set $\{(x, y) \in X \times X\colon d(x, y) \geqslant \gamma\}$ is compact, there is $k = k(\gamma, \epsilon) \in \mathbb{N}$ such that if $d(x, y) \geqslant \gamma$, then $d(f^i(x), f^i(y)) > \epsilon$ for some $|i| \leqslant k$. Thus if A is an (n, γ)-separated set, then $f^{-k}(A)$ is $(n + 2k, \epsilon)$-separated. Hence, by Lemma 2.5.1, $h_\epsilon(f) \geqslant h_{2\gamma}(f)$. □

REMARK 2.5.8. *The topological entropy of a continuous (semi)flow can be defined as the entropy of the time 1 map. Alternatively, it can be defined using the analog d_T, $T > 0$, of the metrics d_n. The two definitions are equivalent because of the equicontinuity of the family of time-t maps, $t \in [0, 1]$.*

Exercise 2.5.1. Let (X, d) be a compact metric space. Show that the metrics d_i all induce the same topology on X.

Exercise 2.5.2. Prove the remaining inequalities in Lemma 2.5.1.

Exercise 2.5.3. Let $\{a_n\}$ be a subadditive sequence of non-negative real numbers, i.e. $0 \leqslant a_{m+n} \leqslant a_m + a_n$ for all $m, n \geqslant 0$. Show that $\lim_{n\to\infty} a_n/n = \inf_{n\geqslant 0} a_n/n$.

Exercise 2.5.4. Show that the topological entropy of an isometry is zero.

Exercise 2.5.5. Let $g\colon Y \to Y$ be a factor of $f\colon X \to X$. Prove that $h(f) \geqslant h(g)$.

Exercise 2.5.6. Let Y and Z be compact metric spaces, $X = Y \times Z$, and π be the projection to Y. Suppose $f\colon X \to X$ is an isometric extension of a continuous map $g\colon Y \to Y$, i.e. $\pi \circ f = g \circ \pi$ and $d(f(x_1), f(x_2)) = d((x_1), (x_2))$ for all $x_1, x_2 \in Y$ with $\pi(x_1) = \pi(x_2)$. Prove that $h(f) = h(g)$.

Exercise 2.5.7. Prove that the topological entropy of a continuously differentiable map of a compact manifold is finite.

2.6 Topological entropy for some examples

In this section, we compute the topological entropy for some of the examples from Chapter 1.

PROPOSITION 2.6.1. *Let $\tilde{A}$ be a 2×2 integer matrix with determinant 1 and eigenvalues λ, λ^{-1}, with $|\lambda| > 1$, and let $A\colon \mathbb{T}^2 \to \mathbb{T}^2$ be the associated hyperbolic toral automorphism. Then $h(A) = \log|\lambda|$.*

Proof. The natural projection $\pi : \mathbb{R}^2 \to \mathbb{R}^2/\mathbb{Z}^2 = \mathbb{T}^2$ is a local homeomorphism and $\pi\tilde{A} = A\pi$. Any metric $\tilde{d}$ on $\mathbb{R}^2$ invariant under integer translations induces a metric d on $\mathbb{T}^2$, where $d(x, y)$ is the $\tilde{d}$-distance between the sets $\pi^{-1}(x)$ and $\pi^{-1}(y)$. For these metrics π is a local isometry.

Let v_1, v_2 be eigenvectors of A with (Euclidean) length 1 corresponding to the eigenvalues λ, λ^{-1}. For $x, y \in \mathbb{R}^2$, write $x - y = a_1v_1 + a_2v_2$ and define $\tilde{d}(x, y) = \max(|a_1|, |a_2|)$. This is a translation-invariant metric on $\mathbb{R}^2$. A $\tilde{d}$-ball of radius ϵ is a parallelogram whose sides are of (Euclidean) length 2ϵ and parallel to v_1 and v_2. In the metric $\tilde{d}_n$ (defined for $\tilde{A}$) a ball of radius ϵ is a parallelogram with side length $2\epsilon|\lambda|^{-n}$ in the v_1-direction and 2ϵ in the v_2-direction. In particular, the Euclidean area of a $\tilde{d}_n$-ball of radius ϵ is not greater than $4\epsilon^2|\lambda|^{-n}$. Since the induced metric d on $\mathbb{T}^2$ is locally isometric to $\tilde{d}$, we conclude that for sufficiently small ϵ, the Euclidean area of a d_n-ball of radius ϵ in $\mathbb{T}^2$ is at most $4\epsilon^2|\lambda|^{-n}$. It follows that the minimal number of balls of d_n-radius ϵ needed to cover $\mathbb{T}^2$ is at least

$$\frac{\text{area}(\mathbb{T}^2)}{4\epsilon^2|\lambda|^{-n}} = \frac{|\lambda|^n}{4\epsilon^2}.$$

Since a set of diameter ϵ is contained in an open ball of radius ϵ, we conclude that $\text{cov}(n, \epsilon, A) \geqslant |\lambda|^n/4\epsilon^2$. Thus $h(A) \geqslant \log|\lambda|$.

Conversely, since the closed $\tilde{d}_n$-balls are parallelograms, there is a tiling of the plane by ϵ-balls whose interiors are disjoint. The Euclidean area of such a ball is $C\epsilon^2|\lambda|^{-n}$, where C depends on the angle between v_1 and v_2. For small enough ϵ, any ϵ-ball that intersects the unit square $[0,1] \times [0,1]$ is entirely contained in the larger square $[-1,2] \times [-1,2]$. Therefore the number of the balls that intersect the unit square does not exceed the area of the larger square divided by the area of a $\tilde{d}_n$-ball of radius ϵ. Thus the torus can be covered by $9\lambda^n/(C\epsilon^2)$ closed d_n-balls of radius ϵ. It follows that $\text{cov}(n, 2\epsilon, A) \leqslant 9\lambda^n/(C\epsilon^2)$, so $h(A) \leqslant \log|\lambda|$. □

To establish the corresponding result in higher dimensions, we need some results from linear algebra. Let B be a $k \times k$ complex matrix. If λ is an eigenvalue of B, let

$$V_\lambda = \{v \in \mathbb{C}^k \colon (B - \lambda I)^i v = 0 \text{ for some } i \in \mathbb{N}\}.$$

If B is real and γ is a real eigenvalue, let

$$V_\gamma^{\mathbb{R}} = \mathbb{R}^k \cap V_\gamma = \{v \in \mathbb{R}^k \colon (B - \gamma I)^i v = 0 \text{ for some } i \in \mathbb{N}\}.$$

If B is real and $\lambda, \bar{\lambda}$ is a pair of complex eigenvalues, let

$$V_{\lambda,\bar{\lambda}}^{\mathbb{R}} = \mathbb{R}^k \cap (V_\lambda \oplus V_{\bar{\lambda}}).$$

These spaces are called *generalized eigenspaces.*

LEMMA 2.6.2. *Let B be a $k \times k$ complex matrix and λ an eigenvalue of B. Then for every $\delta > 0$ there is $C(\delta) > 0$ such that*

$$C(\delta)^{-1}(|\lambda| - \delta)^n \|v\| \leqslant \|B^n v\| \leqslant C(\delta)(|\lambda| + \delta)^n \|v\|$$

for every $n \in \mathbb{N}$ and every $v \in V_\lambda$.

Proof. It suffices to prove the lemma for a Jordan block. Thus without loss of generality, we assume that B has λ's on the diagonal, 1's above and 0's elsewhere. In this setting, $V_\lambda = \mathbb{C}^k$ and in the standard basis $e_1, \ldots, e_k$, we have $Be_1 = \lambda e_1$ and $Be_i = \lambda e_i + e_{i-1}$, for $i = 2, \ldots, k$. For $\delta > 0$, consider the basis $e_1, \delta e_2, \delta^2 e_3, \dots, \delta^{k-1} e_k$. In this basis the linear map B is represented by the matrix

$$B_\delta = \begin{pmatrix} \lambda & \delta & & & \\ & \lambda & \delta & & \\ & & \ddots & \ddots & \\ & & & \lambda & \delta \\ & & & & \lambda \end{pmatrix}.$$

Observe that $B_\delta = \lambda I + \delta A$ with $\|A\| \leqslant 1$, where $\|A\| = \sup_{v \neq 0} \|Av|/\|v\|$. Therefore

$$(|\lambda| - \delta)^n \|v\| \leqslant \|B_\delta^n v\| \leqslant (|\lambda| + \delta)^n \|v\|.$$

Since B_δ is conjugate to B, there is a constant $C(\delta) > 0$ that bounds the distortion of the change of basis. □

LEMMA 2.6.3. *Let B be a $k \times k$ real matrix and λ an eigenvalue of B. Then for every $\delta > 0$ there is $C(\delta) > 0$ such that*

$$C(\delta)^{-1}(|\lambda| - \delta)^n \|v\| \leqslant \|B^n v\| \leqslant C(\delta)(|\lambda| + \delta)^n \|v\|$$

for every $n \in \mathbb{N}$ and every $v \in V_\lambda$ (if $\lambda \in \mathbb{R}$) or every $v \in V_{\lambda,\bar{\lambda}}$ (if $\lambda \notin \mathbb{R}$).

Proof. If λ is real, then the result follows from Lemma 2.6.2. If λ is complex, then the estimates for V_λ and $V_{\bar{\lambda}}$ from Lemma 2.6.2 imply a similar estimate for $V_{\lambda,\bar{\lambda}}$, with a new constant $C(\delta)$ depending on the angle between V_λ and $V_{\bar{\lambda}}$ and the constants in the estimates for V_λ and $V_{\bar{\lambda}}$ (since $|\lambda| = |\bar{\lambda}|$). □

PROPOSITION 2.6.4. *Let $\tilde{A}$ be a $k \times k$ integer matrix with determinant 1 and with eigenvalues $\alpha_1, \dots, \alpha_k$, where $|\alpha_1| \geqslant |\alpha_2| \geqslant \dots \geqslant |\alpha_m| > 1 > |\alpha_{m+1}| \geqslant \dots \geqslant |\alpha_k|$. Let $A\colon \mathbb{T}^k \to \mathbb{T}^k$ be the associated hyperbolic toral automorphism. Then*

$$h(A) = \sum_{i=1}^{m} \log |\alpha_i|.$$

Proof. Let $\gamma_1, \dots, \gamma_j$ be the distinct real eigenvalues of $\tilde{A}$, and $\lambda_1, \bar{\lambda}_1, \dots, \lambda_l, \bar{\lambda}_l$ be the distinct complex eigenvalues of $\tilde{A}$. Then

$$\mathbb{R}^k = \bigoplus_{i=1}^{j} V_{\gamma_i} \oplus \bigoplus_{i=1}^{l} V_{\lambda_i,\bar{\lambda}_i},$$

so any vector $v \in \mathbb{R}^k$ can be decomposed uniquely as $v = v_1 + \cdots + v_{j+l}$ with v_i in the corresponding generalized eigenspace. Given $x, y \in \mathbb{R}^k$, let $v = x - y$, and define $\tilde{d}(x, y) = \max(|v_1|, \ldots, |v_{j+l}|)$. This is a translation invariant metric on $\mathbb{R}^k$, and therefore descends to a metric on $\mathbb{T}^k$. Now, using Lemma 2.6.3, the proposition follows by an argument similar to the one in the proof of Proposition 2.6.1 (Exercise 2.6.3). □

The next example we consider is the solenoid from §1.9.

PROPOSITION 2.6.5. *The topological entropy of the solenoid map $F: S \to S$ is* $\log 2$.

Proof. Recall from §1.9 that F is topologically conjugate to the automorphism $\alpha: \Phi \to \Phi$, where

$$\Phi = \left\{(\phi_i)_{i=0}^{\infty} : \phi_i \in [0, 1),\ \phi_i = 2\phi_{i+1} \bmod 1\right\},$$

and α is coordinatewise multiplication by 2 (mod 1). Thus $h(F) = h(\alpha)$. Let $|x - y|$ denote the distance on $S^1 = [0, 1] \bmod 1$. The distance function

$$d(\phi, \phi') = \sum_{n=0}^{\infty} \frac{1}{2^n} |\phi_n - \phi'_n|$$

generates the topology in Φ introduced in §1.9.

The map $\pi: \Phi \to S^1$, $(\phi_i)_{i=0}^{\infty} \mapsto \phi_0$, is a semiconjugacy from α to E_2. Hence $h(\alpha) \geqslant h(E_2) = \log 2$ (Exercise 2.6.1). We will establish the inequality $h(\alpha) \leqslant \log 2$ by constructing an (n, ϵ)-spanning set.

Fix $\epsilon > 0$ and choose $k \in \mathbb{N}$ such that $2^{-k} < \epsilon/2$. For $n \in \mathbb{N}$, let $A_n \subset \Phi$ consist of the 2^{n+2k} sequences $\psi^j = (\psi_i^j)$, where $\psi_i^j = j \cdot 2^{-(n+k+i)} \bmod 1$, $j = 0, \ldots, 2^{n+2k} - 1$. We claim that A_n is (n, ϵ)-spanning. Let $\phi = (\phi_i)$ be a point in Φ. Choose $j \in \{0, \ldots, 2^{n+2k} - 1\}$ so that $|\phi_k - j \cdot 2^{-(n+2k)}| \leqslant 2^{-(n+2k+1)}$. Then $|\phi_i - \psi_i^j| \leqslant 2^{k-i} 2^{-(n+2k+1)}$, for $0 \leqslant i \leqslant k$. It follows that for $0 \leqslant m \leqslant n$,

$$\begin{aligned} d(\alpha^m \phi, \alpha^m \psi^j) &= \sum_{i=0}^{\infty} \frac{|2^m \phi_i - 2^m \psi_i^j|}{2^i} < \sum_{i=0}^{k} \frac{2^m |\phi_i - \psi_i^j|}{2^i} + \frac{1}{2^k} \\ &< 2^m \sum_{i=0}^{k} \frac{2^{k-i} 2^{-(n+2k+1)}}{2^i} + \frac{1}{2^k} < \frac{1}{2^{k-1}} < \epsilon. \end{aligned}$$

Thus $d_n(\phi, \psi^j) < \epsilon$, so A_n is (n, ϵ)-spanning. Hence

$$h(\alpha) \leqslant \lim_{n \to \infty} \frac{1}{n} \log \operatorname{card} A_n = \log 2 .$$ □

Note that $\alpha\colon \Phi \to \Phi$ is expansive with expansiveness constant 1/3 (Exercise 2.6.4), so by Proposition 2.5.7, $h_\epsilon(\alpha) = h(\alpha)$ for any $\epsilon < 1/3$.

Exercise 2.6.1. Compute the topological entropy of an expanding endomorphism $E_m\colon S^1 \to S^1$.

Exercise 2.6.2. Compute the topological entropy of the full one- and two-sided m-shifts.

Exercise 2.6.3. Finish the proof of Proposition 2.6.4.

Exercise 2.6.4. Prove that the solenoid map (§1.9) is expansive.

2.7 Equicontinuity, distality and proximality[1]

In this section we describe a number of properties related to the asymptotic behavior of the distance between corresponding points on pairs of orbits.

Let $f\colon X \to X$ be a homeomorphism of a compact Hausdorff space. Points $x, y \in X$ are called *proximal* if the closure $\overline{\mathcal{O}((x, y))}$ of the orbit of (x, y) under $f \times f$ intersects the diagonal $\Delta = \{(z, z) \in X \times X : z \in X\}$. Every point is proximal to itself. If two points x and y are not proximal, i.e. if $\overline{\mathcal{O}((x, y))} \cap \Delta = \emptyset$, they are called *distal*. A homeomorphism $f\colon X \to X$ is *distal* if every pair of distinct points $x, y \in X$ is distal. If (X, d) is a compact metric space, then $x, y \in X$ are proximal if there is a sequence $n_k \in \mathbb{Z}$ such that $d(f^{n_k}(x), f^{n_k}(y)) \to 0$ as $k \to \infty$; points $x, y \in X$ are distal if there is $\epsilon > 0$ such that $d(f^n(x), f^n(y)) > \epsilon$ for all $n \in \mathbb{Z}$ (Exercise 2.7.2)

A homeomorphism f of a compact metric space (X, d) is said to be *equicontinuous* if the family of all iterates of f is an equicontinuous family, i.e. for any $\epsilon > 0$, there exists $\delta > 0$ such that $d(x, y) < \delta$ implies that $d(f^n(x), f^n(y)) < \epsilon$ for all $n \in \mathbb{Z}$. An isometry preserves distances and is therefore equicontinuous. Equicontinuous maps share many of the dynamical properties of isometries. The only examples from Chapter 1 that are equicontinuous are the group translations, including circle rotations.

We denote by $f \times f$ the induced action of f in $X \times X$, defined by $f \times f(x, y) = (f(x), f(y))$.

PROPOSITION 2.7.1. *An expansive homeomorphism of an infinite compact metric space is not distal.*

Proof. Exercise 2.7.1. □

[1] Several arguments of this section were conveyed to us by J. Auslander.

PROPOSITION 2.7.2. *Equicontinuous homeomorphisms are distal.*

Proof. Suppose the equicontinuous homeomorphism $f: X \to X$ is not distal. Then there is a pair of proximal points $x, y \in X$, so $d(f^{n_k}(x), f^{n_k}(y)) \to 0$ for some sequence $n_k \in \mathbb{Z}$. Let $x_k = f^{n_k}(x)$ and $y_k = f^{n_k}(y)$. Let $\epsilon = d(x, y)$. Then for any $\delta > 0$, there is some $k \in \mathbb{N}$ such that $d(x_k, y_k) < \delta$, but $d(f^{-n_k}(x_k), f^{-n_k}(y_k)) = \epsilon$, so f is not equicontinuous. □

Distal homeomorphisms are not necessarily equicontinuous. Consider the map $F: \mathbb{T}^2 \to \mathbb{T}^2$ defined by

$$x \mapsto x + \alpha \mod 1$$
$$y \mapsto x + y \mod 1.$$

We view $\mathbb{T}^2$ as the unit square with opposite sides identified and use the metric inherited from the Euclidean metric. To see that this map is distal, let $(x, y), (x', y')$ be distinct points in $\mathbb{T}^2$. If $x \neq x'$, then $d(F^n(x, y), F^n(x', y'))$ is at least $d((x, 0), (x', 0))$, which is constant. If $x = x'$, then $d(F^n(x, y), F^n(x', y')) = d((x, y), (x', y'))$. Therefore the pair $(x, y), (x', y')$ is distal. To see that F is not equicontinuous, let $p = (0, 0)$ and $q = (\delta, 0)$. Then for all n, the difference between the first coordinates of $F^n(p)$ and $F^n(q)$ is δ. The difference between the second coordinates of $F^n(p)$ and $F^n(q)$ is $n\delta$ as long as $n\delta < 1/2$. Therefore there are points that are arbitrarily close together that are moved at least $1/4$ apart, so F is not equicontinuous.

The preceding map is an example of a *distal extension*. Suppose a homeomorphism $g: Y \to Y$ is an extension of a homeomorphism $f: X \to X$ with projection $\pi: Y \to X$. We say that the extension is *distal* if any pair of distinct points $y, y' \in Y$ with $\pi(y) = \pi(y')$ is distal. The map $F: \mathbb{T}^2 \to \mathbb{T}^2$ in the preceding paragraph is a distal extension of a circle rotation, with projection on the first factor as the factor map. A straightforward generalization of the argument in the previous paragraph shows that a distal extension of a distal homeomorphism is distal. Moreover, as we show later in this section, any factor of a distal map is distal. Thus (X_1, f_1) and (X_2, f_2) are distal if and only if $(X_1 \times X_2, f_1 \times f_2)$ is distal.

Similarly, $\pi: Y \to X$ is an *isometric extension* if $d(g(y), g(y')) = d(y, y')$ whenever $\pi(y) = \pi(y')$. The extension $\pi: Y \to X$ is an *equicontinuous extension* if for any $\epsilon > 0$, there exists $\delta > 0$ such that if $\pi(y) = \pi(y')$ and $d(y, y') < \delta$, then $d(g^n(y), g^n(y')) < \epsilon$, for all n. An isometric extension is an equicontinuous extension; an equicontinuous extension is a distal extension.

To prove Theorem 2.7.4 we need the following notion. For a subset $A \subset X$ and a homeomorphism $f: X \to X$, denote by f_A the induced action of

f in the product space X^A (an element z of X^A is a function $z\colon A \to X$, and $f_A(z) = f \circ z$). We say that $A \subset X$ is *almost periodic* if every $z \in X^A$ with range A is an almost periodic point of (X^A, f_A). That is, A is almost periodic if for every finite subset $a_1, \dots, a_n \in A$, and neighborhoods $U_1 \ni a_1, \dots, U_n \ni a_n$, the set $\{k \in \mathbb{Z}\colon f^k(a_i) \in U_i,\ 1 \leqslant i \leqslant n\}$ is syndetic in $\mathbb{Z}$. Every subset of an almost periodic set is an almost periodic set. Note that if x is an almost periodic point of f, then $\{x\}$ is an almost periodic set.

LEMMA 2.7.3. *Every almost periodic set is contained in a maximal almost periodic set.*

Proof. Let A be an almost periodic set, and $\mathcal{C}$ be a collection, totally ordered by inclusion, of almost periodic sets containing A. The set $\bigcup_{C \in \mathcal{C}} C$ is an almost periodic set and a maximal element of $\mathcal{C}$. By Zorn's Lemma there is a maximal almost periodic set containing A. □

THEOREM 2.7.4. *Let f be a homeomorphism of a compact Hausdorff space X. Then every $x \in X$ is proximal to an almost periodic point.*

Proof. If x is an almost periodic point, then we are done, since x is proximal to itself. Suppose x is not almost periodic, and let A be a maximal almost periodic set. By definition, $x \notin A$. Let $z \in X^A$ have range A, and consider $(x, z) \in (X \times X^A)$. Let (x_0, z_0) be an almost periodic point (of $(f \times f_A)$) in $\overline{\mathcal{O}(x, z)}$. Since z is almost periodic, $z \in \overline{\mathcal{O}(z_0)}$. Hence there is $x' \in X$ such that (x', z) is almost periodic and $(x', z) \in \overline{\mathcal{O}(x, z)}$ (Proposition 2.1.1). Therefore $\{x'\} \cup \text{range}(z) = \{x'\} \cup A$ is an almost periodic set. Since A is maximal, $x' \in A$, i.e. x' appears as one of the coordinates of z. It follows that $(x', x') \in \overline{\mathcal{O}(x, x')}$ and x is proximal to x'. □

A homeomorphism f of a compact Hausdorff space X is called *pointwise almost periodic* if every point is almost periodic. By Proposition 2.1.3, this happens if and only if X is a union of minimal sets.

PROPOSITION 2.7.5. *Let f be a distal homeomorphism of a compact Hausdorff space X. Then f is pointwise almost periodic.*

Proof. Let $x \in X$. Then, by Theorem 2.7.4, x is proximal to an almost periodic $y \in X$. Since f is distal, $x = y$ and x is almost periodic. □

PROPOSITION 2.7.6. *A homeomorphism of a compact Hausdorff space is distal if and only if the product system $(X \times X, f \times f)$ is pointwise almost periodic.*

Proof. If f is distal, so is $f \times f$, and hence $f \times f$ is pointwise almost periodic. Conversely, assume that $f \times f$ is pointwise almost periodic, and let $x, y \in X$ be distinct points. If x and y are proximal, then there is z with $(z, z) \in \overline{\mathcal{O}(x, y)}$. Recall that $\overline{\mathcal{O}(x, y)}$ is minimal (Proposition 2.1.3). Since $(x, y) \notin \overline{\mathcal{O}(z, z)}$, we obtain a contradiction. □

COROLLARY 2.7.7. *A factor of a distal homeomorphism f of a compact Hausdorff space X is distal.*

Proof. Let $g \colon Y \to Y$ be a factor of f. Then $f \times f$ is pointwise almost periodic by Proposition 2.7.6. Since $(g \times g)$ is a factor of $f \times f$, it is pointwise almost periodic (Exercise 2.7.5), and hence is distal. □

The class of distal dynamical systems is of special interest because it is closed under factors and isometric extensions. The class of minimal distal systems is the smallest such class of minimal systems: according to Furstenberg's structure theorem (Furstenberg, 1963), every minimal distal homeomorphism (or flow) can be obtained by a (possibly transfinite) sequence of isometric extensions starting with the one-point dynamical system.

Exercise 2.7.1. Prove Proposition 2.7.1.

Exercise 2.7.2. Prove the equivalence of the topological and metric definitions of distal and proximal points at the beginning of this section.

Exercise 2.7.3. Give a non-trivial example of a homeomorphism f of a compact metric space (X, d) such that $d(f^n(x), f^n(y)) \to 0$ as $n \to \infty$ for every pair $x, y \in X$.

Exercise 2.7.4. Show that any infinite closed shift-invariant subset of Σ_m contains a proximal pair of points.

Exercise 2.7.5. Prove that a factor of a pointwise almost periodic system is pointwise almost periodic.

2.8 Applications of topological recurrence to Ramsey Theory[2]

In this section, we establish several Ramsey-type results to illustrate how topological dynamics is applied in combinatorial number theory. One of the main principles of the Ramsey Theory is that a sufficiently rich structure is indestructible by finite partitioning (see Bergelson (1996) for more information on Ramsey Theory). An example of such a statement is van

[2] The exposition in this section follows to a large extent Bergelson (2000).

der Waerden's theorem, which we prove later in this section. We conclude this section by proving a result in Ramsey Theory about infinite-dimensional vector spaces over finite fields.

THEOREM 2.8.1 (van der Waerden). *For each finite partition* $\mathbb{Z} = \bigcup_{k=1}^{m} S_k$, *one of the sets* S_k *contains arbitrarily long (finite) arithmetic progressions.*

We will obtain van der Waerden's theorem as a consequence of a general recurrence property in topological dynamics.

Recall from §1.4 that $\Sigma_m = \{1, 2, \ldots, m\}^{\mathbb{Z}}$ with metric $d(\omega, \omega') = 2^{-k}$, where $k = \min\{|i| : \omega_i \neq \omega'_i\}$, is a compact metric space. The shift $\sigma : \Sigma_m \to \Sigma_m$, $(\sigma\omega)_i = \omega_{i+1}$, is a homeomorphism. A finite partition $\mathbb{Z} = \bigcup_{k=1}^{m} S_k$ can be viewed as a sequence $\xi \in \Sigma_m$ for which $\xi_i = k$ if $i \in S_k$. Let $X = \overline{\bigcup_{i=-\infty}^{\infty} \sigma^i \xi}$ be the orbit closure of ξ under σ, and let $A_k = \{\omega \in X : \omega_0 = k\}$. If $\omega \in A_k$, $\omega' \in X$ and $d(\omega', \omega) < 1$, then $\omega' \in A_k$. Hence if there are integers $p, q \in \mathbb{N}$ and $\omega \in X$ such that $d(\sigma^{ip}\omega, \omega) < 1$ for $0 \leqslant i \leqslant q - 1$, then there is $r \in \mathbb{Z}$ such that $\xi_j = \omega_0$ for $j = r, r + p, \ldots, r + (q-1)p$. Therefore Theorem 2.8.1 follows from the following multiple recurrence property (Exercise 2.8.1).

PROPOSITION 2.8.2. *Let* T *be a homeomorphism of a compact metric space* X. *Then for every* $\epsilon > 0$ *and* $q \in \mathbb{N}$ *there are* $p \in \mathbb{N}$ *and* $x \in X$ *such that* $d(T^{jp}(x), x) < \epsilon$ *for* $0 \leqslant j \leqslant q$.

We will obtain Proposition 2.8.2 as a consequence of a more general statement (Theorem 2.8.3), which has other corollaries useful in combinatorial number theory.

Let $\mathcal{F}$ be the collection of all finite non-empty subsets of $\mathbb{N}$. For $\alpha, \beta \in \mathcal{F}$, we write $\alpha < \beta$ if each element of α is less than each element of β. For a commutative group G, a map $T : \mathcal{F} \to G, \alpha \mapsto T_\alpha$, defines an *IP-system* in G if

$$T_{\{i_1,\ldots,i_k\}} = T_{\{i_1\}} \cdot \ \ldots \ \cdot T_{\{i_k\}}$$

for every $\{i_1, \ldots, i_k\} \in \mathcal{F}$; in particular, if $\alpha, \beta \in \mathcal{F}$ and $\alpha \cap \beta = \emptyset$, then $T_{\alpha\cup\beta} = T_\alpha T_\beta$. Every IP-system T is generated by the elements $T_{\{n\}} \in G$, $n \in \mathbb{N}$.

Let G be a group of homeomorphisms of a topological space X. For $x \in X$, denote by Gx the orbit of x under G. We say that *G acts minimally on X* if for each $x \in X$ the orbit Gx is dense in X.

THEOREM 2.8.3 (Furstenberg–Weiss (Furstenberg and Weiss, 1978)). *Let G be a commutative group acting minimally on a compact topological space X. Then for every non-empty open set $U \subset X$, every $n \in \mathbb{N}$, every $\alpha \in \mathcal{F}$, and any IP-systems $T^{(1)}, \ldots, T^{(n)}$ in G, there is $\beta \in \mathcal{F}$ such that $\alpha < \beta$ and*

$$U \cap T_\beta^{(1)}(U) \cap \cdots \cap T_\beta^{(n)}(U) \neq \emptyset.$$

Proof (Bergelson, 2000). Since G acts minimally and X is compact, there are elements $g_1, \ldots, g_m \in G$ such that $\bigcup_{i=1}^m g_i(U) = X$ (Exercise 2.8.2).

We argue by induction on n. For $n = 1$, let T be an IP-system and $U \subset X$ be open and not empty. Set $V_0 = U$. Define recursively $V_k = T_{\{k\}}(V_{k-1}) \cap g_{i_k}(U)$, where i_k is chosen so that $1 \leqslant i_k \leqslant m$ and $T_{\{k\}}(V_{k-1}) \cap g_{i_k}(U) \neq \emptyset$. By construction, $T_{\{k\}}^{-1}(V_k) \subset V_{k-1}$ and $V_k \subset g_{i_k}(U)$. In particular, by the pigeonhole principle, there are $1 \leqslant i \leqslant m$ and arbitrarily large $p < q$ such that $V_p \cup V_q \subset g_i(U)$. Choose p so that $\beta = \{p+1, p+2, \ldots, q\} > \alpha$. Then the set $W = g_i^{-1}(V_q) \subset U$ is not empty and

$$T_\beta^{-1}(W) = g_i^{-1}\left(T_{\{p+1\}}^{-1} \cdots T_{\{q\}}^{-1}(V_q)\right) \subset g_i^{-1}\left(T_{\{p+1\}}^{-1}(V_{p+1})\right) \subset g_i^{-1}(V_p) \subset U.$$

Therefore $U \cap T_\beta(U) \supset W \neq \emptyset$.

Assume that the theorem holds true for any n IP-systems in G. Let $U \subset X$ be open and not empty. Let $T^{(1)}, \ldots, T^{(n+1)}$ be IP-systems in G. We will construct a sequence of non-empty open subsets $V_k \subset X$ and an increasing sequence $\alpha_k \in \mathcal{F}$, $\alpha_k > \alpha$, such that $V_0 = U$, $\bigcup_{j=1}^{n+1} (T_{\alpha_k}^{(j)})^{-1}(V_k) \subset V_{k-1}$, and $V_k \subset g_{i_k}(U)$ for some $1 \leqslant i_k \leqslant m$.

By the inductive assumption applied to $V_0 = U$ and the n IP-systems $(T^{(n+1)})^{-1}T^{(j)}$, $j = 1, \ldots, n$, there is $\alpha_1 > \alpha$ such that

$$V_0 \cap \left(T_{\alpha_1}^{(n+1)}\right)^{-1} T_{\alpha_1}^{(1)}(V_0) \cap \cdots \cap \left(T_{\alpha_1}^{(n+1)}\right)^{-1} T_{\alpha_1}^{(n)}(V_0) \neq \emptyset.$$

Apply $T_{\alpha_1}^{(n+1)}$ and, for an appropriate $1 \leqslant i_1 \leqslant m$, set

$$V_1 = g_{i_1}(V_0) \cap T_{\alpha_1}^{(1)}(V_0) \cap T_{\alpha_1}^{(2)}(V_0) \cap \cdots \cap T_{\alpha_1}^{(n+1)}(V_0) \neq \emptyset.$$

If V_{k-1} and α_{k-1} have been constructed, apply the inductive assumption to V_{k-1} and the IP-systems $(T^{(n+1)})^{-1}T^{(j)}$, $j = 1, \ldots, n$, to get $\alpha_k > \alpha_{k-1}$ such that

$$V_{k-1} \cap \left(T_{\alpha_k}^{(n+1)}\right)^{-1} T_{\alpha_k}^{(1)}(V_{k-1}) \cap \cdots \cap \left(T_{\alpha_k}^{(n+1)}\right)^{-1} T_{\alpha_k}^{(n)}(V_{k-1}) \neq \emptyset.$$

Apply $T^{(n+1)}_{\alpha_k}$ and, for an appropriate $1 \leqslant i_k \leqslant m$, set

$$V_k = g_{i_k}(V_0) \cap T^{(1)}_{\alpha_k}(V_{k-1}) \cap T^{(2)}_{\alpha_k}(V_{k-1}) \cap \cdots \cap T^{(n+1)}_{\alpha_k}(V_{k-1}) \neq \emptyset.$$

By construction, the sequences α_k and V_k have the desired properties. Since $V_k \subset g_{i_k}(U)$, there is $1 \leqslant i \leqslant m$ such that $V_k \subset g_i(U)$ for infinitely many k's. Hence there are arbitrarily large $p < q$ such that $V_p \cup V_q \subset g_i(U)$. Let $W = g_i^{-1}(V_q) \subset U$ and $\beta = \alpha_{p+1} \cup \cdots \cup \alpha_q$. Then $W \neq \emptyset$, and for each $1 \leqslant j \leqslant n+1$,

$$\left(T^{(j)}_\beta\right)^{-1}(W) = g_i^{-1}\left(\prod_{s=p}^{q-1}\left(T^{(j)}_{\alpha_{s+1}}\right)^{-1}(V_q)\right)$$

$$\subset g_i^{-1}\left(\prod_{s=p}^{q-2}\left(T^{(j)}_{\alpha_{s+1}}\right)^{-1}(V_{q-1})\right) \subset \cdots \subset g_i^{-1}(V_p) \subset U.$$

Therefore $\bigcup_{j=1}^{n+1}(T^{(j)}_\beta)^{-1}W \subset U$ and hence $\bigcup_{j=1}^{n+1}(T^{(j)}_\beta)^{-1}U \neq \emptyset$. □

COROLLARY 2.8.4. *Let G be a commutative group of homeomorphisms of a compact metric space X and let $T^{(1)}, \ldots, T^{(n)}$ be IP-systems in G. Then for every $\alpha \in \mathcal{F}$ and every $\epsilon > 0$ there are $x \in X$ and $\beta > \alpha$ such that $d(x, T^{(i)}_\beta(x)) < \epsilon$ for each $1 \leqslant i \leqslant n$.*

Proof. Similarly to Proposition 2.1.2, there is a non-empty closed G-invariant subset $X' \subset X$ on which G acts minimally (Exercise 2.8.3). Thus the corollary follows from Theorem 2.8.3. □

Proof. *Proof of Proposition 2.8.2.* Let $G = \{T^k\}_{k\in\mathbb{Z}}$. For $\alpha \in \mathcal{F}$, denote by $|\alpha|$ the sum of the elements in α. Apply Corollary 2.8.4 to G, X and the IP-systems $T^{(j)}_\alpha = T^{j|\alpha|}, 1 \leqslant j \leqslant q-1$.

The following generalization of Theorem 2.8.1 also follows from Corollary 2.8.4.

THEOREM 2.8.5. *Let $d \in \mathbb{N}$, and let A be a finite subset of $\mathbb{Z}^d$. Then for each finite partition $\mathbb{Z}^d = \bigcup_{k=1}^m S_k$, there are $k \in \{1, \ldots, m\}$, $z_0 \in \mathbb{Z}^d$, and $n \in \mathbb{N}$ such that $z_0 + na \in S_k$ for each $a \in A$, i.e. $z_0 + nA \subset S_k$.*

Proof. Exercise 2.8.5. □

Let V_F be an infinite vector space over a finite field F. A subset $A \subset V_F$ is a d-dimensional *affine subspace* if there are $v \in V_F$ and linearly independent $x_1, \ldots, x_d \in V_F$ such that $A = v + \text{Span}(x_1, \ldots, x_d)$.

THEOREM 2.8.6 (Graham *et al.*, 1972, 1973). *For each finite partition $V_F = \bigcup_{k=1}^m S_k$, one of the sets S_k contains affine subspaces of arbitrarily large (finite) dimension.*

Proof (Bergelson, 2000; see Theorem 2.8.3). We say that a subset $L \subset V_F$ is *monochromatic* of *color* j if $L \subset S_j$.

Since V_F is infinite, it contains a countable subspace isomorphic to the abelian group

$$F_\infty = \left\{\mathbf{a} = (a_i)_{i=1}^\infty \in F^{\mathbb{N}} : a_i = 0 \text{ for all but finitely many } i \in \mathbb{N}\right\}.$$

Without loss of generality we assume that $V_F = F_\infty$. The set $\Omega = \{1, \dots, m\}^{F_\infty}$ of all functions $F_\infty \to \{1, \dots, m\}$ is naturally identified with the set of all partitions of F_∞ into m subsets. The discrete topology on $\{1, \dots, m\}$ and product topology on Ω make it a compact Hausdorff space.

Let $\xi \in \Omega$ correspond to a partition $F_\infty = \bigcup_{k=1}^m S_k$, i.e. $\xi\colon F_\infty \to \{1, \dots, m\}$, $\xi(\mathbf{a}) = k$ if and only if $\mathbf{a} \in S_k$. Each $\mathbf{b} \in F_\infty$ induces a homeomorphism $T_\mathbf{b}\colon \Omega \to \Omega, (T_\mathbf{b}\eta)(\mathbf{a}) = \eta(\mathbf{a} + \mathbf{b})$. Denote by $X \subset \Omega$ the orbit closure of ξ, $X = \overline{\{\bigcup_{\mathbf{b} \in F_\infty} T_\mathbf{b}\xi\}}$. Similarly to the argument in the proof of Proposition 2.1.2, Zorn's Lemma implies that there is a non-empty closed subset $X' \subset X$ on which the group F_∞ acts minimally.

Let $g\colon \mathcal{F} \to F_\infty$ be an IP-system in F_∞ such that the elements $g_n, n \in \mathbb{N}$ are linearly independent. Define an IP-system T of homeomorphisms of X by setting $T_\alpha = T_{g_\alpha}$. For each $f \in F$, set $T_\alpha^{(f)} = T_{fg_\alpha}$ to get $|F| = \operatorname{card} F$ IP-systems of commuting homeomorphisms of X. Let $\mathbf{0} = (0, 0, \dots)$ be the zero element of F_∞ and $A_i = \{\eta \in \Omega\colon \eta(\mathbf{0}) = i\}$. Then each A_i is open and $\cup_{i=1}^m A_i = \Omega$. Therefore there is $j \in \{1, \dots, m\}$ such that $U = A_j \bigcap X' \neq \emptyset$. By Theorem 2.8.3, there is $\beta_1 \in \mathcal{F}$ such that $U_1 = \cap_{f \in F} T_{\beta_1}^{(f)}(U) \neq \emptyset$. If $\eta \in U_1$, then $\eta(fg_{\beta_1}) = j$ for each $f \in F$. In other words, η contains a monochromatic affine line of color j. Since the orbit of ξ is dense in X', there is $\mathbf{b}_1 \in F_\infty$ such that $\xi(fg_{\beta_1} + \mathbf{b}_1) = \eta(fg_{\beta_1}) = j$. Thus S_j contains an affine line.

To obtain a two-dimensional affine subspace in S_j apply Theorem 2.8.3 to U_1, β_1 and the same collection of IP-systems to get $\beta_2 > \beta_1$ such that $U_2 = \bigcap_{f \in F} T_{\beta_2}^{(f)}(U_1) \neq \emptyset$. Since g_{β_2} is linearly independent with every $g_\alpha, \alpha < \beta_2$, each $\eta \in U_2$ contains a monochromatic two-dimensional affine subspace of color j. Since η can be arbitrarily approximated by the shifts of ξ, the latter also contains a monochromatic two-dimensional affine subspace of color j.

Proceeding in this manner, we obtain a monochromatic subspace of arbitrarily large dimension. □

Exercise 2.8.1. Prove Theorem 2.8.1 using Proposition 2.8.2.

Exercise 2.8.2. Prove that a group G acts minimally on a compact topological space X if and only if for every non-empty open set $U \subset X$ there are elements $g_1, \dots, g_n \in G$ such that $\bigcup_{i=1}^{n} g_i(U) = X$.

Exercise 2.8.3. Prove the following generalization of Proposition 2.1.2. If a commutative group G acts by homeomorphisms on a compact metric space X, then there is a non-empty closed G-invariant subset X' on which G acts minimally.

Exercise 2.8.4. Prove that van der Waerden's Theorem 2.8.1 is equivalent to the following finite version: For each $m, n \in \mathbb{N}$ there is $k \in \mathbb{N}$ such that if the set $\{1, 2, \dots, k\}$ is partitioned into m subsets, then one of them contains an arithmetic progression of length n.

***Exercise 2.8.5.** For $z \in \mathbb{Z}^d$, the translation by z in $\mathbb{Z}^d$ induces a homeomorphism (shift) T_z in $\Sigma = \{1, \dots, m\}^{\mathbb{Z}^d}$. Prove Theorem 2.8.5 by considering the orbit closure under the group of shifts of the element $\xi \in \Sigma$ corresponding to the partition of $\mathbb{Z}^d$ and the IP-systems in $\mathbb{Z}^d$ generated by the translations T_f, $f \in A$.

CHAPTER THREE

Symbolic dynamics

In §1.4, we introduced the symbolic dynamical systems (Σ_m, σ) and (Σ_m^+, σ), and we showed by example throughout Chapter 1 how these shift spaces arise naturally in the study of other dynamical systems. In all of those examples, we encoded an orbit of the dynamical system by its *itinerary* through a finite collection of disjoint subsets. Specifically, following the idea that goes back to Hadamard, suppose $f: X \to X$ is a discrete dynamical system. Consider a partition $\mathcal{P} = \{P_1, P_2, \ldots, P_m\}$ of X, i.e. $P_1 \cup P_2 \cup \cdots \cup P_m = X$ and $P_i \cap P_j = \emptyset$ for $i \neq j$. For each $x \in X$, let $\psi_i(x)$ be the index of the element of $\mathcal{P}$ containing $f^i(x)$. The sequence $(\psi_i(x))_{i \in \mathbb{N}_0}$ is called the *itinerary* of x. This defines a map

$$\psi: X \to \Sigma_m^+ = \{1, 2, \ldots, m\}^{\mathbb{N}_0}, \quad x \mapsto \{\psi_i(x)\}_{i=0}^{\infty},$$

which satisfies $\psi \circ f = \sigma \circ \psi$. The space Σ_m^+ is totally disconnected, and the map ψ usually is not continuous. If f is invertible, then positive and negative iterates of f define a similar map $X \to \Sigma_m = \{1, 2, \ldots, m\}^{\mathbb{Z}}$. The image of ψ in Σ_m or Σ_m^+ is shift-invariant, and ψ semiconjugates f to the shift on the image of ψ. The indices $\psi_i(x)$ are *symbols* – hence the name *symbolic dynamics*. Any finite set can serve as the *symbol set*, or *alphabet*, of a symbolic dynamical system. Throughout this chapter we identify every finite alphabet with $\{1, 2, \ldots, m\}$.

Recall that the *cylinder* sets

$$C_{j_1, \ldots, j_k}^{n_1, \ldots, n_k} = \{\omega = (\omega_l) : \omega_{n_i} = j_i, \ i = 1, \ldots, k\},$$

form a basis for the product topology of Σ_m and Σ_m^+, and that the metric

$$d(\omega, \omega') = 2^{-l}, \quad \text{where } l = \min\{|i| : \omega_i \neq \omega_i'\}$$

generates the product topology.

3.1 Subshifts and codes[1]

In this section we concentrate on two-sided shifts. The case of one-sided shifts is similar.

A *subshift* is a closed subset $X \subset \Sigma_m$ invariant under the shift σ and its inverse. We refer to Σ_m as the *full m-shift*.

Let $X_i \subset \Sigma_{m_i}$, $i = 1, 2$, be two subshifts. A continuous map $c\colon X_1 \to X_2$ is a *code* if it commutes with the shifts, i.e. $\sigma \circ c = c \circ \sigma$ (here and later, σ denotes the shift in any sequence space). Note that a surjective code is a factor map; an injective code is called an *embedding*; a bijective code gives a topological conjugacy of the subshifts and is called an *isomorphism* (since Σ_m is compact, a bijective code is a homeomorphism).

For a subshift $X \subset \Sigma_m$, denote by $W_n(X)$ the set of words of length n that occur in X, and by $|W_n(X)|$ its cardinality. Since different elements of X differ in at least one position, the restriction $\sigma|_X$ is expansive. Therefore Proposition 2.5.7 allows us to compute the topological entropy of $\sigma|_X$ through the asymptotic growth rate of $|W_n(X)|$.

PROPOSITION 3.1.1. *Let $X \subset \Sigma_m$ be a subshift.*

Then $h(\sigma|_X) = \lim_{n\to\infty} \frac{1}{n} \log |W_n(X)|$.

Proof. Exercise 3.1.1. □

Let X be a subshift, $k, l \in \mathbb{N}_0$, $n = k + l + 1$, and let α be a map from $W_n(X)$ to an *alphabet* $\mathcal{A}_{m'}$. The *(k, l)-block code* c_α from X to the full shift $\Sigma_{m'}$ assigns to a sequence $x = (x_i) \in X$ the sequence $c_\alpha(x)$ with $c_\alpha(x)_i = \alpha(x_{i-k} \cdots x_i \cdots x_{i+l})$. Any block code is a code, since it is continuous and commutes with the shift.

PROPOSITION 3.1.2 (Curtis–Lyndon–Hedlund). *Every code $c\colon X \to Y$ is a block code.*

Proof. Let $\mathcal{A}$ be the symbol set of Y, and define $\tilde{\alpha}\colon X \to \mathcal{A}$ by $\tilde{\alpha}(x) = c(x)_0$. Since X is compact, $\tilde{\alpha}$ is uniformly continuous, so there is a $\delta > 0$ such that $\tilde{\alpha}(x) = \tilde{\alpha}(x')$ whenever $d(x, x') < \delta$. Choose $k \in \mathbb{N}$ so that $2^{-k} < \delta$. Then $\tilde{\alpha}(x)$ depends only on $x_{-k} \cdots x_0 \cdots x_k$, and therefore defines a map $\alpha\colon W_{2k+1} \to \mathcal{A}$ satisfying $c(x)_0 = \alpha(x_{-k} \cdots x_0 \cdots x_k)$. Since c commutes with the shift, we conclude that $c = c_\alpha$. □

[1] The exposition of this section as well as §§3.2, 3.4, and 3.5 follows in part the lectures of Boyle (1993).

There is a canonical class of block codes obtained by taking the alphabet of the target shift to be the set of words of length n in the original shift. Specifically, let $k, l \in \mathbb{N}, l < k$, and let X be a subshift. For $x \in X$ set

$$c(x)_i = x_{i-k+l+1} \cdots x_i \cdots x_{i+l}, \ i \in \mathbb{Z}.$$

This defines a block code c from X to the full shift on the alphabet $W_k(X)$ which is an isomorphism onto its image (Exercise 3.1.2). Such a code (or sometimes its image) is called a *higher block presentation* of X.

Exercise 3.1.1. Prove Proposition 3.1.1.

Exercise 3.1.2. Prove that a higher block presentation of X is an isomorphism.

Exercise 3.1.3. Use a higher block presentation to prove that for any block code $c\colon\ X \to Y$, there is a subshift Z and an isomorphism $f\colon\ Z \to X$ such that $c \circ f\colon\ Z \to Y$ is a $(0, 0)$-block code.

Exercise 3.1.4. Show that the full shift has points whose full orbit is dense but whose forward orbit is nowhere dense.

3.2 Subshifts of finite type

The complement of a subshift $X \subset \Sigma_m$ is open and hence is a union of at most countably many cylinders. By shift invariance, if C is a cylinder and $C \subset \Sigma_m \setminus X$, then $\sigma^n(C) \subset \Sigma_m \setminus X$ for all $n \in \mathbb{Z}$, i.e. there is a countable list of forbidden words such that no sequence in X contains a forbidden word and each sequence in $\Sigma_m \setminus X$ contains at least one forbidden word. If there is a finite list of finite words such that X consists of precisely the sequences in Σ_m that do not contain any of these words, then X is called a *subshift of finite type* (SFT); X is a k-step SFT if it is defined by a set of words of length at most $k + 1$. A 1-step SFT is called a *topological Markov chain*.

In §1.4 we introduced a vertex shift Σ_A^v determined by an adjacency matrix A of zeros and ones. A *vertex shift* is an example of an SFT. The forbidden words have length 2 and are precisely those that are not allowed by A, i.e. a word uv is forbidden if there is no edge from u to v in the graph Γ_A determined by A. Since the list of forbidden words is finite, Σ_A^v is an SFT. A sequence in Σ_A^v can be viewed as an infinite path in the directed graph Γ_A, labeled by the vertices.

An infinite path in the graph Γ_A can also be specified by a sequence of edges (rather than vertices). This gives a subshift Σ_A^e whose alphabet is the set of edges in Γ_A. More generally, a finite directed graph Γ, possibly with multiple directed edges connecting pairs of vertices, corresponds to an adjacency matrix B whose i, jth entry is a non-negative integer specifying the number of directed edges in $\Gamma = \Gamma_B$ from the ith vertex to the jth vertex. The set Σ_B^e of infinite directed paths in Γ_B, labeled by the edges, is closed and shift-invariant and is called the *edge shift* determined by B. Any edge shift is a subshift of finite type (Exercise 3.2.3).

For any matrix A of zeros and ones, the map $uv \mapsto e$, where e is the edge from u to v, defines a 2-block isomorphism from Σ_A^v to Σ_A^e. Conversely, any edge shift is naturally isomorphic to a vertex shift (Exercise 3.2.4).

PROPOSITION 3.2.1. *Every SFT is isomorphic to a vertex shift.*

Proof. Let X be a k-step SFT with $k > 0$. Let $W_k(X)$ be the set of words of length k that occur in X. Let Γ be the directed graph whose set of vertices is $W_k(X)$; a vertex $x_1 \cdots x_k$ is connected to a vertex $x_1' \cdots x_k'$ by a directed edge if $x_1 \cdots x_k x_k' = x_1 x_1' \cdots x_k' \in W_{k+1}(X)$. Let A be the adjacency matrix of Γ. The code $c(x)_i = x_i \cdots x_{i+k-1}$ gives an isomorphism from X to Σ_A^v. □

COROLLARY 3.2.2. *Every SFT is isomorphic to an edge shift.*

The last proposition implies that "the future is independent of the past" in an SFT; i.e. with appropriate one-step coding, if the sequences $\cdots x_{-2}x_{-1}x_0$ and $x_0x_1x_2\cdots$ are allowed, then $\cdots x_{-2}x_{-1}x_0x_1x_2\cdots$ is allowed.

Exercise 3.2.1. Show that the collection of all isomorphism classes of subshifts of finite type is countable.

***Exercise 3.2.2.** Show that the collection of all subshifts of Σ_2 is uncountable.

Exercise 3.2.3. Show that every edge shift is an SFT.

Exercise 3.2.4. Show that every edge shift is naturally isomorphic to a vertex shift. What are the vertices?

3.3 The Perron–Frobenius Theorem

The Perron–Frobenius Theorem guarantees the existence of special invariant measures – *Markov measures* – for subshifts of finite type.

A vector or matrix all of whose coordinates are positive (non-negative), is called *positive* (*non-negative*). Let A be a square non-negative matrix. If for any i, j there is $n \in \mathbb{N}$ such that $(A^n)_{ij} > 0$, then A is called *irreducible*; otherwise A is called *reducible*. If some power of A is positive, A is called *primitive*.

An integer non-negative square matrix A is primitive if and only if the directed graph Γ_A has the property that there is $n \in \mathbb{N}$ such that for every pair of vertices u and v there is a directed path from u to v of length n (see Exercise 1.4.2). An integer non-negative square matrix A is irreducible if and only if the directed graph Γ_A has the property that for every pair of vertices u and v there is a directed path from u to v (see Exercise 1.4.2).

A real non-negative $m \times m$ matrix is *stochastic* if the sum of the entries in each row is 1 or, equivalently, the column vector with all entries 1 is an eigenvector with eigenvalue 1.

THEOREM 3.3.1 (Perron). *Let A be a primitive $m \times m$ matrix. Then A has a positive eigenvalue λ with the following properties:*

1. *λ is a simple root of the characteristic polynomial of A;*
2. *λ has a positive eigenvector v;*
3. *any other eigenvalue of A has modulus strictly less than λ;*
4. *any non-negative eigenvector of A is a positive multiple of v.*

Proof. Denote by $\mathrm{int}(W)$ the interior of a set W. We will need the following lemma.

LEMMA 3.3.2. *Let $L\colon \mathbb{R}^k \to \mathbb{R}^k$ be a linear operator, and assume that there is a non-empty compact set P such that $0 \in \mathrm{int}(P)$ and $L^i(P) \subset \mathrm{int}(P)$ for some $i > 0$. Then the modulus of any eigenvalue of L is strictly less than 1.*

Proof. If the conclusion holds for L^i with some $i > 0$, then it holds for L. Hence we may assume that $L(P) \subset \mathrm{int}(P)$. It follows that $L^n(P) \subset \mathrm{int}(P)$ for all $n > 0$. The matrix L cannot have an eigenvalue of modulus greater than 1, since otherwise the iterates of L would move some vector in the open set $\mathrm{int}(P)$ off to ∞.

Suppose that σ is an eigenvalue of L and $|\sigma| = 1$. If $\sigma^j = 1$, then L^j has a fixed point on ∂P, a contradiction.

If σ is not a root of unity, there is a 2-dimensional subspace U on which L acts as an irrational rotation and any point $p \in \partial P \cap U$ is a limit point of $\bigcup_{n>0} L^n(P)$, a contradiction. □

Since A is non-negative, it induces a continuous map f from the unit simplex $S = \{x \in \mathbb{R}^m\colon\ \sum x_j = 1,\ x_j \geqslant 0,\ j = 1, \dots, m\}$ into itself; $f(x)$ is the radial projection of Ax onto S. By the Brouwer Fixed Point Theorem, there is a fixed point $v \in S$ of f, which is a non-negative eigenvector of A with eigenvalue $\lambda > 0$. Since some power of A is positive, all coordinates of v are positive.

Let V be the diagonal matrix that has the entries of v on the diagonal. The matrix $M = \lambda^{-1}V^{-1}AV$ is primitive, and the column vector $\mathbf{1}$ with all entries 1 is an eigenvector of M with eigenvalue 1, i.e. M is a stochastic matrix. To prove statements (1) and (3) of Theorem 3.3.1, it suffices to show that 1 is a simple root of the characteristic polynomial of M and that all other eigenvalues of M have moduli strictly less than 1. Consider the action of M on row vectors. Since M is stochastic and non-negative, the row action preserves the unit simplex S. By the Brouwer Fixed Point Theorem, there is a fixed row vector $w \in S$ all of whose coordinates are positive. Let $P = S - w$ be the translation of S by $-w$. Since for some $j > 0$ all entries of M^j are positive, $M^j(P) \subset \text{int}(P)$ and, by Lemma 3.3.2, the modulus of any eigenvalue of the row action of M in the $(m-1)$-dimensional invariant subspace spanned by P is strictly less than 1.

The last statement of the theorem follows from the fact that the codimension-one subspace spanned by P is M^t-invariant and its intersection with the cone of non-negative vectors in $\mathbb{R}^n$ is $\{0\}$. □

COROLLARY 3.3.3. *Let A be a primitive stochastic matrix. Then* 1 *is a simple root of the characteristic polynomial of A, both A and the transpose of A have positive eigenvectors with eigenvalue* 1, *and any other eigenvalue of A has modulus strictly less than* 1.

Frobenius extended Theorem 3.3.1 to irreducible matrices.

THEOREM 3.3.4 (Frobenius). *Let A be a non-negative irreducible square matrix. Then there exists an eigenvalue λ of A with the following properties: (i) $\lambda > 0$; (ii) λ is a simple root of the characteristic polynomial; (iii) λ has a positive eigenvector; (iv) if μ is any other eigenvalue of A, then $|\mu| \leqslant \lambda$; (v) if k is the number of eigenvalues of modulus $|\lambda|$, then the spectrum of A (with multiplicity) is invariant under the rotation of the complex plane by angle $2\pi/k$.*

A proof of Theorem 3.3.4 is outlined in Exercise 3.3.3. A complete argument can be found in Gantmacher (1959) or Berman and Plemmons (1994).

Exercise 3.3.1. Show that if A is a primitive integral matrix, then the edge shift Σ_A^e is topologically mixing.

Exercise 3.3.2. Show that if A is an irreducible integral matrix, then the edge shift Σ_A^e is topologically transitive.

Exercise 3.3.3. This exercise outlines the main steps in the proof of Theorem 3.3.4. Let A be a non-negative irreducible matrix and let B be the matrix with entries $b_{ij} = 0$ if $a_{ij} = 0$ and $b_{ij} = 1$ if $a_{ij} > 0$. Let Γ be the graph whose adjacency matrix is B. For a vertex v in Γ, let $d = d(v)$ be the greatest common divisor of the lengths of closed paths in Γ starting from v. Let V_k, $k = 0, 1, \ldots, d-1$, be the set of vertices of Γ that can be connected to v by a path whose length is congruent to $k \mod d$.

(a) Prove that d does not depend on v.

(b) Prove that any path of length l starting in V_k ends in V_m with m congruent to $k+l \mod d$.

(c) Prove that there is a permutation of the vertices that conjugates B^d to a block-diagonal matrix with square blocks B_k, $k = 0, 1, \ldots, d-1$, along the diagonal and 0's elsewhere, each B_k being a primitive matrix whose size equals the cardinality of V_k.

(d) What are the implications for the spectrum of A?

(e) Deduce Theorem 3.3.4.

3.4 Topological entropy and the zeta function of an SFT

For an edge or vertex shift, dynamical invariants can be computed from the adjacency matrix. In this section we compute the topological entropy of an edge shift and introduce the zeta function, an invariant that collects combinatorial information about the periodic points.

PROPOSITION 3.4.1. *Let A be a square non-negative integer matrix. Then the topological entropy of both the edge shift Σ_A^e and the vertex shift Σ_A^v is equal to the logarithm of the largest eigenvalue of A.*

Proof. We consider only the edge shift. By Proposition 3.1.1, it suffices to compute the cardinality of $W_n(\Sigma_A)$ (the words of length n in Σ_A), which is the sum S_n of all entries of A^n (Exercise 1.4.2). The proposition now follows from Exercise 3.4.1. □

For a discrete dynamical system f, denote by $\mathrm{Fix}(f)$ the set of fixed points of f and by $|\mathrm{Fix}(f)|$ its cardinality. If $|\mathrm{Fix}(f^n)|$ is finite for every n, we define the *zeta function* $\zeta_f(z)$ of f to be the formal power series

$$\zeta_f(z) = \exp \sum_{n=1}^{\infty} \frac{1}{n} |\mathrm{Fix}(f^n)| z^n.$$

The zeta function can also be expressed by the *product formula*:

$$\zeta_f(z) = \prod_{\gamma} \left(1 - z^{|\gamma|}\right)^{-1},$$

where the product is taken over all periodic orbits γ of f, and $|\gamma|$ is the number of points in γ (Exercise 3.4.4). The *generating function* $g_f(z)$ is another way to collect information about the periodic points of f:

$$g_f(z) = \sum_{n=1}^{\infty} |\operatorname{Fix}(f^n)| z^n.$$

The generating function is related to the zeta function by $\zeta_f'(z) = \zeta_f(z) \times g_f(z)/z$.

The zeta function of the edge shift determined by an adjacency matrix A is denoted by ζ_A. *A priori*, the zeta function is merely a formal power series. The next proposition shows that the zeta function of an SFT is a rational function.

PROPOSITION 3.4.2. $\zeta_A(z) = (\det(I - zA))^{-1}$.

Proof. Observe that

$$\exp\left(\sum_{n=1}^{\infty} \frac{x^n}{n}\right) = \exp(-\log(1-x)) = \frac{1}{1-x},$$

and that $|\operatorname{Fix}(\sigma^n|\Sigma_A)| = \operatorname{tr}(A^n) = \sum_\lambda \lambda^n$, where the sum is over the eigenvalues of A, repeated with the proper multiplicity (see Exercise 1.4.2). Therefore, if A is $N \times N$,

$$\zeta_A(z) = \exp\left(\sum_{n=1}^{\infty}\sum_{\lambda} \frac{(\lambda z)^n}{n}\right) = \prod_{\lambda} \exp\left(\sum_{n=1}^{\infty} \frac{(\lambda z)^n}{n}\right) = \prod_{\lambda}(1-\lambda z)^{-1}$$

$$= \frac{1}{z^N}\prod_{\lambda}\left(\frac{1}{z} - \lambda\right)^{-1} = \left(z^N \det\left(\frac{1}{z}I - A\right)\right)^{-1} = (\det(I - zA))^{-1}.$$

□

The following theorem (which we do not prove) addresses the rationality of the zeta function for a general subshift.

THEOREM 3.4.3 (Bowen–Lanford (Bowen and Lanford, 1970)). *The zeta function of a subshift $X \subset \Sigma_m$ is rational if and only if there are matrices A and B such that $\left|\operatorname{Fix}\left(\sigma^n|_X\right)\right| = \operatorname{tr} A^n - \operatorname{tr} B^n$ for all $n \in \mathbb{N}_0$.*

Exercise 3.4.1. Let A be a non-negative, non-zero, square matrix, S_n the sum of entries of A^n, and λ the eigenvalue of A with largest modulus. Prove that $\lim_{n\to\infty} \frac{\log S_n}{n} = \log \lambda$.

Exercise 3.4.2. Calculate the zeta and generating functions of the full 2-shift.

Exercise 3.4.3. Let $A = \begin{pmatrix} 1 & 1 \\ 1 & 0 \end{pmatrix}$. Calculate the zeta function of Σ_A^e.

Exercise 3.4.4. Prove the product formula for the zeta function.

Exercise 3.4.5. Calculate the generating function of an edge shift with adjacency matrix A.

Exercise 3.4.6. Calculate the zeta function of a 2-dimensional hyperbolic toral automorphism (see Exercise 1.7.4).

Exercise 3.4.7. Prove that if the zeta function is rational, then so is the generating function.

3.5 Strong shift equivalence and shift equivalence

We saw in §3.2 that any subshift of finite type is isomorphic to an edge shift Σ_A^e for some adjacency matrix A. In this section we give an algebraic condition on pairs of adjacency matrices that is equivalent to topological conjugacy of the corresponding edge shifts.

Square matrices A and B are *elementary strong shift equivalent* if there are (not necessarily square) non-negative integer matrices U and V such that $A = UV$ and $B = VU$. Matrices A and B are *strong shift equivalent* if there are (square) matrices $A_1, \ldots, A_n$ such that $A_1 = A$, $A_n = B$, and the matrices A_i and A_{i+1} are elementary strong shift equivalent. For example, the matrices $\begin{pmatrix} 1 & 1 & 0 \\ 1 & 1 & 1 \\ 2 & 2 & 1 \end{pmatrix}$ and (3) are strong shift equivalent but not elementary strong shift equivalent (Exercise 3.5.1).

THEOREM 3.5.1 (Williams, 1973). *The edge shifts Σ_A^e and Σ_B^e are topologically conjugate if and only if the matrices A and B are strong shift equivalent.*

Proof. We show here only that strong shift equivalence gives an isomorphism of the edge shifts. The other direction is much more difficult (see Lind and Marcus (1995)).

It is sufficient to consider the case when A and B are elementary strong shift equivalent. Let $A = UV$, $B = VU$, and Γ_A, Γ_B be the (disjoint) directed graphs with adjacency matrices A and B. If A is $k \times k$ and B is $l \times l$, then U is $k \times l$ and V is $l \times k$. We interpret the entry U_{ij} as the number of (additional) edges from vertex i of Γ_A to vertex j of Γ_B, and similarly we interpret V_{ji} as the number of edges from vertex j of Γ_B to vertex i of Γ_A. Since $A_{pq} = \sum_{j=1}^{l} U_{pj} V_{jq}$, the number of edges in Γ_A from vertex p to vertex q is the same as the number of paths of length 2 from vertex p to vertex q through a vertex in Γ_B. Therefore we can choose a one-to-one correspondence ϕ between the edges a of Γ_A and pairs uv of edges determined by U and V, i.e. $\phi(a) = uv$, so that the starting vertex of u is the starting vertex of a, the terminal vertex of u is the starting vertex of v, and the terminal vertex of v is the terminal

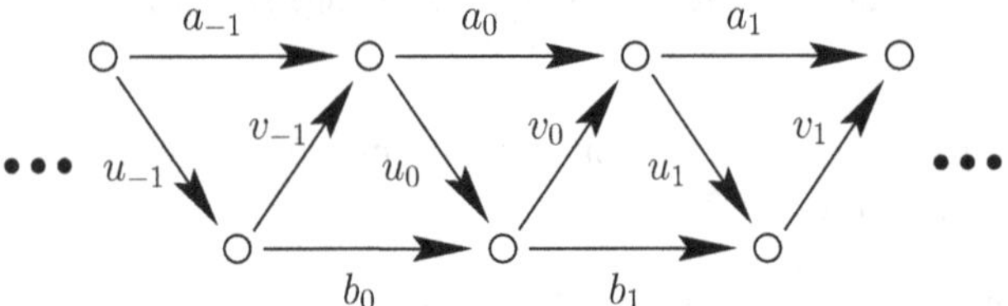

Figure 3.1. A graph constructed from an elementary strong shift equivalence.

vertex of a. Similarly, there is a bijection ψ from the edges b of Γ_B to pairs vu of edges determined by V and U. For each sequence $\cdots a_{-1}a_0a_1\cdots \in \Sigma_A^e$ apply ϕ to get

$$\cdots\phi(a_{-1})\phi(a_0)\phi(a_1)\cdots = \cdots u_{-1}v_{-1}u_0v_0u_1v_1\cdots$$

and then apply ψ^{-1} to get $\cdots b_{-1}b_0b_1\cdots \in \Sigma_B^e$ with $b_i = \psi^{-1}(v_iu_{i+1})$ (see Figure 3.1). This gives an isomorphism from Σ_A^e to Σ_B^e. □

Square matrices A and B are *shift equivalent* if there are (not necessarily square) non-negative integer matrices U, V and a positive integer k (called the *lag*) such that

$$A^k = UV, \quad B^k = VU, \quad AU = UB, \quad BV = VA.$$

The notion of shift equivalence was introduced by R. Williams, who conjectured that if two primitive matrices are shift equivalent, then they are strong shift equivalent, or, in view of Theorem 3.5.1, that shift equivalence classifies subshifts of finite type. Kim and Roush (1999) constructed a counterexample to this conjecture.

For other notions of equivalence for SFTs see Boyle (1993).

Exercise 3.5.1. Show that the matrices $A = \begin{pmatrix} 1&1&0\\ 1&1&1\\ 2&2&1 \end{pmatrix}$ and $B = (3)$ are strong shift equivalent but not elementary strong shift equivalent. Write down an explicit isomorphism from (Σ_A, σ) to (Σ_B, σ).

Exercise 3.5.2. Show that strong shift equivalence and shift equivalence are equivalence relations and elementary strong shift equivalence is not.

3.6 Substitutions[2]

For an alphabet $\mathcal{A}_m = \{0, 1, \ldots, m-1\}$, denote by $\mathcal{A}_m^*$ the collection of all finite words in $\mathcal{A}_m$, and by $|w|$ the length of $w \in \mathcal{A}_m^*$. A *substitution*

[2] Several arguments in this section follow in part those of Queffélec (1987).

$s: \mathcal{A}_m \to \mathcal{A}_m^*$ assigns to every symbol $a \in \mathcal{A}_m$ a finite word $s(a) \in \mathcal{A}_m^*$. We assume throughout this section that $|s(a)| > 1$ for some $a \in \mathcal{A}_m$ and that $|s^n(b)| \to \infty$ for every $b \in \mathcal{A}_m$. Applying the substitution to each element of a sequence or a word gives maps $s: \mathcal{A}_m^* \to \mathcal{A}_m^*$ and $s: \Sigma_m^+ \to \Sigma_m^+$,

$$x_0 x_1 \cdots \overset{s}{\mapsto} s(x_0)s(x_1)\cdots .$$

These maps are continuous but not surjective. If $s(a)$ has the same length for all $a \in \mathcal{A}_m$, then s is said to have *constant length*.

Consider the example $m = 2$, $s(0) = 01$, $s(1) = 10$. We have: $s^2(0) = 0110$, $s^3(0) = 01101001$, $s^4(0) = 0110100110010110, \ldots$. If $\overline{w}$ is the word obtained from w by interchanging 0 and 1, then $s^{n+1}(0) = s^n(0)\overline{s^n(0)}$. The sequence of finite words $s^n(0)$ stabilizes to an infinite sequence

$$\mathcal{M} = 01101001100101101001011001101001\ldots$$

called the *Morse sequence*. The sequences $\mathcal{M}$ and $\overline{\mathcal{M}}$ are the only fixed points of s in Σ_m^+.

PROPOSITION 3.6.1. *Every substitution s has a periodic point in* Σ_m^+.

Proof. Consider the map $a \mapsto s(a)_0$. Since $\mathcal{A}_m$ contains m elements, there are $n \in \{1, \ldots, m\}$ and $a \in \mathcal{A}_m$ such that $s^n(a)_0 = a$. If $|s^n(a)| = 1$, then the sequence $aaa\ldots$ is a fixed point of s^n. Otherwise, $|s^{ni}(a)| \to \infty$, and the sequence of finite words $s^{ni}(a)$ stabilizes to a fixed point of s^n in Σ_m^+. □

If a substitution s has a fixed point $x = x_0 x_1 \cdots \in \Sigma_m^+$ and $|s(x_0)| > 1$, then $s(x_0)_0 = x_0$ and the sequence $s^n(x_0)$ stabilizes to x; we write $x = s^\infty(x_0)$. If $|s(a)| > 1$ for every $a \in \mathcal{A}_m$, then s has at most m fixed points in Σ_m.

The closure $\Sigma_s(a)$ of the (forward) orbit of a fixed point $s^\infty(a)$ under the shift σ is a subshift.

We call a substitution $s: \mathcal{A}_m \to \mathcal{A}_m^*$ *irreducible* if for any $a, b \in \mathcal{A}_m$ there is $n(a, b) \in \mathbb{N}$ such that $s^{n(a,b)}(a)$ contains b; s is *primitive* if there is $n \in \mathbb{N}$ such that $s^n(a)$ contains b for all $a, b \in \mathcal{A}_m$.

We assume from now on that $|s^n(b)| \to \infty$ for every $b \in \mathcal{A}_m$.

PROPOSITION 3.6.2. *Let s be an irreducible substitution over* $\mathcal{A}_m$. *If* $s(a)_0 = a$ *for some* $a \in \mathcal{A}_m$, *then s is primitive and the subshift* $(\Sigma_s(a), \sigma)$ *is minimal.*

Proof. Observe that $s^n(a)_0 = a$ for all $n \in \mathbb{N}$. Since s is irreducible, for every $b \in \mathcal{A}_m$ there is $n(b)$ such that b appears in $s^{n(b)}(a)$ and therefore appears in $s^n(a)$ for all $n \geqslant n(b)$. Hence $s^n(a)$ contains all symbols from $\mathcal{A}_m$ if $n \geqslant N = \max n(b)$. Since s is irreducible, for every $b \in \mathcal{A}_m$ there is $k(b)$ such that a

appears in $s^{k(b)}(b)$ and hence in $s^n(b)$ with $n \geqslant k(b)$. It follows that for every $c \in \mathcal{A}_m$, $s^n(c)$ contains all symbols from $\mathcal{A}_m$ if $n \geqslant 2(N + \max k(b))$, so s is primitive.

Recall (Proposition 2.1.3) that $(\Sigma_s(a), \sigma)$ is minimal if and only if $s^\infty(a)$ is almost periodic, i.e. for every $n \in \mathbb{N}$ the word $s^n(a)$ occurs in $s^\infty(a)$ infinitely often, and the gaps between successive occurrences are bounded. This happens if and only if a recurs in $s^\infty(a)$ with bounded gaps, which holds true because s is primitive (Exercise 3.6.1). □

For two words $u, v \in \mathcal{A}_m^*$ denote by $N_u(v)$ the number of times u occurs in v. The *composition matrix* $M = M(s)$ of a substitution s is the non-negative integer matrix with entries $M_{ij} = N_i(s(j))$. The matrix $M(s)$ is primitive (respectively, irreducible) if and only if the substitution s is primitive (respectively, irreducible). For a word $w \in \mathcal{A}_m^*$, the numbers $N_i(w)$, $i \in \mathcal{A}_m$, form a vector $N(w) \in \mathbb{R}^m$. Observe that $M(s^n) = (M(s))^n$ for all $n \in \mathbb{N}$ and $N(s(w)) = M(s)N(w)$. If s has constant length l, then the sum of every column of M is l and the transpose of $l^{-1}M$ is a stochastic matrix.

PROPOSITION 3.6.3. *Let $s\colon \mathcal{A}_m \to \mathcal{A}_m^*$ be a primitive substitution and let λ be the largest in modulus eigenvalue of $M(s)$. Then for every $a \in \mathcal{A}_m$,*

1. *$\lim_{n\to\infty} \lambda^{-n} N(s^n(a))$ is an eigenvector of $M(s)$ with eigenvalue λ;*
2. *$\displaystyle\lim_{n\to\infty} \frac{|s^{n+1}(a)|}{|s^n(a)|} = \lambda$;*
3. *$v = \lim_{n\to\infty} |s^n(a)|^{-1} N(s^n(a))$ is an eigenvector of $M(s)$ corresponding to λ, and $\sum_{i=0}^{m-1} v_i = 1$.*

Proof. The proposition follows directly from Theorem 3.3.1 (Exercise 3.6.2). □

PROPOSITION 3.6.4. *Let s be a primitive substitution, $s^\infty(a)$ be a fixed point of s, and l_n be the number of different words of length n occurring in $s^\infty(a)$. Then there is a constant C such that $l_n \leqslant C \cdot n$ for all $n \in \mathbb{N}$. Consequently, the topological entropy of $(\Sigma_s(a), \sigma)$ is 0.*

Proof. Let $\underline{\nu}_k = \min_{a\in\mathcal{A}_m} |s^k(a)|$ and $\overline{\nu}_k = \max_{a\in\mathcal{A}_m} |s^k(a)|$, and note that $\underline{\nu}_k, \overline{\nu}_k \to \infty$ monotonically in k. Hence for every $n \in \mathbb{N}$ there is $k = k(n) \in \mathbb{N}$ such that $\underline{\nu}_{k-1} \leqslant n \leqslant \underline{\nu}_k$. Therefore every word of length n occurring in x is contained in $s^k(ab)$ for a pair of consecutive symbols ab from x. Let λ be the maximal-modulus eigenvalue λ of the primitive composition matrix $M = M(s)$. Then for every non-zero vector v with non-negative components there are constants $C_1(v)$ and $C_2(v)$ such that for all $k \in \mathbb{N}$,

$$C_1(v)\lambda^k \leqslant \|M^k v\| \leqslant C_2(v)\lambda^k,$$

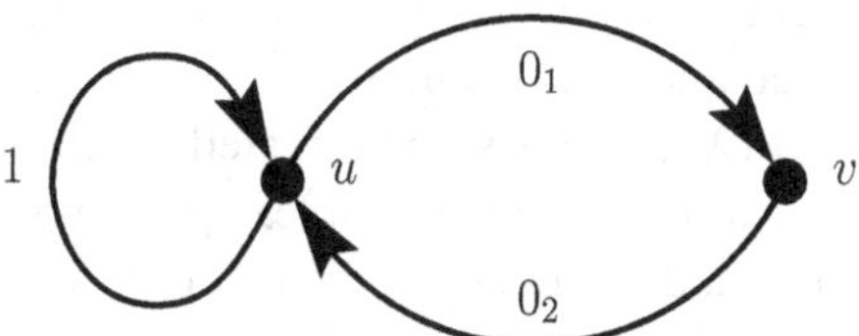

Figure 3.2. The directed graph used to construct the even system of Weiss.

where $\|\cdot\|$ is the Euclidean norm. Hence, by Proposition 3.6.3(1), there are positive constants C_1 and C_2 such that for all $k \in \mathbb{N}$

$$C_1 \cdot \lambda^k \leqslant \underline{\nu}_k \leqslant \overline{\nu}_k \leqslant C_2 \cdot \lambda^k.$$

Since for every $a \in \mathcal{A}_m$ there are at most $\overline{\nu}_k$ different words of length n in $s^k(ab)$ with initial symbol in $s^k(a)$, we have

$$l_n \leqslant m^2\overline{\nu}_k \leqslant C_2\lambda^k m^2 = \left(\frac{C_2}{C_1}m^2\lambda\right) C_1\lambda^{k-1} \leqslant \left(\frac{C_2}{C_1}m^2\lambda\right) \underline{\nu}_{k-1} \leqslant \left(\frac{C_2}{C_1}m^2\lambda\right) n.$$

□

Exercise 3.6.1. Prove that if s is primitive and $s(a)_0 = a$, then each symbol $b \in \mathcal{A}_m$ appears in $s^\infty(a)$ infinitely often and with bounded gaps.

Exercise 3.6.2. Prove Proposition 3.6.3.

3.7 Sofic shifts

A subshift $X \subset \Sigma_m$ is called *sofic* if it is a factor of a subshift of finite type, i.e. there is an adjacency matrix A and a code $c\colon\ \Sigma_A^e \to X$ such that $c \circ \sigma = \sigma \circ c$. Sofic shifts have applications in finite-state automata and data transmission and storage (Marcus *et al.*, 1995).

A simple example of a sofic shift is the following subshift of (Σ_2, σ), called the *even system of Weiss* (Weiss, 1973). Let A be the adjacency matrix of the graph Γ_A consisting of two vertices u and v, an edge from u to itself labeled 1, an edge from u to v labeled 0_1, and an edge from v to u labeled 0_2 (see Figure 3.2). Let X be the set of sequences of 0's and 1's such that there is an even number of 0's between every two 1's. The surjective code $c\colon\ \Sigma_A \to X$ replaces both 0_1 and 0_2 by 0.

As Proposition 3.7.1 shows, every sofic shift can be obtained by the following construction. Let Γ be a finite directed labeled graph, i.e. the edges of Γ are labeled by an alphabet $\mathcal{A}_m$. Note that we do *not* assume that different edges of Γ are labeled differently. The subset $X_\Gamma \subset \Sigma_m$ consisting of all infinite directed paths in Γ is closed and shift-invariant.

If a subshift (X, σ) is isomorphic to (X_Γ, σ) for some directed labeled graph Γ, then we say that Γ is a *presentation* of X. For example, a presentation for the even system of Weiss is obtained by replacing the labels 0_1 and 0_2 with 0 in Figure 3.2.

PROPOSITION 3.7.1. *A subshift $X \subset \Sigma_m$ is sofic if and only if it admits a presentation by a finite directed labeled graph.*

Proof. Since X is sofic, there is a matrix A and a code $c\colon \Sigma_A^e \to X$ (see Corollary 3.2.2). By Proposition 3.1.2, c is a block code. By passing to a higher block presentation we may assume that c is a 1-block code. Hence X admits a presentation by a finite directed labeled graph. The converse is Exercise 3.7.2. □

Exercise 3.7.1. Prove that the even system of Weiss is not a subshift of finite type.

Exercise 3.7.2. Prove that for any directed labeled graph Γ, the set X_Γ is a sofic shift.

Exercise 3.7.3. Show that there are only countably many non-isomorphic sofic shifts. Conclude that there are subshifts that are not sofic.

3.8 Data storage[3]

Most computer storage devices (floppy disk, hard drive, etc.) store data as a chain of magnetized segments on tracks. A magnetic head can either change or detect the polarity of a segment as it passes the head. Since it is technically much easier to detect a change of polarity than to measure the polarity, a common technique is to record a 1 as a change of polarity and a 0 as no change in polarity. The two major problems that restrict the effectiveness of this method are intersymbol interference and clock drift. Both of these problems can be ameliorated by applying a block code to the data before it is written to the storage device.

Intersymbol interference occurs when two polarity changes are adjacent to each other on the track; the magnetic fields from the adjacent positions partially cancel each other, and the magnetic head may not read the track

[3] The presentation of this section follows in part Berman and Plemmons (1994).

correctly. This effect can be minimized by requiring that in the encoded sequence every two 1's are separated by at least one 0.

A sequence of n 0's with 1's on both ends is read off the track as two pulses separated by n non-pulses. The length n is obtained by measuring the time between the pulses. Every time a 1 is read, the clock is synchronized. However, for a long sequence of 0's, clock error accumulates, which may cause the data to read incorrectly. To counteract this effect the encoded sequence is required to have no long stretches of 0's.

A common coding scheme called *modified frequency modulation* (MFM) inserts a 0 between each two symbols unless they are both 0's, in which case it inserts a 1. For example the sequence

$$10100110001$$

is encoded for storage as

$$100010010010100101001.$$

This requires twice the length of track, but results in fewer read/write errors. The set of sequences produced by the MFM coding is a sofic system (Exercise 3.8.3).

There are other considerations for storage devices that impose additional conditions on the sequences used to encode data. For example, the total magnetic charge of the device should not be too large. This restriction leads to a subset of (Σ_2, σ) that is not of finite type and not sofic.

Recall that the topological entropy of the factor does not exceed the topological entropy of the extension (Exercise 2.5.5). Therefore in any one-to-one coding scheme which increases the length of the sequence by a factor of $n > 1$, the topological entropy of the original subshift must be not more than n times the topological entropy of the target subshift.

Exercise 3.8.1. Prove that the sequences produced by MFM have at least one and at most three 0's between every two 1's.

Exercise 3.8.2. Describe an algorithm to reverse the MFM coding.

Exercise 3.8.3. Prove that the set of sequences produced by the MFM coding is a sofic system.

CHAPTER FOUR

Ergodic theory

Ergodic[1] theory is the study of statistical properties of dynamical systems relative to a measure on the underlying space of the dynamical system. The name comes from classical statistical mechanics, where the "ergodic hypothesis" asserts that, asymptotically, the time average of an observable is equal to the space average. Among the dynamical systems with natural invariant measures that we have encountered before are circle rotations §1.2 and toral automorphisms §1.7. Unlike topological dynamics, which studies the behavior of individual orbits (e.g. periodic orbits), ergodic theory is concerned with the behavior of the system on a set of full measure and with the induced action in spaces of measurable functions such as L^p (especially L^2).

The proper setting for ergodic theory is a dynamical system on a *measure space*. Most natural (non-atomic) measure spaces are measure-theoretically isomorphic to an interval $[0, a]$ with Lebesgue measure, and the results in this chapter are most important in that setting. The first section of this chapter recalls some notation, definitions and facts from measure theory. It is not intended to serve as a complete exposition of measure theory (for a full introduction see, for example, Halmos (1950) or Rudin (1987)).

4.1 Measure theory preliminaries

A non-empty collection $\mathfrak{A}$ of subsets of a set X is called a σ-algebra if $\mathfrak{A}$ is closed under complements and countable unions (and hence countable intersections). A *measure* μ on $\mathfrak{A}$ is a non-negative (possibly infinite) function on $\mathfrak{A}$ that is σ-additive, i.e. $\mu(\bigcup_i A_i) = \sum_i \mu(A_i)$ for any countable collection of disjoint sets $A_i \in \mathfrak{A}$. A set of measure 0 is called a *null set*. A set

[1] From the Greek words $\epsilon\rho\gamma o\nu$ (*ergon* – "work") and $o\delta o\varsigma$ (*odos* – "path").

whose complement is a null set is said to have *full measure*. The σ-algebra is *complete* (relative to μ) if it contains every subset of every null set. Given a σ-algebra $\mathfrak{A}$, and a measure μ, the *completion* $\overline{\mathfrak{A}}$ is the smallest σ-algebra containing $\mathfrak{A}$ and all subsets of null sets in $\mathfrak{A}$; the σ-algebra $\overline{\mathfrak{A}}$ is complete.

A *measure space* is a triple $(X, \mathfrak{A}, \mu)$, where X is a set, $\mathfrak{A}$ is a σ-algebra of subsets of X, and μ is a σ-additive measure. We always assume that $\mathfrak{A}$ is complete, and that μ is σ-finite, i.e. that X is a countable union of subsets of finite measure. The elements of $\mathfrak{A}$ are called *measurable sets*.

If $\mu(X) = 1$, then $(X, \mathfrak{A}, \mu)$ is called a *probability space* and μ is a *probability measure*. If $\mu(X)$ is finite, then we can rescale μ by the factor $1/\mu(X)$ to obtain a probability measure.

Let $(X, \mathfrak{A}, \mu)$ and $(Y, \mathfrak{B}, \nu)$ be measure spaces. The product measure space is the triple $(X \times Y, \mathfrak{C}, \mu \times \nu)$, where $\mathfrak{C}$ is the completion relative to $\mu \times \nu$ of the σ-algebra generated by $\mathfrak{A} \times \mathfrak{B}$.

Let $(X, \mathfrak{A}, \mu)$ and $(Y, \mathfrak{B}, \nu)$ be measure spaces. A map $T\colon X \to Y$ is called *measurable* if the preimage of any measurable set is measurable. A measurable map T is *non-singular* if the preimage of every set of measure 0 has measure 0 and is *measure-preserving* if $\mu(T^{-1}(B)) = \nu(B)$ for every $B \in \mathfrak{B}$. A non-singular map from a measure space into itself is called a *non-singular transformation* (or simply a *transformation*). If a transformation T preserves a measure μ, then μ is called T*-invariant*. If T is an invertible measurable transformation, and its inverse is measurable and non-singular, then the iterates T^n, $n \in \mathbb{Z}$, form a group of measurable transformations. Measure spaces $(X, \mathfrak{A}, \mu)$ and $(Y, \mathfrak{B}, \nu)$ are *isomorphic* if there is a subset X' of full measure in X, a subset Y' of full measure in Y, and an invertible bijection $T : X' \to Y'$ such that T and T^{-1} are measurable and measure-preserving with respect to $(\mathfrak{A}, \mu)$ and $(\mathfrak{B}, \nu)$. An isomorphism from a measure space into itself is an *automorphism*.

Denote by λ the Lebesgue measure on $\mathbb{R}$. A flow T^t on a measure space $(X, \mathfrak{A}, \mu)$ is *measurable* if the map $T : X \times \mathbb{R} \to X, (x, t) \mapsto T^t(x)$, is measurable with respect to the product measure on $X \times \mathbb{R}$, and $T^t\colon X \to X$ is a non-singular measurable transformation for each $t \in \mathbb{R}$. A measurable flow T^t is a *measure-preserving flow* if each T^t is a measure-preserving transformation.

Let T be a measure-preserving transformation of a measure space $(X, \mathfrak{A}, \mu)$, and S a measure-preserving transformation of a measure space $(Y, \mathfrak{B}, \nu)$. We say that T is an *extension* of S if there are sets $X' \subset X$ and $Y' \subset Y$ of full measure and a measure-preserving map $\psi : X' \to Y'$ such that $\psi \circ T = S \circ \psi$. A similar definition holds for measure-preserving flows.

If ψ is an isomorphism, then T and S are called *isomorphic*. The product $T \times S$ is a measure-preserving transformation of $(X \times Y, \mathfrak{C}, \mu \times \nu)$, where $\mathfrak{C}$ is the completion of the σ-algebra generated by $\mathfrak{A} \times \mathfrak{B}$.

Let X be a topological space. The smallest σ-algebra containing all the open subsets of X is called the *Borel σ-algebra* of X. If $\mathfrak{A}$ is the Borel σ-algebra, then a measure μ on $\mathfrak{A}$ is a *Borel measure* if the measure of any compact set is finite. A Borel measure is *regular* in the sense that the measure of any set is the infimum of measures of open sets containing it, and the supremum of measures of compact sets contained in it.

A one-point subset with positive measure is called an *atom*. A finite measure space is a *Lebesgue space* if it is isomorphic to the union of an interval $[0, a]$ (with Lebesgue measure) and at most countably many atoms. Most natural measure spaces are Lebesgue spaces. For example, if X is a complete separable metric space, μ a finite Borel measure on X, and $\mathfrak{A}$ the completion of the Borel σ-algebra with respect to μ, then $(X, \mathfrak{A}, \mu)$ is a Lebesgue space. In particular, the unit square $[0, 1] \times [0, 1]$ with Lebesgue measure is (measure theoretically) isomorphic to the unit interval $[0, 1]$ with Lebesgue measure (Exercise 4.1.1).

A Lebesgue space without atoms is called *non-atomic*, and is isomorphic to an interval $[0, a]$ with Lebesgue measure.

A set has full measure if its complement has measure 0. We say a property holds mod 0 in X, or holds for *μ-almost every (a.e.) x*, if it holds on a subset of full μ-measure in X. We also use the word *essentially* to indicate that a property holds mod 0.

Let $(X, \mathfrak{A}, \mu)$ be a measure space. Two measurable functions are equivalent if they coincide on a set of full measure. For $p \in (0, \infty)$, the space $L^p(X, \mu)$ consists of equivalence classes mod 0 of measurable functions $f\colon X \to \mathbb{C}$ such that $\int |f|^p \, d\mu < \infty$. As a rule, if there is no ambiguity, we identify the function with its equivalence class. For $p \geqslant 1$, the L^p norm is defined by $\|f\|_p = \left(\int |f|^p \, d\mu\right)^{1/p}$. The space $L^2(X, \mu)$ is a Hilbert space with inner product $\langle f, g \rangle = \int f \cdot \bar{g} \, d\mu$. The space $L^\infty(X, \mu)$ consists of equivalence classes of essentially bounded measurable functions. If μ is finite, then $L^\infty(X, \mu) \subset L^p(X, \mu)$ for all $p > 0$. If X is a topological space and μ is a Borel measure on X, then the space $C_0(X, \mathbb{C})$ of continuous, complex-valued, compactly-supported functions on X is dense in $L^p(X, \mu)$ for all $p > 0$.

Exercise 4.1.1. Prove that the unit square $[0, 1] \times [0, 1]$ with Lebesgue measure is (measure theoretically) isomorphic to the unit interval $[0, 1]$ with Lebesgue measure.

4.2 Recurrence

The following famous result of Poincaré implies that recurrence is a generic property of orbits of measure-preserving dynamical systems.

THEOREM 4.2.1 (Poincaré Recurrence Theorem). *Let T be a measure-preserving transformation of a probability space $(X, \mathfrak{A}, \mu)$. If A is a measurable set, then for a.e. $x \in A$, there is some $n \in \mathbb{N}$ such that $T^n(x) \in A$. Consequently, for a.e. $x \in A$, there are infinitely many $k \in \mathbb{N}$ for which $T^k(x) \in A$.*

Proof. Let

$$B = \{x \in A : T^k(x) \notin A \text{ for all } k \in \mathbb{N}\} = A \setminus \bigcup_{k \in \mathbb{N}} T^{-k}(A).$$

Then $B \in \mathfrak{A}$, and all the preimages $T^{-k}(B)$ are disjoint, measurable, and have the same measure as B. Since X has finite total measure, it follows that B has measure 0. Since every point in $A \backslash B$ returns to A, this proves the first assertion. The proof of the second assertion is Exercise 4.2.1. □

For continuous maps of topological spaces, there is a connection between measure-theoretic recurrence and the topological recurrence introduced in Chapter 2. If X is a topological space, and μ is a Borel measure on X, then $\operatorname{supp} \mu$ (the *support* of μ) is the complement of the union of all open sets with measure 0 or, equivalently, the intersection of all closed sets with full measure. Recall from §2.1 that the set of recurrent points of a continuous map $T\colon X \to X$ is $\mathcal{R}(T) = \{x \in X : x \in \omega(x)\}$.

PROPOSITION 4.2.2. *Let X be a separable metric space, μ a Borel probability measure on X, and $f\colon X \to X$ a continuous measure-preserving transformation. Then almost every point is recurrent and hence* $\operatorname{supp} \mu \subset \overline{\mathcal{R}(f)}$.

Proof. Since X is separable, there is a countable basis $\{U_i\}_{i \in \mathbb{N}}$ for the topology of X. A point $x \in X$ is recurrent if it returns (in the future) to every basis element containing it. By the Poincaré Recurrence Theorem, for each i, there is a subset $\widetilde{U}_i$ of full measure in U_i such that every point of $\widetilde{U}_i$ returns to U_i. Then $X_i = \widetilde{U}_i \cup (X \backslash U_i)$ has full measure in X, so $\widetilde{X} = \bigcap_{i \in \mathbb{Z}} X_i = \mathcal{R}(T)$ has full measure in X. □

We will discuss some applications of measure-theoretic recurrence in §4.11.

Given a measure-preserving transformation T in a finite measure space $(X, \mathfrak{A}, \mu)$ and a measurable subset $A \in \mathfrak{A}$ of positive measure, the *derivative transformation* $T_A\colon A \to A$ is defined by $T_A(x) = T^k(x)$, where $k \in \mathbb{N}$ is the smallest natural number for which $T^k(x) \in A$. The derivative transformation is often called the *first return map*, or the *Poincaré map*. By Theorem 4.2.1, T_A is defined on a subset of full measure in A.

Let T be a transformation on a measure space $(X, \mathfrak{A}, \mu)$ and $f\colon X \to \mathbb{N}$ a measurable function. Let $X_f = \{(x, k) : x \in X,\ 1 \leqslant k \leqslant f(x)\} \subset X \times \mathbb{N}$. Let $\mathfrak{A}_f$ be the σ-algebra generated by the sets $A \times \{k\}, A \in \mathfrak{A}, k \in \mathbb{N}$, and define $\mu_f(A \times \{k\}) = \mu(A)$. Define the *primitive transformation* $T_f\colon X_f \to X_f$ by $T_f(x, k) = (x, k+1)$ if $k < f(x)$ and $T_f(x, f(x)) = (T(x), 1)$. If $\mu(X) < \infty$ and $f \in L^1(X, \mathfrak{A}, \mu)$, then $\mu_f(X_f) = \int_X f(x)\, d\mu$. Note that the derivative transformation of T_f on the set $X \times \{1\}$ is just the original transformation T.

Primitive and derivative transformations are both referred to as *induced transformations*; we will encounter them later.

Exercise 4.2.1. Prove the second assertion of Theorem 4.2.1.

Exercise 4.2.2. Suppose $T\colon X \to X$ is a continuous transformation of a second countable topological space X, and μ is a finite T-invariant Borel measure on X with $\operatorname{supp} \mu = X$. Show that every point is non-wandering and μ-a.e. point is recurrent.

Exercise 4.2.3. Prove that if T is a measure-preserving transformation, then so are the induced transformations.

4.3 Ergodicity and mixing

A dynamical system induces an action on functions: T acts on a function f by $(T_* f)(x) = f(T(x))$. The ergodic properties of a dynamical system correspond to the degree of statistical independence between f and $T_*^n f$. The strongest possible dependence happens for an invariant function $f(T(x)) = f(x)$. The strongest possible independence happens when a non-zero L^2 function is orthogonal to its images.

Let T be a measure-preserving transformation (or flow) on a measure space $(X, \mathfrak{A}, \mu)$. A measurable function $f\colon X \to \mathbb{R}$ is *essentially T-invariant* if $\mu(\{x \in X : f(T^t x) \neq f(x)\}) = 0$ for every t. A measurable set A is *essentially T-invariant* if its characteristic function $\mathbf{1}_A$ is essentially T-invariant; equivalently, if $\mu(T^{-1}(A) \Delta A) = 0$ (we denote by Δ the symmetric difference, $A \Delta B = (A \setminus B) \cup (B \setminus A)$).

A measure-preserving transformation (or flow) T is *ergodic* if any essentially T-invariant measurable set has either measure 0 or full measure. Equivalently (Exercise 4.3.1), T is ergodic if any essentially T-invariant measurable function is constant mod 0.

PROPOSITION 4.3.1. *Let T be a measure-preserving transformation or flow on a finite measure space $(X, \mathfrak{A}, \mu)$ and let $p \in (0, \infty]$. Then T is ergodic if and only if every essentially invariant function $f \in L^p(X, \mu)$ is constant mod 0.*

Proof. If T is ergodic, then every essentially invariant function is constant mod 0.

To prove the converse, let f be an essentially invariant measurable function on X. Then for every $M > 0$, the function

$$f_M(x) = \begin{cases} f(x) & \text{if } f(x) \leqslant M \\ 0 & \text{if } f(x) > M \end{cases}$$

is bounded, essentially invariant, and belongs to $L^p(X, \mu)$. Therefore it is constant mod 0. It follows that f itself is constant mod 0. □

As the following proposition shows, any essentially invariant set or function is equal mod 0 to a strictly invariant set or function.

PROPOSITION 4.3.2. *Let $(X, \mathfrak{A}, \mu)$ be a measure space, and suppose $f\colon X \to \mathbb{R}$ is essentially invariant for a measurable transformation or flow T on X. Then there is a strictly invariant measurable function $\tilde{f}$ such that $f(x) = \tilde{f}(x)$ mod 0.*

Proof. We prove the proposition for a measurable flow. The case of a measurable transformation follows by a similar but easier argument, and is left as an exercise.

Consider the measurable map $\Phi\colon X \times \mathbb{R} \to \mathbb{R}$, $\Phi(x, t) = f(T^t x) - f(x)$, and the product measure $\nu = \mu \times \lambda$ in $X \times \mathbb{R}$, where λ is Lebesgue measure on $\mathbb{R}$. The set $A = \Phi^{-1}(0)$ is a measurable subset of $X \times \mathbb{R}$. Since f is essentially T-invariant, for each $t \in \mathbb{R}$ the set

$$A_t = \{(x, t) \in (X \times \mathbb{R}) : f(T^t x) = f(x)\}$$

has full μ-measure in $X \times \{t\}$. By the Fubini Theorem, the set

$$A_f = \{x \in X \ : \ f(T^t x) = f(x) \ \text{ for a.e. } t \in \mathbb{R}\}$$

has full μ-measure in X. Set

$$\tilde{f}(x) = \begin{cases} f(y) & \text{if } T^t x = y \in A_f \text{ for some } t \in \mathbb{R} \\ 0 & \text{otherwise.} \end{cases}$$

If $T^t x = y \in A_f$ and $T^s x = z \in A_f$, then y and z lie on the same orbit, and the value of f along this orbit is equal λ-almost everywhere to $f(y)$ and to $f(z)$, so $f(y) = f(z)$. Therefore $\tilde{f}$ is well-defined and strictly T-invariant. □

A measure-preserving transformation (or flow) T on a probability space $(X, \mathfrak{A}, \mu)$ is called *(strong) mixing* if

$$\lim_{t\to\infty} \mu(T^{-t}(A) \cap B) = \mu(A) \cdot \mu(B)$$

for any two measurable sets $A, B \in \mathfrak{A}$. Equivalently (Exercise 4.3.3), T is mixing if

$$\lim_{t\to\infty} \int_X f(T^t(x)) \cdot g(x)\, d\mu = \int_X f(x)\, d\mu \cdot \int_X g(x)\, d\mu$$

for any bounded measurable functions f, g.

A measure-preserving transformation T of a probability space $(X, \mathfrak{A}, \mu)$ is called *weak mixing* if for all $A, B \in \mathfrak{A}$,

$$\lim_{n\to\infty} \frac{1}{n} \sum_{i=0}^{n-1} \left|\mu(T^{-i}(A) \cap B) - \mu(A) \cdot \mu(B)\right| = 0$$

or, equivalently (Exercise 4.3.3), if for all bounded measurable functions f, g,

$$\lim_{n\to\infty} \frac{1}{n} \sum_{i=0}^{n-1} \left| \int_X f(T^i(x))g(x)\, d\mu - \int_X f\, d\mu \cdot \int_X g\, d\mu \right| = 0.$$

A measure-preserving flow T^t on $(X, \mathfrak{A}, \mu)$ is *weak mixing* if for all $A, B \in \mathfrak{A}$

$$\lim_{t\to\infty} \frac{1}{t} \int_0^t \left|\mu(T^{-s}(A) \cap B) - \mu(A) \cdot \mu(B)\right| ds = 0,$$

or, equivalently (Exercise 4.3.3), if for all bounded measurable functions f, g,

$$\lim_{t\to\infty} \frac{1}{t} \int_0^t \left| \int_X f(T^s(x))g(x)\, d\mu\, ds - \int_X f\, d\mu \cdot \int_X g\, d\mu \right| = 0.$$

In practice, the definitions of ergodicity and mixing in terms of L^2 functions are often easier to work with than the definitions in terms of measurable sets. For example, to establish a certain property for each L^2 function on a separable topological space with Borel measure it suffices to do it for a countable set of continuous functions that is dense in L^2 (Exercise 4.3.5). If the property is "linear", it is enough to check it for a basis in L^2, e.g. for the exponential functions $e^{2\pi ikx}$ on the circle $[0, 1)$.

PROPOSITION 4.3.3. *Mixing implies weak mixing, and weak mixing implies ergodicity.*

Proof. Suppose T is a measure-preserving transformation of the probability space $(X, \mathfrak{A}, \mu)$. Let A and B be measurable subsets of X. If T is mixing, then $\left|\mu(T^{-i}(A) \cap B) - \mu(A) \cdot \mu(B)\right|$ converges to 0, so the averages do as well; thus T is weak mixing.

Let A be an invariant measurable set. Then applying the definition of weak mixing with $B = A$, we conclude that $\mu(A) = \mu(A)^2$, so either $\mu(A) = 1$ or $\mu(A) = 0$. □

For continuous maps, ergodicity and mixing have the following topological consequences.

PROPOSITION 4.3.4. *Let X be a compact metric space, $T\colon X \to X$ a continuous map, and μ a T-invariant Borel measure on X.*

1. *If T is ergodic, then the orbit of μ-almost every point is dense in* $\operatorname{supp} \mu$.
2. *If T is mixing, then T is topologically mixing on* $\operatorname{supp} \mu$.

Proof. Suppose T is ergodic. Let U be a non-empty open set in $\operatorname{supp} \mu$. Then $\mu(U) > 0$. By ergodicity, the backward invariant set $\bigcup_{k \in \mathbb{N}} T^{-k}(U)$ has full measure. Thus the forward orbit of almost every point visits U. It follows that the set of points whose forward orbit visits every element of a countable open basis has full measure in X. This proves the first assertion.

The proof of the second assertion is Exercise 4.3.4. □

Exercise 4.3.1. Show that a measurable transformation is ergodic if and only if every essentially invariant measurable function is constant mod 0 (see the remark after Corollary 4.5.7).

Exercise 4.3.2. Let T be an ergodic measure-preserving transformation in a finite measure space $(X, \mathfrak{A}, \mu)$, $A \in \mathfrak{A}$, $\mu(A) > 0$ and $f \in L^1(X, \mathfrak{A}, \mu)$, $f\colon X \to \mathbb{N}$. Prove that the induced transformations T_A and T_f are ergodic.

Exercise 4.3.3. Show that the two definitions of strong and weak mixing given in terms of sets and bounded measurable functions are equivalent.

Exercise 4.3.4. Prove the second statement of Proposition 4.3.4.

Exercise 4.3.5. Let T be a measure-preserving transformation of $(X, \mathfrak{A}, \mu)$ and let $f \in L^1(X, \mu)$ satisfy $f(T(x)) \leqslant f(x)$ for a.e. x. Prove that $f(T(x)) = f(x)$ for a.e. x.

Exercise 4.3.6. Let X be a compact topological space, μ a Borel measure and $T\colon X \to X$ a transformation preserving μ. Suppose that for every continuous f and g with 0 integrals,

$$\int_X f(T^n(x)) \cdot g(x) d\mu \to 0 \text{ as } n \to \infty.$$

Prove that T is mixing.

Exercise 4.3.7. Show that if $T\colon X \to X$ is mixing, then $T \times T : X \times X \to X \times X$ is mixing.

4.4 Examples

We now prove ergodicity or mixing for some of the examples from Chapter 1.

PROPOSITION 4.4.1. *The circle rotation R_α is ergodic with respect to Lebesgue measure if and only if α is irrational.*

Proof. Suppose α is irrational. By Proposition 4.3.1, it is enough to prove that any bounded R_α-invariant function $f\colon S^1 \to \mathbb{R}$ is constant mod 0. Since $f \in L^2(S^1, \lambda)$, the Fourier series $\sum_{n=-\infty}^{\infty} a_n e^{2n\pi ix}$ of f converges to f in the L^2 norm. The series $\sum_{n=-\infty}^{\infty} a_n e^{2n\pi i(x+\alpha)}$ converges to $f \circ R_\alpha$. Since $f = f \circ R_\alpha$ mod 0, uniqueness of Fourier coefficients implies that $a_n = a_n e^{2n\pi i\alpha}$ for all $n \in \mathbb{Z}$. Since $e^{2n\pi i\alpha} \neq 1$ for $n \neq 0$, we conclude that $a_n = 0$ for $n \neq 0$, so f is constant mod 0.

The proof of the converse is left as an exercise. □

PROPOSITION 4.4.2. *An expanding endomorphism $E_m\colon S^1 \to S^1$ is mixing with respect to Lebesgue measure.*

Proof. Since any measurable subset of S^1 can be approximated by a finite union of intervals, it is sufficient to consider two intervals $A = [p/m^i, (p+1)/m^i]$, $p \in \{0, \ldots, m^i - 1\}$ and $B = [q/m^j, (q+1)/m^j]$, $q \in \{0, \ldots, m^j - 1\}$. Recall that $E_m^{-1}(B)$ is the union of m uniformly spaced intervals of length $1/m^{j+1}$:

$$E_m^{-1}(B) = \bigcup_{k=0}^{m-1} [(km^j + q)/m^{j+1}, (km^j + q + 1)/m^{j+1}].$$

Similarly, $E_m^{-n}(B)$ is the union of m^n uniformly spaced intervals of length $1/m^{j+n}$. Thus for $n > i$, the intersection $A \cap E_m^{-n}(B)$ consists of m^{n-i} intervals of length $m^{-(n+j)}$. Thus

$$\mu(A \cap E_m^{-n}(B)) = m^{n-i}(1/m^{n+j}) = m^{-i-j} = \mu(A) \cdot \mu(B).$$ □

PROPOSITION 4.4.3. *Any hyperbolic toral automorphism $A\colon \mathbb{T}^n \to \mathbb{T}^n$ is ergodic with respect to Lebesgue measure.*

Proof. We consider here only the case $A = \begin{pmatrix} 2 & 1 \\ 1 & 1 \end{pmatrix} : \mathbb{T}^2 \to \mathbb{T}^2$; the argument in the general case is similar. Let $f\colon \mathbb{T}^2 \to \mathbb{R}$ be a bounded A-invariant measurable function. The Fourier series $\sum_{m,n=-\infty}^{\infty} a_{mn} e^{2\pi i(mx+ny)}$ of f converges to f in L^2. The series

$$\sum_{m,n=-\infty}^{\infty} a_{mn} e^{2\pi i(m(2x+y)+n(x+y))}$$

converges to $f \circ A$. Since f is invariant, uniqueness of Fourier coefficients implies that $a_{mn} = a_{(2m+n)(m+n)}$ for all m, n. Since A does not have eigenvalues on the unit circle, if $a_{mn} \neq 0$ for some $(m, n) \neq (0, 0)$, then $a_{ij} = a_{mn} \neq 0$ with arbitrarily large $|i| + |j|$, and the Fourier series diverges. □

A toral automorphism of $\mathbb{T}^n$ corresponding to an integer matrix A is ergodic if and only if no eigenvalue of A is a root of unity; for a proof see, for example, Petersen (1989). A hyperbolic toral automorphism is mixing (Exercise 4.4.3).

Let A be an $m \times m$ stochastic matrix, i.e. A has non-negative entries and the sum of every row is 1. Suppose A has a non-negative left eigenvector q with eigenvalue 1 and the sum of entries equal to 1 (recall that if A is irreducible, then by Corollary 3.3.3, q exists and is unique). We define a Borel probability measure $P = P_{A,q}$ on Σ_m (and Σ_m^+) as follows. For a cylinder C_j^n of length 1, we define $P(C_j^n) = q_j$; for a cylinder $C_{j_0,j_1,\dots,j_k}^{n,n+1,\dots,n+k} \subset \Sigma_m$ (or Σ_m^+) with $k+1 > 1$ consecutive indices,

$$P\left(C_{j_0,j_1,\dots,j_k}^{n,n+1,\dots,n+k}\right) = q_{j_0} \prod_{i=0}^{k-1} A_{j_i j_{i+1}}.$$

In other words, we interpret q as an initial probability distribution on the set $\{1, \dots, m\}$ and A as the matrix of transition probabilities. The number $P(C_j^n)$ is the probability of observing symbol j in the nth place, and A_{ij} is the probability of passing from i to j. The fact that $qA = q$ means that the probability distribution q is invariant under transition probabilities A, i.e.

$$q_j = P(C_j^{n+1}) = \sum_{i=0}^{m-1} P(C_i^n) A_{ij}.$$

The pair (A, q) is called a *Markov chain* on the set $\{1, \dots, m\}$.

It can be shown that P extends uniquely to a shift-invariant σ-additive measure defined on the completion $\mathfrak{C}$ of the Borel σ-algebra generated by the cylinders (Exercise 4.4.5); it is called the *Markov measure* corresponding to A and q. The measure space $(\Sigma_m, \mathfrak{C}, P)$ is a non-atomic Lebesgue probability space. If A is irreducible, this measure is uniquely determined by A.

A very important particular case of this situation arises when the transition probabilities do not depend on the initial state. In this case each row of A is the left eigenvector q, the shift-invariant measure P is called a *Bernoulli measure*, and the shift is called a *Bernoulli automorphism*.

Let A' be the adjacency matrix defined by $A'_{ij} = 0$ if $A_{ij} = 0$ and $A'_{ij} = 1$ if $A_{ij} > 0$. Then the support of P is precisely $\Sigma^v_{A'} \subset \Sigma_m$ (Exercise 4.4.6).

PROPOSITION 4.4.4. *If A is a primitive stochastic $m \times m$ matrix, then the shift σ is mixing in Σ_m with respect to the Markov measure $P(A)$.*

Proof. Exercise 4.4.7 □

Markov chains can be generalized to the class of *stationary (discrete) stochastic processes*, dynamical systems with invariant measures on shift spaces with a *continuous* alphabet. Let $(\Omega, \mathfrak{A}, P)$ be a probability space. A *random variable* on Ω is a measurable real-valued function on Ω. A sequence $(f_i)_{i=-\infty}^{\infty}$ of *random variables* is *stationary* if, for any $i_1, \ldots, i_k \in \mathbb{Z}$ and any Borel subsets $B_1, \ldots, B_k \subset \mathbb{R}$

$$P\{\omega \in \Omega : f_{i_j}(\omega) \in B_j, \ j = 1, \ldots, k\}$$
$$= P\{\omega \in \Omega : f_{i_j+n}(\omega) \in B_j, \ j = 1, \ldots, k\}.$$

Define the map $\Phi\colon\ \Omega \to \mathbb{R}^{\mathbb{Z}}$ by

$$\Phi(\omega) = (\ldots, f_{-1}(\omega), f_0(\omega), f_1(\omega), \ldots)$$

and the measure μ on the Borel subsets of $\mathbb{R}^{\mathbb{Z}}$ by $\mu(A) = P(\Phi^{-1}(A))$. Since the sequence (f_i) is stationary, the shift $\sigma : \mathbb{R}^{\mathbb{Z}} \to \mathbb{R}^{\mathbb{Z}}$ defined by $(\sigma x)_n = x_{n+1}$ preserves μ (Exercise 4.4.8).

Exercise 4.4.1. Prove that the circle rotation R_α is not weak mixing.

Exercise 4.4.2. Let $\alpha \in \mathbb{R}$ be irrational and let $F\colon\ \mathbb{T}^2 \to \mathbb{T}^2$ be the map $(x, y) \mapsto (x + \alpha, x + y)$ mod 1 introduced in §2.4. Prove that F preserves the Lebesgue measure and is ergodic but not weak mixing.

Exercise 4.4.3. Prove that any hyperbolic automorphism of $\mathbb{T}^n$ is mixing.

Exercise 4.4.4. Show that an isometry of a compact metric space is not mixing for any invariant Borel measure whose support is not a single point. In particular, circle rotations are not mixing.

Exercise 4.4.5. Prove that any Markov measure is shift invariant.

Exercise 4.4.6. Prove that supp $P_{A,q} = \Sigma^v_{A'}$.

Exercise 4.4.7. Prove Proposition 4.4.4.

Exercise 4.4.8. Prove that the measure μ on $\mathbb{R}^{\mathbb{Z}}$ constructed above for a stationary sequence (f_i) is invariant under the shift σ.

4.5 Ergodic theorems[2]

The collection of all orbits represents a complete evolution of the dynamical system T. The values $f(T^n(x))$ of a (measurable) function f may represent observations such as position or velocity. Long term averages $\frac{1}{n}\sum_{k=0}^{n-1} f(T^k(x))$ of these quantities are important in statistical physics and other areas. A central question in ergodic theory is whether these averages converge as $n \to \infty$ and, if yes, whether the limit depends on x. In the context of statistical physics, the *ergodic hypothesis* states that the asymptotic time average $\lim_{n\to\infty} 1/n \sum_{k=0}^{n-1} f(T^k(x))$ equals the space average $\int_X f\,d\mu$ for a.e. x. We show that this happens if T is ergodic.

Let $(X, \mathfrak{A}, \mu)$ be a measure space and $T\colon X \to X$ a measure preserving transformation. For a measurable function $f\colon X \to \mathbb{C}$ set $(U_T f)(x) = f(T(x))$. The operator U_T is linear and multiplicative: $U_T(f \cdot g) = U_T f \cdot U_T g$. Since T is measure-preserving, U_T is an isometry of $L^p(X, \mathfrak{A}, \mu)$ for any $p \geqslant 1$, i.e. $\|U_T f\|_p = \|f\|_p$ for any $f \in L^p$ (Exercise 4.5.3). If T is an automorphism, then $U_T^{-1} = U_{T^{-1}}$ is also an isometry, and hence U_T is a unitary operator on $L^2(X, \mathfrak{A}, \mu)$. We denote the scalar product on $L^2(X, \mathfrak{A}, \mu)$ by $\langle f, g\rangle$, the norm by $\|\,.\,\|$ and the adjoint operator of U by U^*.

LEMMA 4.5.1. *Let U be a linear isometry of a Hilbert space H. Then $Uf = f$ if and only if $U^*f = f$.*

Proof. For every $f, g \in H$ we have $\langle U^*Uf, g\rangle = \langle Uf, Ug\rangle = \langle f, g\rangle$ and hence $U^*Uf = f$. If $Uf = f$, then (multiplying both sides by U^*) $U^*f = f$. Conversely, if $U^*f = f$, then $\langle f, Uf\rangle = \langle U^*f, f\rangle = \|f\|^2$ and $\langle Uf, f\rangle = \langle f, U^*f\rangle = \|f\|^2$. Therefore $\langle Uf - f, Uf - f\rangle = \|Uf\|^2 - \langle f, Uf\rangle - \langle Uf, f\rangle + \|f\|^2 = 0$. □

THEOREM 4.5.2 (von Neumann Ergodic Theorem). *Let U be a linear isometry of a separable Hilbert space H and let P be orthogonal projection onto the subspace $I = \{f \in H : Uf = f\}$ of U-invariant vectors in H. Then for every $f \in H$*

$$\lim_{n\to\infty} \frac{1}{n}\sum_{i=0}^{n-1} U^i f = Pf.$$

[2] Several proofs in this section are due to F. Riesz, see Halmos (1960).

Proof. Let $U_n = \frac{1}{n}\sum_{i=0}^{n-1} U^i$ and $L = \{g - Ug : g \in H\}$. Note that L and I are U-invariant, and I is closed. If $f = g - Ug \in L$, then $\sum_{i=0}^{n-1} U^i f = g - U^n g$ and hence $U_n f \to 0$ as $n \to \infty$. If $f \in I$, then $U_n f = f$ for all $n \in \mathbb{N}$. We will show that $L \perp I$ and $H = \overline{L} \oplus I$, where $\overline{L}$ is the closure of L.

Let $\{f_k\}$ be a sequence in L, and suppose $f_k \to f \in \overline{L}$. Then $\|U_n f\| \leqslant \|U_n(f - f_k)\| + \|U_n f_k\| \leqslant \|U_n\| \cdot \|f - f_k\| + \|U_n f_k\|$, and hence $U_n f \to 0$ as $n \to \infty$.

Let $\perp$ denote the orthogonal complement and note that $\overline{L}^{\perp} = L^{\perp}$. If $h \in L^{\perp}$, then $0 = \langle h, g - Ug\rangle = \langle h - U^*h, g\rangle$ for all $g \in H$ so that $h = U^*h$, and hence $Uh = h$, by Lemma 4.5.1. Conversely (again using Lemma 4.5.1), if $h \in I$, then $\langle h, g - Ug\rangle = \langle h, g\rangle - \langle U^*h, g\rangle = 0$ for every $g \in H$, and hence $h \in L^{\perp}$.

Therefore, $H = \overline{L} \oplus I$ and $\lim_{n\to\infty} U_n$ is the identity on I and 0 on $\overline{L}$. □

The following theorem is an immediate corollary of the von Neumann Ergodic Theorem.

THEOREM 4.5.3. *Let T be a measure-preserving transformation of a finite measure space $(X, \mathfrak{A}, \mu)$. For $f \in L^2(X, \mathfrak{A}, \mu)$, set*

$$f_N^+(x) = \frac{1}{N}\sum_{n=0}^{N-1} f(T^n(x)).$$

Then f_N^+ converges in $L^2(X, \mathfrak{A}, \mu)$ to a T-invariant function $\overline{f}$.

If T is invertible, then $f_N^-(x) = \frac{1}{N}\sum_{n=0}^{N-1} f(T^{-n}(x))$ also converges in $L^2(X, \mathfrak{A}, \mu)$ to $\overline{f}$.

Similarly, let T be a measure-preserving flow in a finite measure space $(X, \mathfrak{A}, \mu)$. For a function $f \in L^2(X, \mathfrak{A}, \mu)$ set

$$f_\tau^+(x) = \frac{1}{\tau}\int_0^\tau f(T^t(x))\,dt \quad \text{and} \quad f_\tau^-(x) = \frac{1}{\tau}\int_0^\tau f(T^{-t}(x))\,dt.$$

Then f_τ^+ and f_τ^- converge in $L^2(X, \mathfrak{A}, \mu)$ to a T-invariant function $\overline{f}$. □

Our next objective is to prove a pointwise version of the preceding theorem. First, we need a combinatorial lemma. If $a_1, \ldots, a_m$ are real numbers and $1 \leqslant n \leqslant m$, we say that a_k is an *n-leader* if $a_k + \cdots + a_{k+p-1} \geqslant 0$ for some p, $1 \leqslant p \leqslant n$.

LEMMA 4.5.4. *For every n, $1 \leqslant n \leqslant m$, the sum of all n-leaders is non-negative.*

Proof. If there are no n-leaders, the lemma is true. Otherwise, let a_k be the first n-leader and $p \geqslant 1$ be the smallest integer for which $a_k + \cdots + a_{k+p-1} \geqslant 0$. If $k \leqslant j \leqslant k+p-1$, then $a_j + \cdots + a_{k+p-1} \geqslant 0$, by the choice of p, and hence a_j is an n-leader. The same argument can be applied to the sequence $a_{k+p}, \ldots, a_m$, which proves the lemma. □

THEOREM 4.5.5 (Birkhoff Ergodic Theorem). *Let T be a measure-preserving transformation in a finite measure space $(X, \mathfrak{A}, \mu)$ and let $f \in L^1(X, \mathfrak{A}, \mu)$. Then the limit*

$$\overline{f}(x) = \lim_{n\to\infty} \frac{1}{n} \sum_{k=0}^{n-1} f(T^k(x))$$

exists for a.e. $x \in X$, is μ-integrable and T-invariant, and satisfies

$$\int_X \overline{f}(x)\, d\mu = \int_X f(x)\, d\mu.$$

If, in addition, $f \in L^2(X, \mathfrak{A}, \mu)$, then by Theorem 4.5.3, $\overline{f}$ is the orthogonal projection of f to the subspace of T-invariant functions.

If T is invertible, then $\frac{1}{n}\sum_{k=0}^{n-1} f(T^{-k}(x))$ also converges almost everywhere to $\overline{f}$.

Similarly, let T be a measure-preserving flow in a finite measure space $(X, \mathfrak{A}, \mu)$. Then

$$f_\tau^+(x) = \frac{1}{\tau}\int_0^\tau f(T^t(x))\, dt \quad and \quad f_\tau^-(x) = \frac{1}{\tau}\int_0^\tau f(T^{-t}(x))\, dt$$

converge almost everywhere to the same μ-integrable and T-invariant limit function $\overline{f}$, and $\int_X f(x)\, d\mu = \int_X \overline{f}(x)\, d\mu$.

Proof. We consider only the case of a transformation. We assume without loss of generality that f is real-valued. Let

$$A = \{x \in X : f(x) + f(T(x)) + \cdots + f(T^k(x)) \geqslant 0 \text{ for some } k \in \mathbb{N}_0\}.$$

LEMMA 4.5.6 (Maximal Ergodic Theorem). $\int_A f(x)\, d\mu \geqslant 0$.

Proof. Let $A_n = \{x \in X : \sum_{i=0}^k f(T^i(x)) \geqslant 0 \text{ for some } k,\ 0 \leqslant k \leqslant n\}$. Then $A_n \subset A_{n+1}, A = \bigcup_{n\in\mathbb{N}} A_n$ and, by the Dominated Convergence Theorem, it suffices to show that $\int_{A_n} f(x)\, d\mu \geqslant 0$ for each n.

Fix an arbitrary $m \in \mathbb{N}$. Let $s_n(x)$ be the sum of the n-leaders in the sequence $f(x), f(T(x)), \ldots, f(T^{m+n-1}(x))$. For $k \leqslant m-1$, let $B_k \subset X$ be the set of points for which $f(T^k(x))$ is an n-leader of this sequence. By Lemma 4.5.4,

$$0 \leqslant \int_X s_n(x)\, d\mu = \sum_{k=0}^{m+n-1} \int_{B_k} f(T^k(x))\, d\mu. \tag{4.1}$$

Note that $x \in B_k$ if and only if $T(x) \in B_{k-1}$. Therefore, $B_k = T^{-1}(B_{k-1})$ and $B_k = T^{-k}(B_0)$ for $1 \leqslant k \leqslant m-1$, and hence

$$\int_{B_k} f(T^k(x))\, d\mu = \int_{T^{-k}(B_0)} f(T^k(x))\, d\mu = \int_{B_0} f(x)\, d\mu.$$

Thus the first m terms in (4.1) are equal and since $B_0 = A_n$,

$$m \int_{A_n} f(x)\, d\mu + n \int_X |f(x)|\, d\mu \geqslant 0.$$

Since m is arbitrary, the lemma follows. □

Now we can finish the proof of the Birkhoff Ergodic Theorem. For any $a, b \in \mathbb{R},\ a < b$, the set

$$X(a,b) = \left\{ x \in X : \underline{\lim}_{n\to\infty} \frac{1}{n} \sum_{i=0}^{n-1} f(T^i(x)) < a < b < \overline{\lim}_{n\to\infty} \frac{1}{n} \sum_{i=0}^{n-1} f(T^i(x)) \right\}$$

is measurable and T-invariant. We claim that $\mu(X(a,b)) = 0$. Apply Lemma 4.5.6 to $T|_{X(a,b)}$ and $f - b$ to obtain that $\int_{X(a,b)} (f(x) - b)\, d\mu \geqslant 0$. Similarly, $\int_{X(a,b)} (a - f(x))\, d\mu \geqslant 0$, and hence $\int_{X(a,b)} (a - b)\, d\mu \geqslant 0$. Therefore $\mu(X(a,b)) = 0$. Since a and b are arbitrary, we conclude that the averages $\frac{1}{n} \sum_{i=0}^{n-1} f(T^i(x))$ converge for a.e. $x \in X$.

For $n \in \mathbb{N}$, let $f_n(x) = \frac{1}{n} \sum_{i=0}^{n-1} f(T^i(x))$. Define $\overline{f}\colon X \to \mathbb{R}$ by $\overline{f}(x) = \underline{\lim}_{n\to\infty} f_n(x)$. Then $\overline{f}$ is measurable, and f_n converges a.e. to $\overline{f}$. By Fatou's Lemma and invariance of μ,

$$\begin{aligned}\int_X \underline{\lim}_{n\to\infty} |f_n(x)|\, d\mu &\leqslant \underline{\lim}_{n\to\infty} \int_X |f_n(x)|\, d\mu \\ &\leqslant \underline{\lim}_{n\to\infty} \frac{1}{n} \sum_{j=0}^{n-1} \int_X |f(T^j(x))|\, d\mu = \int_X |f(x)|\, d\mu.\end{aligned}$$

Thus $\int_X |\overline{f}(x)|\, d\mu = \int_X \underline{\lim}\, |f_n(x)|\, d\mu$ is finite, so $\overline{f}$ is integrable.

The proof that $\int_X f(x)\, d\mu = \int_X \overline{f}(x)\, d\mu$ is left as an exercise (Exercise 4.5.2). □

The following facts are immediate corollaries of Theorem 4.5.5 (Exercises 4.5.4 and 4.5.5).

COROLLARY 4.5.7. *A measure-preserving transformation T in a finite measure space $(X, \mathfrak{A}, \mu)$ is ergodic if and only if for each $f \in L^1(X, \mathfrak{A}, \mu)$*

$$\lim_{n\to\infty} \frac{1}{n} \sum_{i=0}^{n-1} f(T^i(x)) = \frac{1}{\mu(X)} \int_X f(x)\, d\mu, \tag{4.2}$$

for a.e. x, i.e. if and only if the time average equals the space average for every L^1 function. □

The preceding corollary implies that to check the ergodicity of a measure-preserving transformation, it suffices to verify (4.2) for a dense subset of $L^1(X, \mathfrak{A}, \mu)$, e.g. for all continuous functions if X is a compact topological space and μ is a Borel measure. Moreover, due to linearity it suffices to check the convergence for a countable collection of functions that form a basis.

COROLLARY 4.5.8. *A measure-preserving transformation T of a finite measure space $(X, \mathfrak{A}, \mu)$ is ergodic if and only if for every $A \in \mathfrak{A}$, for a.e. $x \in X$*

$$\lim_{n\to\infty} \frac{1}{n} \sum_{k=0}^{n-1} \chi_A(T^k(x)) = \frac{\mu(A)}{\mu(X)},$$

where χ_A is the characteristic function of A. □

Exercise 4.5.1. Let T be a measure-preserving transformation of a finite measure space $(X, \mathfrak{A}, \mu)$. Prove that T is ergodic if and only if

$$\lim_{n\to\infty} \frac{1}{n} \sum_{k=0}^{n-1} \mu(T^{-k}(A) \cap B) = \mu(A) \cdot \mu(B)$$

for any $A, B \in \mathfrak{A}$.

Exercise 4.5.2. Using the Dominated Convergence Theorem, finish the proof of Theorem 4.5.5 by showing that the averages $\frac{1}{n} \sum_{j=0}^{n-1} f$ converge to $\overline{f}$ in L^1.

Exercise 4.5.3. Prove that if T is a measure-preserving transformation, then U_T is an isometry of $L^p(X, \mathfrak{A}, \mu)$ for any $p \geqslant 1$.

Exercise 4.5.4. Prove Corollary 4.5.7.

Exercise 4.5.5. Prove Corollary 4.5.8.

Exercise 4.5.6. A real number x is said to be *normal in base n* if for any $k \in \mathbb{N}$, every finite word of length k in the alphabet $\{0, \dots, n-1\}$ appears with asymptotic frequency n^{-k} in the base n expansion of x. Prove that almost every real number is normal with respect to every base $n \in \mathbb{N}$.

4.6 Invariant measures for continuous maps

In this section we show that a continuous map T of a compact metric space X into itself has at least one invariant Borel probability measure. Every finite Borel measure μ on X defines a bounded linear functional $L_\mu(f) = \int_X f d\mu$ on the space $C(X)$ of continuous functions on X; moreover, L_μ is positive in the sense that $L_\mu(f) \geqslant 0$ if $f \geqslant 0$. The Riesz Representation Theorem (Rudin, 1987) states that the converse is also true: for every positive bounded linear functional L on $C(X)$, there is a finite Borel measure μ on X such that $L = \int_X f d\mu$.

THEOREM 4.6.1 (Krylov–Bogolubov). *Let X be a compact metric space and $T\colon X \to X$ a continuous map. Then there is a T-invariant Borel probability measure μ on X.*

Proof. Fix $x \in X$. For a function $f\colon X \to \mathbb{R}$ set $S_f^n(x) = \frac{1}{n}\sum_{i=0}^{n-1} f(T^i(x))$. Let $\mathcal{F} \subset C(X)$ be a dense countable collection of continuous functions on X. For any $f \in \mathcal{F}$ the sequence $S_f^n(x)$ is bounded, and hence has a convergent subsequence. Since $\mathcal{F}$ is countable, there is a sequence $n_j \to \infty$ such that the limit

$$S_f^\infty(x) = \lim_{j\to\infty} S_f^{n_j}(x)$$

exists for every $f \in \mathcal{F}$. For any $g \in C(X)$ and any $\epsilon > 0$ there is $f \in \mathcal{F}$ such that $\max_{y\in X} |g(y) - f(y)| < \epsilon/3$. For large enough j and k we have $|S_f^{n_j}(x) - S_f^{n_k}(x)| < \epsilon/3$. Therefore

$$\begin{aligned}&\left|S_g^{n_j}(x) - S_g^{n_k}(x)\right| \\ &\quad \leqslant \left|S_g^{n_j}(x) - S_f^{n_j}(x)\right| + \left|S_f^{n_j}(x) - S_f^{n_k}(x)\right| + \left|S_f^{n_k}(x) - S_g^{n_k}(x)\right| < \epsilon,\end{aligned}$$

so $S_g^{n_j}(x)$ is a Cauchy sequence. Thus, the limit $S_g^\infty(x)$ exists for every $g \in C(X)$ and defines a bounded positive linear functional L_x on $C(X)$. By the Riesz Representation Theorem, there is a Borel probability measure μ such that $L_x(g) = \int_X g\, d\mu$. Note that

$$\left|S_g^{n_j}(T(x)) - S_g^{n_j}(x)\right| = \frac{1}{n_j}|g(T^{n_j}(x)) - g(x)|.$$

Therefore $S_g^{\infty}(T(x)) = S_g^{\infty}(x)$ and μ is T-invariant. □

Let $\mathcal{M} = \mathcal{M}(X)$ denote the set of all Borel probability measures on X. A sequence of measures $\mu_n \in \mathcal{M}$ converges in the *weak* topology* to a measure $\mu \in \mathcal{M}$ if $\int_X f\, d\mu_n \to \int_X f\, d\mu$ for every $f \in C(X)$. If μ_n is any sequence in $\mathcal{M}$ and $F \subset C(X)$ is a dense countable subset, then, by a diagonal process, there is a subsequence μ_{n_j} such that $\int_X f\, d\mu_{n_j}$ converges for every $f \in F$, and hence the sequence $\int_X g\, d\mu_{n_j}$ converges for every $g \in C(X)$. Therefore $\mathcal{M}$ is compact in the weak* topology. It is also convex: $t\mu + (1-t)\nu \in \mathcal{M}$ for any $t \in [0, 1]$ and $\mu, \nu \in \mathcal{M}$. A point in a convex set is *extreme* if it cannot be represented as a non-trivial convex combination of two other points. The extreme points of $\mathcal{M}$ are the probability measures supported on points; they are called *Dirac measures.*

Let $\mathcal{M}_T \subset \mathcal{M}$ denote the set of all T-invariant Borel probability measures on X. Then $\mathcal{M}_T$ is closed, and therefore compact, in the weak* topology, and convex.

Recall that if μ and ν are finite measures on a space X with σ-algebra $\mathfrak{A}$, then ν is *absolutely continuous* with respect to μ if $\nu(A) = 0$ whenever $\mu(A) = 0$, for $A \in \mathfrak{A}$. If ν is absolutely continuous with respect to μ, then the Radon–Nikodym Theorem asserts that there is an L^1 function $d\nu/d\mu$, called the *Radon–Nikodym derivative*, such that $\nu(A) = \int_A (d\nu/d\mu)(x)\, d\mu$ for every $A \in \mathfrak{A}$ (Royden, 1988).

PROPOSITION 4.6.2. *Ergodic T-invariant measures are precisely the extreme points of $\mathcal{M}_T$.*

Proof. If μ is not ergodic, then there is a T-invariant measurable subset $A \subset X$ with $0 < \mu(A) < 1$. Let $\mu_A(B) = \mu(B \cap A)/\mu(A)$ and $\mu_{X\setminus A}(B) = \mu(B \cap (X \setminus A))/\mu(X \setminus A)$ for any measurable set B. Then μ_A and $\mu_{X\setminus A}$ are T-invariant and $\mu = \mu(A)\mu_A + \mu(X \setminus A)\mu_{X\setminus A}$, so μ is not an extreme point.

Conversely, assume that $\mu = t\nu + (1-t)\kappa$ with $\nu, \kappa \in \mathcal{M}_T$ and $t \in (0, 1)$. Then ν is absolutely continuous with respect to μ and $\nu(A) = \int_A r\, d\mu$, where $r = d\nu/d\mu \in L^1(X, \mu)$ is the Radon–Nikodym derivative. Since μ and ν are T-invariant, so is r. If μ is ergodic, then r is essentially constant, so $\mu = \nu = \kappa$. □

By the Krein–Milman Theorem (Royden, 1988; Rudin, 1991), $\mathcal{M}_T$ is the closed convex hull of its extreme points. Therefore, the set $\mathcal{M}_T^e$ of all T-invariant, ergodic, Borel, probability measures is not empty. However, $\mathcal{M}_T^e$ may be rather complicated; for example, it may be dense in $\mathcal{M}_T$ in the weak* topology (Exercise 4.6.5).

Exercise 4.6.1. Describe $\mathcal{M}_T$ and $\mathcal{M}_T^e$ for the homeomorphism of the circle $T(x) = x + a \sin 2\pi x \bmod 1,\ 0 < a \leqslant \frac{1}{2\pi}$.

Exercise 4.6.2. Describe $\mathcal{M}_T$ and $\mathcal{M}_T^e$ for the homeomorphism of the torus $T(x, y) = (x, x + y) \bmod 1$.

Exercise 4.6.3

(a) Give an example of a map of the circle that is discontinuous at exactly one point and does not have non-trivial finite invariant Borel measures.

(b) Give an example of a continuous map of the real line that does not have non-trivial finite invariant Borel measures.

Exercise 4.6.4. Let X and Y be compact metric spaces and $T\colon X \to Y$ a continuous map. Show that T induces a natural map $\mathcal{M}(X) \to \mathcal{M}(Y)$, and that this map is continuous in the weak* topology.

***Exercise 4.6.5.** Prove that if σ is the two-sided 2-shift, then $\mathcal{M}_\sigma^e$ is dense in $\mathcal{M}_\sigma$ in the weak* topology.

4.7 Unique ergodicity and Weyl's Theorem[3]

In this section T is a continuous map of a compact metric space X. By §4.6, there are T-invariant Borel probability measures. If there is only one such measure, then T is said to be *uniquely ergodic*. Note that this unique invariant measure is necessarily ergodic by Proposition 4.6.2.

An irrational circle rotation is uniquely ergodic (Exercise 4.7.1). Moreover, any topologically transitive translation on a compact abelian group is uniquely ergodic (Exercise 4.7.2). On the other hand, unique ergodicity does not imply topological transitivity (Exercise 4.7.3).

PROPOSITION 4.7.1. *Let X be a compact metric space. A continuous map $T\colon X \to X$ is uniquely ergodic if and only if $S_f^n = \frac{1}{n}\sum_{i=0}^{n-1} f \circ T^i$ converges uniformly to a constant function S_f^∞ for any continuous function $f \in C(X)$.*

[3] The arguments of this section follow in part those of Furstenberg (1981a) and Cornfeld *et al.* (1982).

Proof. Suppose first that T is uniquely ergodic and μ is the unique T-invariant Borel probability measure. We will show that

$$\lim_{n\to\infty} \max_{x\in X} \left| S_f^n(x) - \int_X f d\mu \right| \to 0.$$

Assume, for a contradiction, that there are $f \in C(X)$ and sequences $x_k \in X$ and $n_k \to \infty$ such that $\lim_{k\to\infty} S_f^{n_k}(x_k) = c \neq \int_X f\, d\mu$. As in the proof of Proposition 4.6.1, there is a subsequence $n_{k_i} \to \infty$ such that the limit $L(g) = \lim_{i\to\infty} S_f^{n_{k_i}}(x_{k_i})$ exists for any $g \in C(X)$. As in Proposition 4.6.1, L defines a T-invariant, positive, bounded linear functional on $C(X)$. By the Riesz Representation Theorem, $L(g) = \int_X g\, d\nu$ for some $\nu \in \mathcal{M}_T$. Since $L(f) = c \neq \int_X f\, d\mu$, the measures μ and ν are different, which contradicts unique ergodicity.

The proof of the converse is left as an exercise (Exercise 4.7.4). □

Uniform convergence of the time averages of continuous functions does not, by itself, imply unique ergodicity. For example, if (X, T) is uniquely ergodic and $I = [0, 1]$, then $(X \times I, T \times \mathrm{Id})$ is not uniquely ergodic, but the time averages converge uniformly for all continuous functions.

PROPOSITION 4.7.2. *Let T be a topologically transitive continuous map of a compact metric space X. Suppose that the sequence of time averages S_f^n converges uniformly for every continuous function $f \in C(X)$. Then T is uniquely ergodic.*

Proof. Since the convergence is uniform, $S_f^\infty = \lim_{n\to\infty} S_f^n$ is a continuous function. As in the proof of Proposition 4.6.1, $S_f^\infty(T(x)) = S_f^\infty(x)$ for every x. Since T is topologically transitive, S_f^∞ is constant. As in previous arguments, the linear functional $f \mapsto S_f^\infty$ defines a measure $\mu \in \mathcal{M}_T$ with $\int_X f\, d\mu = S_f^\infty$. Let $\nu \in \mathcal{M}_T$. By the Birkhoff Ergodic Theorem (Theorem 4.5.5), $S_f^\infty(x) = \int_X f\, d\nu$ for every $f \in C(X)$ and ν a.e. $x \in X$. Therefore $\nu = \mu$. □

Let X be a compact metric space with a Borel probability measure μ. Let $T\colon X \to X$ be a homeomorphism preserving μ. A point $x \in X$ is called *generic* for (X, μ, T) if for every continuous function f

$$\lim_{n\to\infty} \frac{1}{n} \sum_{k=0}^{n-1} f(T^k(x)) = \int_X f d\mu.$$

If T is ergodic, then by Corollary 4.5.8, μ-a.e. x is generic.

For a compact topological group G, the Haar measure on G is the unique Borel probability measure invariant under all left and right translations. Let $T\colon X \to X$ be a homeomorphism of a compact metric space, G a compact group, and $\phi\colon X \to G$ a continuous function. The homeomorphism $S : X \times G \to X \times G$ given by $S(x, g) = (T(x), \phi(x)g)$ is a *group extension* (or *G-extension*) of T. Observe that S commutes with the right translations $R_g(x, h) = (x, hg)$. If μ is a T-invariant measure on X and m is the Haar measure on G, then the product measure $\mu \times m$ is S-invariant (Exercise 4.7.7).

PROPOSITION 4.7.3 (Furstenberg). *Let G be a compact group with Haar measure m, X a compact metric space with a Borel probability measure μ, $T\colon X \to X$ a homeomorphism preserving μ, $Y = X \times G$, $\nu = \mu \times m$, and $S\colon Y \to Y$ a G-extension of T. If T is uniquely ergodic and S is ergodic, then S is uniquely ergodic.*

Proof. Since ν is R_g-invariant for every $g \in G$, if (x, h) is generic for ν, then (x, hg) is generic for ν. Since S is ergodic, ν-a.e. (x, h) is ν-generic. Therefore for μ-a.e. $x \in X$, the point (x, h) is ν-generic for every h. If a measure $\nu' \neq \nu$ is S-invariant and ergodic, then ν'-a.e. (x, h) is ν'-generic. The points that are ν'-generic cannot be ν-generic. Hence there is a subset $N \subset X$ such that $\mu(N) = 0$ and the first coordinate x of every ν'-generic point (x, h) lies in N. However, the projection of ν' to X is T-invariant and therefore is μ. This is a contradiction. □

PROPOSITION 4.7.4. *Let $\alpha \in (0, 1)$ be irrational and let $T\colon \mathbb{T}^k \to \mathbb{T}^k$ be defined by*

$$T(x_1, \ldots, x_k) = (x_1 + \alpha, x_2 + a_{21}x_1, \ldots, x_k + a_{k1}x_1 + \cdots a_{k\,k-1}x_{k-1}),$$

where the coefficients a_{ij} are integers and $a_{i\,i-1} \neq 0$, $i = 2, \ldots, k$. Then T is uniquely ergodic.

Proof. By Exercise 4.7.8, T is ergodic with respect to Lebesgue measure on $\mathbb{T}^k$. An inductive application of Proposition 4.7.3 yields the result. □

Let X be a compact topological space with a Borel probability measure μ. A sequence $(x_i)_{i\in\mathbb{N}}$ in X is *uniformly distributed* if for any continuous function f on X,

$$\lim_{n\to\infty} \frac{1}{n} \sum_{k=1}^{n} f(x_k) = \int_X f d\mu.$$

THEOREM 4.7.5 (Weyl). *If* $P(x) = b_k x^k + \cdots + b_0$ *is a real polynomial such that at least one of the coefficients* b_i, $i > 0$, *is irrational, then the sequence* $(P(n) \bmod 1)_{n\in\mathbb{N}}$ *is uniformly distributed in* $[0, 1]$.

Proof. Assume first that $b_k = \alpha/k!$ with α irrational. Consider the map $T\colon \mathbb{T}^k \to \mathbb{T}^k$ given by

$$T(x_1, \ldots, x_k) = (x_1 + \alpha, x_2 + x_1, \ldots, x_k + x_{k-1}).$$

Let $\pi : \mathbb{R}^k \to \mathbb{T}^k$ be the projection. Let $P_k(x) = P(x)$ and $P_{i-1}(x) = P_i(x+1) - P_i(x)$, $i = k, \ldots, 1$. Then $P_1(x) = \alpha x + \beta$. Observe that $T^n(\pi(P_1(0), \ldots, P_k(0))) = \pi(P_1(n), \ldots, P_k(n))$. Since T is uniquely ergodic by Proposition 4.7.4, this orbit (and any other orbit) is uniformly distributed on $\mathbb{T}^k$. It follows that the last coordinate $P_k(n) = P(n)$ is uniformly distributed on S^1.

Exercise 4.7.9 finishes the proof. □

Exercise 4.7.1. Prove that an irrational circle rotation is uniquely ergodic.

Exercise 4.7.2. Prove that any topologically transitive translation on a compact abelian group is uniquely ergodic.

Exercise 4.7.3. Prove that the diffeomorphism $T\colon S^1 \to S^1$ defined by $T(x) = x + a\sin^2(\pi x)$, $a < 1/\pi$, is uniquely ergodic but not topologically transitive.

Exercise 4.7.4. Prove the remaining statement of Proposition 4.7.1.

Exercise 4.7.5. Prove that the subshift defined by a fixed point a of a primitive substitution s is uniquely ergodic.

Exercise 4.7.6. Let T be a uniquely ergodic continuous transformation of a compact metric space X, and μ the unique invariant Borel probability measure. Show that $\operatorname{supp}\mu$ is a minimal set for T.

Exercise 4.7.7. Let $S : X \times G \to X \times G$ be a G-extension of $T : (X, \mu) \to (X, \mu)$ and let m be the Haar measure on G. Prove that the product measure $\mu \times m$ is S-invariant.

Exercise 4.7.8. Use Fourier series on $\mathbb{T}^k$ to prove that T from Proposition 4.7.4 is ergodic with respect to Lebesgue measure.

Exercise 4.7.9. Reduce the general case of Theorem 4.7.5 to the case where the leading coefficient is irrational.

4.8 The Gauss transformation revisited[4]

Recall that the Gauss transformation (§1.6) is the map of the unit interval to itself defined by

$$\phi(x) = \frac{1}{x} - \left[\frac{1}{x}\right] \quad \text{for } x \in (0,1],\ \phi(0) = 0.$$

The Gauss measure μ defined by

$$\mu(A) = \frac{1}{\log 2} \int_A \frac{dx}{1+x} \tag{4.3}$$

is a ϕ-invariant probability measure on $[0, 1]$.

For an irrational $x \in (0, 1]$, the nth entry $a_n(x) = \left[1/\phi^{n-1}(x)\right]$ of the continued fraction representing x is called the nth *quotient*, and we write $x = [a_1(x), a_2(x), \dots]$. The irreducible fraction $p_n(x)/q_n(x)$ that is equal to the truncated continued fraction $[a_1(x), \dots, a_n(x)]$ is called the nth *convergent* of x. The numerators and denominators of the convergents satisfy the following relations:

$$p_0(x) = 0, \quad p_1(x) = 1, \quad p_n(x) = a_n(x)p_{n-1}(x) + p_{n-2}(x), \tag{4.4}$$

$$q_0(x) = 1,\ q_1(x) = a_1(x),\ q_n(x) = a_n(x)q_{n-1}(x) + q_{n-2}(x), \tag{4.5}$$

for $n > 1$. We have

$$x = \frac{p_n(x) + (\phi^n(x))\, p_{n-1}(x)}{q_n(x) + (\phi^n(x))\, q_{n-1}(x)}.$$

By an inductive argument

$$p_n(x) \geqslant 2^{(n-2)/2} \text{ and } q_n(x) \geqslant 2^{(n-1)/2} \text{ for } n \geqslant 2$$

and

$$p_{n-1}(x)q_n(x) - p_n(x)q_{n-1}(x) = (-1)^n, \quad n \geqslant 1. \tag{4.6}$$

For positive integers $b_k, k = 1, \dots, n,$ let

$$\Delta_{b_1,\dots,b_n} = \{x \in (0,1] : a_k(x) = b_k, k = 1, \dots, n\}.$$

The interval $\Delta_{b_1,\dots,b_n}$ is the image of the interval $[0, 1)$ under the map $\psi_{b_1,\dots,b_n}$ defined by

$$\psi_{b_1,\dots,b_n}(t) = [b_1, \dots, b_{n-1}, b_n + t].$$

[4] The arguments of this section follow in part those of Billingsley (1965).

If n is odd, $\psi_{b_1,\dots,b_n}$ is decreasing; if n is even, it is increasing. For $x \in \Delta_{b_1,\dots,b_n}$,

$$x = \psi_{b_1,\dots,b_n}(t) = \frac{p_n + tp_{n-1}}{q_n + tq_{n-1}}, \tag{4.7}$$

where p_n and q_n are given by the recursive relations (4.4) and (4.5) with $a_n(x)$ replaced by b_n. Therefore

$$\Delta_{b_1,\dots,b_n} = \left[\frac{p_n}{q_n}, \frac{p_n + p_{n-1}}{q_n + q_{n-1}}\right), \text{ if } n \text{ is even}$$

and

$$\Delta_{b_1,\dots,b_n} = \left(\frac{p_n + p_{n-1}}{q_n + q_{n-1}}, \frac{p_n}{q_n}\right], \text{ if } n \text{ is odd.}$$

If λ is Lebesgue measure, then $\lambda(\Delta_{b_1,\dots,b_n}) = (q_n(q_n + q_{n-1}))^{-1}$.

PROPOSITION 4.8.1. *The Gauss transformation is ergodic for the Gauss measure μ.*

Proof. For a measure ν and measurable sets A and B with $\nu(B) \neq 0$, let $\nu(A \mid B) = \nu(A \cap B)/(\nu(B))$ denote the conditional measure. Fix $b_1, \dots, b_n$ and let $\Delta_n = \Delta_{b_1,\dots,b_n}$, $\psi_n = \psi_{b_1,\dots,b_n}$. The length of Δ_n is $\pm(\psi_n(1) - \psi_n(0))$, and for $0 \leqslant x < y \leqslant 1$,

$$\lambda(\{z : x \leqslant \phi^n(z) < y\} \cap \Delta_n) = \pm(\psi_n(y) - \psi_n(x)),$$

where the sign depends on the parity of n. Therefore

$$\lambda(\phi^{-n}([x, y)) \mid \Delta_n) = \frac{\psi_n(y) - \psi_n(x)}{\psi_n(1) - \psi_n(0)}$$

and by (4.6) and (4.7),

$$\lambda(\phi^{-n}([x, y)) \mid \Delta_n) = (y - x) \cdot \frac{q_n(q_n + q_{n-1})}{(q_n + xq_{n-1})(q_n + yq_{n-1})}.$$

The second factor in the right-hand side is between $1/2$ and 2. Hence

$$\frac{1}{2}\lambda([x, y)) \leqslant \lambda(\phi^{-n}([x, y)) \mid \Delta_n) \leqslant 2\lambda([x, y)).$$

Since the intervals $[x, y)$ generate the σ-algebra,

$$\frac{1}{2}\lambda(A) \leqslant \lambda(\phi^{-n}(A) \mid \Delta_n) \leqslant 2\lambda(A) \tag{4.8}$$

for any measurable set $A \subset [0, 1]$.

Because the density of the Gauss measure μ is between $1/(2\log 2)$ and $1/(\log 2)$,

$$\frac{1}{2\log 2}\lambda(A) \leqslant \mu(A) \leqslant \frac{1}{\log 2}\lambda(A).$$

By (4.8),

$$\frac{1}{4}\mu(A) \leqslant \mu\big(\phi^{-n}(A) \mid \Delta_n\big) \leqslant 4\mu(A)$$

for any measurable $A \subset [0,1]$.

Let A be a measurable ϕ-invariant set with $\mu(A) > 0$. Then $\frac{1}{4}\mu(A) \leqslant \mu\big(A \mid \Delta_n\big)$, or, equivalently, $\frac{1}{4}\mu(\Delta_n) \leqslant \mu\big(\Delta_n \mid A\big)$. Since the intervals Δ_n generate the σ-algebra,

$$\frac{1}{4}\mu(B) \leqslant \mu(B|A)$$

for any measurable set B. By choosing $B = [0,1]\backslash A$ we obtain that $\mu(A) = 1$. □

The ergodicity of the Gauss transformation has the following number-theoretic consequences.

PROPOSITION 4.8.2. *For almost every $x \in [0,1]$ (with respect to μ measure or Lebesgue measure), we have the following:*

1. *Each integer $k \in \mathbb{N}$ appears in the sequence $a_1(x), a_2(x), \dots$ with asymptotic frequency* $\dfrac{1}{\log 2}\log\left(\dfrac{k+1}{k}\right)$;
2. $\lim\limits_{n\to\infty} \dfrac{1}{n}\big(a_1(x) + \cdots + a_n(x)\big) = \infty$;
3. $\lim\limits_{n\to\infty} \sqrt[n]{a_1(x)a_2(x)\cdots a_n(x)} = \prod\limits_{k=1}^{\infty}\left(1 + \dfrac{1}{k^2+2k}\right)^{\log k/\log 2}$;
4. $\lim\limits_{n\to\infty} \dfrac{\log q_n(x)}{n} = \dfrac{\pi^2}{12\log 2}$

Proof

1. Let f be the characteristic function of the interval $[1/k, 1/(k+1))$. Then $a_n(x) = k$ if and only if $f(\phi^n(x)) = 1$. By the Birkhoff Ergodic Theorem, for almost every x,

$$\lim_{n\to\infty}\frac{1}{n}\sum_{i=0}^{n-1} f(\phi^i(x)) = \int_0^1 f\,d\mu = \mu\left(\left[\frac{1}{k}, \frac{1}{k+1}\right)\right) = \frac{1}{\log 2}\log\left(\frac{k+1}{k}\right),$$

which proves the first assertion.

2. Let $f(x) = [1/x]$, *i.e.* $f(x) = a_1(x)$. Note that $\int_0^1 f(x)/(1+x)\,dx = \infty$ since $f(x) > (1-x)/x$ and $\int_0^1 \frac{(1-x)}{x(1+x)}\,dx = \infty$. For $N > 0$, define

$$f_N(x) = \begin{cases} f(x) & \text{if } f(x) \leqslant N \\ 0 & \text{otherwise.} \end{cases}$$

Then, for any $N > 0$, for almost every x,

$$\begin{aligned} \varliminf_{n\to\infty} \frac{1}{n}\sum_{k=0}^{n-1} f(\phi^k(x)) &\geqslant \varliminf_{n\to\infty} \frac{1}{n}\sum_{k=0}^{n-1} f_N(\phi^k(x)) \\ &= \lim_{n\to\infty} \frac{1}{n}\sum_{k=0}^{n-1} f_N(\phi^k(x)) \\ &= \frac{1}{\log 2}\int_0^1 \frac{f_N(x)}{1+x}\,dx. \end{aligned}$$

Since $\lim_{N\to\infty} \int_0^1 \frac{f_N(x)}{1+x}\,dx \to \infty$, the conclusion follows.

3. Let $f(x) = \log a_1(x) = \log([\frac{1}{x}])$. Then $f \in L^1([0,1])$ with respect to the Gauss measure μ (Exercise 4.8.1). By the Birkhoff Ergodic Theorem,

$$\begin{aligned} \lim_{n\to\infty} \frac{1}{n}\sum_{k=1}^{n} \log a_k(x) &= \frac{1}{\log 2}\int_0^1 \frac{f(x)}{1+x}\,dx \\ &= \frac{1}{\log 2}\sum_{k=1}^{\infty} \int_{\frac{1}{k+1}}^{\frac{1}{k}} \frac{\log k}{1+x}\,dx \\ &= \sum_{k=1}^{\infty} \frac{\log k}{\log 2}\cdot \log\left(1 + \frac{1}{k^2+2k}\right). \end{aligned}$$

Exponentiating this expression gives part (3).

4. Note that $p_n(x) = q_{n-1}(\phi(x))$ (Exercise 4.8.2), so

$$\frac{1}{q_n(x)} = \frac{p_n(x)}{q_n(x)}\frac{p_{n-1}(\phi(x))}{q_{n-1}(\phi(x))}\cdots\frac{p_1(\phi^{n-1}(x))}{q_1(\phi^{n-1}(x))}.$$

Thus

$$\begin{aligned} -\frac{1}{n}\log q_n(x) = \frac{1}{n}\sum_{k=0}^{n-1} \log\left(\frac{p_{n-k}(\phi^k(x))}{q_{n-k}(\phi^k(x))}\right) &= \frac{1}{n}\sum_{k=0}^{n-1} \log(\phi^k(x)) \\ + \frac{1}{n}\sum_{k=0}^{n-1}\left(\log\frac{p_{n-k}(\phi^k(x))}{q_{n-k}(\phi^k(x))} - \log(\phi^k(x))\right). \end{aligned} \tag{4.9}$$

It follows from the Birkhoff Ergodic Theorem that the first term of (4.9) converges a.e. to $(1/\log 2)\int_0^1 \log x/(1+x)\,dx = -\pi^2/12$. The second term converges to 0 (Exercise 4.8.2). □

Exercise 4.8.1. Show that $\log([1/x]) \in L^1([0,1])$ with respect to the Gauss measure μ.

Exercise 4.8.2. Show that $p_n(x) = q_n(\phi(x))$ and that

$$\lim_{n\to\infty} \frac{1}{n} \sum_{k=0}^{n-1} \left(\log(\phi^k(x)) - \log \frac{p_{n-k}(\phi^k(x))}{q_{n-k}(\phi^k(x))} \right) = 0.$$

4.9 Discrete spectrum

Let T be an automorphism of a probability space $(X, \mathfrak{A}, \mu)$. The operator $U_T : L^2(X, \mathfrak{A}, \mu) \to L^2(X, \mathfrak{A}, \mu)$ is unitary and each of its eigenvalues is a complex number of absolute value 1. Denote by Σ_T the set of all eigenvalues of U_T. Since constant functions are T-invariant, 1 is an eigenvalue of U_T. Any T-invariant function is an eigenfunction of U_T with eigenvalue 1. Therefore, T is ergodic if and only if 1 is a simple eigenvalue of U_T. If f, g are two eigenfunctions with different eigenvalues $\sigma \neq \kappa$, then $\langle f, g\rangle = 0$ since $\langle f, g\rangle = \langle U_T f, U_T g\rangle = \sigma\overline{\kappa}\langle f, g\rangle$. Note that U_T is a multiplicative operator, i.e. $U_T(f \cdot g) = U_T(f) \cdot U_T(g)$, which has important implications for its spectrum.

PROPOSITION 4.9.1. *Σ_T is a subgroup of the unit circle $S^1 = \{z \in \mathbb{C} : |z| = 1\}$. If T is ergodic, then every eigenvalue of U_T is simple.*

Proof. If $\sigma \in \Sigma_T$ and $f(T(x)) = \sigma f(x)$, then $\overline{f}(T(x)) = \overline{\sigma}\overline{f}(x)$, and hence $\overline{\sigma} = \sigma^{-1} \in \Sigma_T$. If $\sigma_1, \sigma_2 \in \Sigma_T$ and $f_1(T(x)) = \sigma_1 f_1(x)$, $f_2(T(x)) = \sigma_2 f_2(x)$, then $f = f_1 f_2$ has eigenvalue $\sigma_1\sigma_2$, and hence $\sigma_1\sigma_2 \in \Sigma_T$. Therefore, Σ_T is a subgroup of S^1.

If T is ergodic, the absolute value of any eigenfuction f is essentially constant (and non-zero). Thus, if f and g are eigenfuctions with the same eigenvalue σ, then f/g is in L^2, and is an eigenfunction with eigenvalue 1, so it is essentially constant by ergodicity. Therefore every eigenvalue is simple. □

An ergodic automorphism T has *discrete spectrum* if the eigenfunctions of U_T span $L^2(X, \mathfrak{A}, \mu)$. An automorphism T has *continuous spectrum* if 1 is a simple eigenvalue of U_T and U_T has no other eigenvalues.

Consider a circle rotation $R_\alpha(x) = x + \alpha \bmod 1, x \in [0, 1)$. For each $n \in \mathbb{Z}$, the function $f_n(x) = \exp(2\pi inx)$ is an eigenfunction of U_{R_α} with eigenvalue $\exp(2\pi in\alpha)$. If α is irrational, the eigenfunctions f_n span L^2, and hence R_α has discrete spectrum. On the other hand, every weak mixing transformation has continuous spectrum (Exercise 4.9.1).

Let G be an abelian topological group. A *character* is a continuous homomorphism $\chi: G \to S^1$. The set of characters of G with the compact–open topology form a topological group $\hat{G}$ called the *group of characters* (or the *dual group*). For every $g \in G$, the evaluation map $\chi \mapsto \chi(g)$ is a character $\iota_g \in \hat{\hat{G}}$, the dual of $\hat{G}$, and the map $\iota: G \to \hat{\hat{G}}$ is a homomorphism. If $\iota_g(\chi) \equiv 1$, then $\chi(g) = 1$ for every $\chi \in \hat{G}$, and hence ι is injective. By the Pontryagin Duality Theorem (Helson, 1995), ι is also surjective and $\hat{\hat{G}} \cong G$. Moreover, if G is discrete then $\hat{G}$ is a compact abelian group, and conversely.

For example, each character $\chi \in \hat{\mathbb{Z}}$ is completely determined by the value $\chi(1) \in S^1$. Therefore $\hat{\mathbb{Z}} \cong S^1$. On the other hand, if $\lambda \in \hat{S}^1$, then $\lambda: S^1 \to S^1$ is a homomorphism, so $\lambda(z) = z^n$ for some $n \in \mathbb{Z}$. Therefore $\hat{S}^1 = \mathbb{Z}$.

On a compact abelian group G with Haar measure λ, every character is in L^∞, and therefore in L^2. The integral of any non-trivial character with respect to Haar measure is 0 (Exercise 4.9.3). If σ and σ' are characters of G, then $\sigma\overline{\sigma'}$ is also a character. If σ and σ' are different, then

$$\langle \sigma, \sigma' \rangle = \int_G \sigma(g)\overline{\sigma'}(g)\, d\lambda(g) = \int_G (\sigma\overline{\sigma'})(g)\, d\lambda(g) = 0.$$

Thus the characters of G are pairwise orthogonal in $L^2(G, \lambda)$.

THEOREM 4.9.2. *For every countable subgroup $\Sigma \subset S^1$ there is an ergodic automorphism T with discrete spectrum such that $\Sigma_T = \Sigma$.*

Proof. The identity character $\mathrm{Id}: \Sigma \to S^1, \mathrm{Id}(\sigma) = \sigma$, is a character of Σ. Let $T: \hat{\Sigma} \to \hat{\Sigma}$ be the translation $\chi \mapsto \chi \cdot \mathrm{Id}$. The normalized Haar measure λ on $\hat{\Sigma}$ is invariant under T. For $\sigma \in \Sigma$, let $f_\sigma \in \hat{\hat{\Sigma}}$ be the character of $\hat{\Sigma}$ such that $f_\sigma(\chi) = \chi(\sigma)$. Since

$$U_T f_\sigma(\chi) = f_\sigma(\chi\, \mathrm{Id}) = f_\sigma(\chi) f_\sigma(\mathrm{Id}) = \sigma f_\sigma(\chi),$$

f_σ is an eigenfunction with eigenvalue σ.

We claim that the linear span $\mathcal{A}$ of the set of characters $\{f_\sigma : \sigma \in \Sigma\}$, is dense in $L^2(\hat{\Sigma}, \lambda)$, which will complete the proof. The set of characters separates points of $\hat{\Sigma}$, is closed under complex conjugation, and contains the constant function 1. Since the set of characters is closed under multiplication,

$\mathcal{A}$ is closed under multiplication, and is therefore an algebra. By the Stone–Weierstrass Theorem (Royden, 1988), $\mathcal{A}$ is dense in $C(\Sigma, \mathbb{C})$, and therefore in $L^2(\hat{\Sigma}, \lambda)$. □

The following theorem (which we do not prove) is a converse to Theorem 4.9.2.

THEOREM 4.9.3 (Halmos–von Neumann). *Let T be an ergodic automorphism with discrete spectrum, and let $\Sigma \subset S^1$ be its spectrum. Then T is isomorphic to the translation on $\hat{\Sigma}$ by the identity character* $\text{Id}: \Sigma \to S^1$.

A measure-preserving transformation $T : (X, \mathfrak{A}, \mu) \to (X, \mathfrak{A}, \mu)$ is *aperiodic* if $\mu(\{x \in X : T^n(x) = x\}) = 0$ for every $n \in \mathbb{N}$. Theorem 4.9.4 (which we do not prove) implies that every aperiodic transformation can be approximated by a periodic transformation with an arbitrary period n. Many of the examples and counterexamples in abstract ergodic theory are constructed using the method of *cutting and stacking* based on this theorem.

THEOREM 4.9.4 (Rokhlin–Halmos). *Let T be an aperiodic automorphism of a Lebesgue probability space $(X, \mathfrak{A}, \mu)$. Then for every $n \in \mathbb{N}$ and $\epsilon > 0$ there is a measurable subset $A = A(n, \epsilon) \subset X$ such that the sets $T^i(A)$, $i = 0, \dots, n-1$, are pairwise disjoint and $\mu\left(X \setminus \bigcup_{i=0}^{n-1} T^i(A)\right) < \epsilon$.*

Exercise 4.9.1. Prove that every weak mixing measure-preserving transformation has continuous spectrum.

Exercise 4.9.2. Suppose that $\alpha, \beta \in (0, 1)$ are irrational and α/β is irrational. Let T be the translation of $\mathbb{T}^2$ given by $T(x, y) = (x + \alpha, y + \beta)$. Prove that T is topologically transitive, ergodic and has discrete spectrum.

Exercise 4.9.3. Show that on a compact topological group G, the integral of any non-trivial character with respect to the Haar measure is 0.

4.10 Weak mixing[5]

The property of weak mixing is typical in the following sense. Since each non-atomic probability Lebesgue space is isomorphic to the unit interval with Lebesgue measure λ, every measure-preserving transformation can be viewed as a transformation of $[0, 1]$ preserving λ. The *weak topology* on the

[5] The presentation of this section to a large extent follows (Krengel, 1985, §2.3).

set of all measure-preserving transformations of $[0, 1]$ is given by $T_n \to T$ if $\lambda(T_n(A) \triangle T(A)) \to 0$ for each measurable $A \subset [0, 1]$. Halmos showed that a residual (in the weak topology) subset of transformations are weak mixing (Halmos, 1944). Rokhlin showed that the set of strong mixing transformations is of first category (in the weak topology) (Rohlin, 1948).

The weak mixing transformations, as Theorem 4.10.6 below shows, are precisely those that have continuous spectrum. To show this we first prove a splitting theorem for isometries in a Hilbert space.

We say that a sequence of complex numbers a_n, $n \in \mathbb{Z}$ is *non-negative definite* if for each $N \in \mathbb{N}$,

$$\sum_{k,m=-N}^{N} z_k \bar{z}_m a_{k-m} \geqslant 0$$

for each finite sequence of complex numbers z_k, $-N \leqslant k \leqslant N$.

For a (linear) isometry U in a separable Hilbert space H denote by U^* the adjoint of U and for $n \geqslant 0$ set $U_n = U^n$ and $U_{-n} = U^{*n}$.

LEMMA 4.10.1. *For every $v \in H$, the sequence $\langle U_n v, v\rangle$ is non-negative definite.*

Proof

$$\sum_{k,m=-N}^{N} z_k \bar{z}_m \langle U_{k-m} v, v\rangle = \sum_{k,m=-N}^{N} z_k \bar{z}_m \langle U_k v, U_m v\rangle = \left\| \sum_{l=-N}^{N} z_l U_l v \right\|^2. \quad \square$$

LEMMA 4.10.2 (Wiener). *For a finite measure ν on $[0, 1)$ set $\hat{\nu}_k = \int_0^1 e^{2\pi i k x} \nu(dx)$. Then $\lim_{n\to\infty} n^{-1} \sum_{k=0}^{n-1} |\hat{\nu}_k| = 0$ if and only if ν has no atoms.*

Proof. Observe that $n^{-1} \sum_{k=0}^{n-1} |\hat{\nu}_k| \to 0$ if and only if $n^{-1} \sum_{k=0}^{n-1} |\hat{\nu}_k|^2 \to 0$. Now

$$\begin{aligned} \frac{1}{n} \sum_{k=0}^{n-1} |\hat{\nu}_k|^2 &= \frac{1}{n} \sum_{k=0}^{n-1} \int_0^1 e^{2\pi i k x} \nu(dx) \int_0^1 e^{-2\pi i k y} \nu(dy) \\ &= \int_0^1 \int_0^1 \left[\frac{1}{n} \sum_{k=0}^{n-1} e^{2\pi i k (x-y)} \right] \nu(dx) \nu(dy). \end{aligned}$$

The functions $n^{-1} \sum_{k=0}^{n-1} \exp(2\pi i k(x-y))$ are bounded in absolute value by 1 and converge to 1 for $x = y$ and to 0 for $x \neq y$. Therefore the last integral tends to the product measure $\nu \times \nu$ of the diagonal of $[0, 1) \times [0, 1)$. It follows that

$$\lim_{n\to\infty}\frac{1}{n}\sum_{k=0}^{n-1}|\nu_k|^2=\sum_{0\leqslant x<1}\left(\nu(\{x\})\right)^2. \qquad \square$$

For a (linear) isometry U of a separable Hilbert space H, set

$$H_w(U)=\left\{v\in H:\lim_{n\to\infty}\frac{1}{n}\sum_{k=0}^{n-1}\left|\langle U^k v,v'\rangle\right|=0 \text{ for each } v'\in H\right\}$$

and denote by $H_e(U)$ the closure of the subspace spanned by the eigenvectors of U. Both $H_w(U)$ and $H_e(U)$ are closed and U-invariant.

PROPOSITION 4.10.3. *Let U be a (linear) isometry of a separable Hilbert space H. Then*

1. *For each $v\in H$, there is a unique finite measure ν_v on the interval* $[0,1)$ *(called the* spectral measure*) such that for every $n\in\mathbb{Z}$*

$$\langle U_n v,v\rangle=\int_0^1 e^{2\pi inx}\nu_v(dx).$$

2. *If v is an eigenvector of U with eigenvalue* $\exp(2\pi i\alpha)$, *then ν_v consists of a single atom at α of measure* 1.
3. *If $v\perp H_e(U)$, then ν_v has no atoms and $v\in H_w(U)$.*

Proof. The first statement follows immediately from Lemma 4.10.1 and the Spectral Theorem for isometries in a Hilbert space (Helson, 1995; Folland, 1995). The second statement follows from the first (Exercise 4.10.3).

To prove the last statement let $v\perp H_e$ and $W=e^{-2\pi ix}U$. Applying the von Neumann Ergodic Theorem 4.5.2, let $u=\lim_{n\to\infty}n^{-1}\sum_{k=0}^{n-1}W^k v$. Then $Wu=u$. By Proposition 4.10.3,

$$\langle u,v\rangle=\lim_{n\to\infty}\frac{1}{n}\sum_{k=0}^{n-1}e^{-2\pi ixk}\langle U^k v,v\rangle=\lim_{n\to\infty}\int_0^1\frac{1}{n}\sum_{k=0}^{n-1}e^{-2\pi i(x-y)k}\nu_v(dy)=\nu_v(x).$$

If $\nu_v(x)>0$, then u is a non-zero eigenvector of U with eigenvalue $e^{2\pi ix}$ and $v\not\perp u$, which is a contradiction. Therefore $\nu_v(x)=0$ for each x, and Lemma 4.10.2 completes the proof. $\square$

For a finite subset $B\subset\mathbb{N}$ denote by $|B|$ the cardinality of B. For a subset $A\subset\mathbb{N}$, define the *upper density* $\bar{d}(A)$ by

$$\bar{d}(A)=\limsup_{n\to\infty}\frac{1}{n}|A\cap[1,n]|.$$

We say that a sequence b_n *converges in density* to b and write d-$\lim_n b_n = b$ if there is a subset $A \subset \mathbb{N}$ such that $\bar{d}(A) = 0$ and $\lim_{n\to\infty,\, n\notin A} b_n = b$.

LEMMA 4.10.4. *If (b_n) is a bounded sequence, then d-$\lim_n b_n = b$ if and only if $\lim_{n\to\infty} \frac{1}{n}\sum_{k=0}^{n-1} |b_n - b| = 0$.*

Proof. Exercise 4.10.1. □

The following splitting theorem is an immediate consequence of Proposition 4.10.3.

THEOREM 4.10.5 (Koopman–von Neumann (Koopman and von Neumann, 1932)). *Let U be a linear isometry in a separable Hilbert space H. Then $H = H_e \oplus H_w$. A vector $v \in H$ lies in $H_w(U)$ iff d-$\lim_n \langle U^n v, v\rangle = 0$ and iff d-$\lim_n \langle U^n v, w\rangle = 0$ for each $w \in H$.*

Proof. The splitting follows from Proposition 4.10.3. To prove the remaining statement that d-$\lim\langle U^n v, v\rangle = 0$ iff d-$\lim\langle U^n v, w\rangle = 0$, observe that $\langle U^n v, w\rangle \equiv 0$ if $w \perp U^k v$ for all $k \in \mathbb{N}$. If $w = U^k v$, then $\langle U^n v, w\rangle = \langle U^n v, U^k v\rangle = \langle U^{n-k} v, v\rangle$. □

Recall that if T and S are measure-preserving transformations in finite measure spaces $(X, \mathfrak{A}, \mu)$ and $(Y, \mathfrak{B}, \nu)$, then $T \times S$ is a measure-preserving transformation in the product space $(X \times Y, \mathfrak{A} \times \mathfrak{B}, \mu \times \nu)$. As in §4.9 we denote by U_T the isometry $U_T f(x) = f(T(x))$ of $L^2(X, \mathfrak{A}, \mu)$.

THEOREM 4.10.6. *Let T be a measure-preserving transformation of a probability space $(X, \mathfrak{A}, \mu)$. Then the following are equivalent:*

1. *T is weak mixing;*
2. *T has continuous spectrum;*
3. *d-$\lim_n \int_X f(T^n(x))\,\overline{f(x)}\,d\mu = 0$ if $f \in L^2(X, \mathfrak{A}, \mu)$ and $\int_X f\,d\mu = 0$;*
4. *d-$\lim_n \int_X f(T^n(x))\,\overline{g(x)}\,d\mu = \int_X f\,d\mu \cdot \int_X g\,d\mu$ for all $f, g \in L^2(X, \mathfrak{A}, \mu)$;*
5. *$T \times T$ is ergodic;*
6. *$T \times S$ is weak mixing for each weak mixing S;*
7. *$T \times S$ is ergodic for each ergodic S.*

Proof. The transformation T is weak mixing if and only if $H_w(U_T)$ is the orthogonal complement of the constants in $L^2(X, \mathfrak{A}, \mu)$. Therefore, by Proposition 4.10.3, (1)⇔(2). By Lemma 4.10.4, (1)⇔(3). Clearly (4) ⇒ (3). Assume that (3) holds. It is enough to show (4) for f with $\int_X f\,d\mu = 0$. Observe that (4) holds for g satisfying $\int_X f(T^k(x))\,\overline{g(x)}\,d\mu = 0$ for all $k \in \mathbb{N}$. Hence it suffices to consider $g(x) = f(T^k(x))$. But

$\int_X f(T^n(x))\overline{f(T^k(x))}d\mu = \int_X f(T^{n-k}(x))\overline{f(x)}d\mu \to 0$ as $n \to \infty$ by (3). Therefore (3)⇔(4).

Assume (5). Observe that T is ergodic and if U_T has an eigenfunction f, then $|f|$ is T-invariant, and hence constant. Therefore $f(x)/f(y)$ is $T \times T$-invariant and (5)⇒(2). Clearly (6)⇒(2) and (7)⇒(5).

Assume (3). To prove (7) observe that $L^2(X \times Y, \mathfrak{A} \times \mathfrak{B}, \mu \times \nu)$ is spanned by functions of the form $f(x)g(y)$. Let $\int_X f d\mu = \int_Y g d\nu = 0$. Then

$$\int_{X\times Y} f(T^n(x))g(S^n(y))f(x)g(y)d\mu \times \nu = \int_X f(T^n(x))f(x)d\mu \cdot \\ \times \int_Y g(S^n(y))g(y)d\nu.$$

The first integral on the right-hand side converges in density to 0 by (3) while the second one is bounded. Therefore the product converges in density to 0 and (7) follows. The proof of (3) ⇒ (6) is similar (Exercise 4.10.4). □

Exercise 4.10.1. Let (b_n) be a bounded sequence. Prove that d-lim $b_n = b$ if and only if $\lim_{n\to\infty} \frac{1}{n}\sum_{k=0}^{n-1} |b_n - b| = 0$.

Exercise 4.10.2. Prove that d-lim has the usual arithmetic properties of limits.

Exercise 4.10.3. Prove the second statement of Proposition 4.10.3.

Exercise 4.10.4. Prove that (3)⇒(6) in Theorem 4.10.6.

Exercise 4.10.5. Let T be a weak mixing measure-preserving transformation and let S be a measure-preserving transformation such that $S^k = T$ for some $k \in \mathbb{N}$ (S is called a kth root of T). Prove that S is weak mixing.

4.11 Applications of measure-theoretic recurrence to number theory

In this section we give highlights of applications of measure-theoretic recurrence to number theory initiated by H. Furstenberg. As an illustration of this approach we prove Sárközy's Theorem (Theorem 4.11.5). Our exposition follows to a large extent Furstenberg (1977) and Furstenberg (1981a).

For a finite subset $F \subset \mathbb{Z}$, denote by $|F|$ the number of elements in F. A subset $D \subset \mathbb{Z}$ has *positive upper density* if there are a_n, $b_n \in \mathbb{Z}$ such that

$b_n - a_n \to \infty$ and for some $\delta > 0$,

$$\frac{|D \cap [a_n, b_n]|}{b_n - a_n + 1} > \delta, \text{ for all } n \in \mathbb{N}.$$

Let $D \subset \mathbb{Z}$ have positive upper density. Let $\omega_D \in \Sigma_2 = \{0, 1\}^{\mathbb{Z}}$ be the sequence for which $(\omega_D)_n = 1$ if $n \in D$ and $(\omega_D)_n = 0$ if $n \notin D$, and let X_D be the closure of its orbit under the shift σ in Σ_2. Set $Y_D = \{\omega \in X_D : \omega_0 = 1\}$.

PROPOSITION 4.11.1 (Furstenberg). *Let $D \subset \mathbb{Z}$ have positive upper density. Then there exists a shift invariant Borel probability measure μ on X_D such that $\mu(Y_D) > 0$.*

Proof. By §4.6, every σ-invariant Borel probability measure on X_D is a linear functional L on the space $C(X_D)$ of continuous functions on X_D such that $L(f) \geqslant 0$ if $f \geqslant 0$, $L(1) = 1$, and $L(f \circ \sigma) = L(f)$.

For a function $f \in C(X_D)$, set

$$L_n(f) = \frac{1}{b_n - a_n + 1} \sum_{i=a_n}^{b_n} f(\sigma^i(\omega_D)),$$

where a_n and b_n are as in the definition above. Observe that $L_n(f) \leqslant \max f$ for each n. Let $(f_j)_{j \in \mathbb{N}}$ be a countable dense subset in $C(X_D)$. By a diagonal process, one can find a sequence $n_k \to \infty$ such that $\lim_{k\to\infty} L_{n_k}(f_j)$ exists for each j. Since $(f_j)_{j \in \mathbb{N}}$ is dense in $C(X_D)$, we have that

$$L(f) = \lim_{k\to\infty} \frac{1}{b_{n_k} - a_{n_k} + 1} \sum_{i=a_{n_k}}^{b_{n_k}} f(\sigma^i(\omega_D))$$

exists for each $f \in C(X_D)$ and determines a σ-invariant Borel probability measure μ.

Let $\chi \in C(X_D)$ be the characteristic function of Y_D. Then

$$L(\chi) = \int \chi \, d\mu = \mu(Y_D) \geqslant \delta > 0. \qquad \square$$

PROPOSITION 4.11.2. *Let $p(k)$ be a polynomial with integer coefficients and $p(0) = 0$. Let U be an isometry of a separable Hilbert space H and $H_{\mathrm{rat}} \subset H$ be the closure of the subspace spanned by the eigenvectors of U whose eigenvalues are roots of 1. Suppose $v \in H$ is such that $\langle U^{p(k)}v, v\rangle = 0$ for all $k \in \mathbb{N}$. Then $v \perp H_{\mathrm{rat}}$.*

Proof. Let $v = v_{\text{rat}} + w$ with $v_{\text{rat}} \in H_{\text{rat}}$ and $w \perp H_{\text{rat}}$. We use the following lemma, whose proof is similar to the proof of Lemma 4.10.2 (Exercise 4.11.1).

LEMMA 4.11.3. $\lim_{n\to\infty} \frac{1}{n}\sum_{k=0}^{n-1} U^{p(k)}w = 0$ *for all* $w \perp H_{\text{rat}}$.

Fix $\epsilon > 0$ and let $v'_{\text{rat}} \in H_{\text{rat}}$ and m be such that $\|v_{\text{rat}} - v'_{\text{rat}}\| < \epsilon$ and $U^m v'_{\text{rat}} = v'_{\text{rat}}$. Then $\left\|U^{mk}v_{\text{rat}} - v_{\text{rat}}\right\| < 2\epsilon$ for each k and, since $p(mk)$ is divisible by m,

$$\left\| \frac{1}{n}\sum_{k=0}^{n-1} U^{p(mk)}v_{\text{rat}} - v_{\text{rat}} \right\| < 2\epsilon.$$

Since $1/n\sum_{k=0}^{n-1} U^{p(mk)}w \to 0$ by Lemma 4.11.3, for n large enough we have

$$\left\| \frac{1}{n}\sum_{k=0}^{n-1} U^{p(mk)}v - v_{\text{rat}} \right\| < 2\epsilon.$$

By assumption, $\langle U^{p(mk)}v, v\rangle = 0$. Hence $|\langle v_{\text{rat}}, v\rangle| < 2\epsilon\|v\|$, so $\langle v_{\text{rat}}, v\rangle = 0$. □

As a corollary of the preceding proposition we obtain Furstenberg's Polynomial Recurrence Theorem[6].

THEOREM 4.11.4 (Furstenberg). *Let $p(t)$ be a polynomial with integer coefficients and $p(0) = 0$. Let T be a measure-preserving transformation of a finite measure space $(X, \mathfrak{A}, \mu)$ and $A \in \mathfrak{A}$ be a set with positive measure. Then there is $n \in \mathbb{N}$ such that $\mu(A \cap T^{p(n)}A) > 0$.*

Proof. Let U be the isometry induced by T in $H = L^2(X, \mathfrak{A}, \mu)$, $(Uh)(x) = h(T^{-1}(x))$. If $\mu(A \cap T^{p(n)}A) = 0$ for each $n \in \mathbb{N}$, then the characteristic function χ_A of A satisfies $\langle U^{p(n)}\chi_A, \chi_A\rangle = 0$ for each n. By Proposition 4.11.2, χ_A is orthogonal to all eigenfunctions of U whose eigenvalues are roots of 1. However $\mathbf{1}(x) \equiv 1$ is an eigenfuction of U with eigenvalue 1 and $\langle \mathbf{1}, \chi_A\rangle = \mu(A) \neq 0$. □

Theorem 4.11.4 and Proposition 4.11.1 imply the following result in combinatorial number theory.

THEOREM 4.11.5 (Sárközy, 1978). *Let $D \subset \mathbb{Z}$ have positive upper density and let p be a polynomial with integer coefficients and $p(0) = 0$. Then there are $x, y \in D$ and $n \in \mathbb{N}$ such that $x - y = p(n)$.*

[6] A slight modification of the arguments above yields Proposition 4.11.2 and Theorem 4.11.4 for polynomials with integer values at integer points (rather than integer coefficients).

The following extension of the Poincaré Recurrence Theorem (whose proof is beyond the scope of this book) was used by Furstenberg to give an ergodic-theoretic proof of the Szemerédi Theorem on arithmetic progressions (Theorem 4.11.7).

THEOREM 4.11.6 (Furstenberg's Multiple Recurrence Theorem (Furstenberg, 1977)). *Let T be an automorphism of a probability space $(X, \mathfrak{A}, \mu)$. Then for every $n \in \mathbb{N}$ and every $A \in \mathfrak{A}$ with $\mu(A) > 0$ there is $k \in \mathbb{N}$ such that*

$$\mu\big(A \cap T^{-k}(A) \cap T^{-2k}(A) \cap \cdots \cap T^{-nk}(A)\big) > 0.$$

THEOREM 4.11.7 (Szemerédi, 1969). *Every subset $D \subset \mathbb{Z}$ of positive upper density contains arbitrarily long arithmetic progressions.*

Proof. Exercise 4.11.3. □

Exercise 4.11.1. Prove Lemma 4.11.3.

Exercise 4.11.2. Use Theorem 4.11.4 and Proposition 4.11.1 to prove Theorem 4.11.5.

Exercise 4.11.3. Use Proposition 4.11.1 and Theorem 4.11.6 to prove Theorem 4.11.7.

4.12 Internet search[7]

In this section we describe a surprising application of ergodic theory to the problem of searching the Internet. This approach was pioneered by the Internet search engine *Google*™, http://www.google.com/.

The Internet offers enormous amounts of information. Looking for information on the Internet is analogous to looking for a book in a huge library without a catalog. The task of locating information on the web is performed by search engines. The first search engines appeared in the early 1990s. The most popular engines handle billions of searches per day.

The main tasks performed by a search engine are: gathering information from web pages; processing and storing this information in a database; and producing from this database a list of web pages relevant to a query consisting of one or more words. The gathering of information is performed by robot programs called *crawlers* that "crawl" the web by following links embedded in web pages. Raw information collected by the crawlers is parsed

[7] The exposition in this section follows to a certain extent that of Brin and Page (1998).

and coded by the *indexer*, which produces, for each web page, a set of word occurrences (including word position, font type and capitalization) and records all links from this web page to other pages, thus creating the *forward index*. The *sorter* rearranges information by words (rather than web pages), thus creating the *inverted index*. The *searcher* uses the inverted index to answer the query, i.e. to compile a list of documents relevant to the keywords and phrases of the query.

The order of the documents on the list is extremely important. A typical list may contain tens of thousands of web pages, but at best only the first several dozen may be reviewed by the user. Google relies primarily on two characteristics of the web page to determine the order of the returned pages – the relevance of the document to the query and the *PageRank* of the web page. The relevance is based on the relative position, fontification and frequency of the keyword(s) in the document. This factor by itself often does not produce good search results. For example, a query on the word "Internet" in one of the early search engines returned a list whose first entry was a web page in Chinese containing no English words other than "Internet".

Google uses Markov chains to rank web pages. The collection of all web pages and links between them is viewed as a directed graph G in which the web pages serve as vertices and the links as directed edges (from the web page on which they appear to the web page to which they point). At the moment there are billions of web pages with tens of billions links. We number the vertices by positive integers $i = 1, 2, \dots, N$. Let $\widetilde{G}$ be the graph obtained from G by adding a vertex 0 with edges to and from all other vertices. Let $b_{ij} = 1$ if there is an edge from vertex i to vertex j in $\widetilde{G}$ and let $O(i)$ be the number of outgoing edges adjacent to vertex i in $\widetilde{G}$. Note that $O(i) > 0$ for all i. Fix a damping parameter $p \in (0, 1)$ (for example, $p = .75$). Set $B_{ii} = 0$ for $i \geqslant 0$. For $i, j > 0$ and $i \neq j$ set

$$B_{0i} = 1/N, \quad B_{i0} = \begin{cases} 1 \text{ if } O(i) = 1 \\ 1 - p \text{ if } O(i) \neq 1 \end{cases}, \quad B_{ij} = \begin{cases} 0 \text{ if } b_{ij} = 0 \\ \frac{p}{O(i)} \text{ if } b_{ij} = 1 \end{cases}.$$

The matrix B is stochastic and primitive. Therefore, by Corollary 3.3.3, it has a unique positive left eigenvector q with eigenvalue 1 whose entries add up to 1. The pair (B, q) is a Markov chain on the vertices of $\widetilde{G}$. Google interprets q_i as the PageRank of web page i and uses it together with the relevance factor of the page to determine how high on the return list this page should be.

For any initial probability distribution p on the vertices of $\widetilde{G}$, the sequence pB^n converges exponentially to q. Thus one can find an approximation for

q by computing $q'B^n$ where q' is the uniform distribution. This approach to finding q is computationally much easier than trying to find an eigenvector for a matrix with billions rows and columns.

Exercise 4.12.1. Let A be an $N \times N$ stochastic matrix and let A^n_{ij} be the entries of A^n, i.e. A^n_{ij} is the probability of going from i to j in exactly n steps, §4.4. Suppose q is an invariant probability distribution, $qA = q$.

(a) Suppose that for some j, we have $A_{ij} = 0$ for all $i \neq j$, and $A^n_{jk} > 0$ for some $k \neq j$ and some $n \in \mathbb{N}$. Show that $q_j = 0$.

(b) Prove that if $A_{ij} > 0$ for some $j \neq i$ and $A^n_{ji} = 0$ for all $n \in \mathbb{N}$, then $q_i = 0$.

CHAPTER FIVE

Hyperbolic dynamics

In Chapter 1, we saw several examples of dynamical systems that were locally linear and had complementary expanding and/or contracting directions: expanding endomorphisms of S^1, hyperbolic toral automorphisms, the horseshoe, and the solenoid. In this chapter, we develop the theory of *hyperbolic differentiable dynamical systems*, which include these examples. Locally, a differentiable dynamical system is well-approximated by a linear map – its derivative. Hyperbolicity means that the derivative has complementary expanding and contracting directions.

The proper setting for a differentiable dynamical system is a *differentiable manifold* with a differentiable map or flow. A detailed introduction to the theory of differentiable manifolds is beyond the scope of this book. For the convenience of the reader, we give a brief formal introduction to differentiable manifolds in §5.13, and an even briefer informal introduction here.

For the purposes of this book, and without loss of generality (see the embedding theorems in Hirsch (1994)), it suffices to think of a differentiable manifold M^n as an n-dimensional differentiable surface, or *submanifold*, in $\mathbb{R}^N, N > n$. The Implicit Function Theorem implies that each point in M has a *local coordinate system* that identifies a neighborhood of the point with a neighborhood of 0 in $\mathbb{R}^n$. For each point x on such a surface $M \subset \mathbb{R}^N$, the tangent space $T_xM \subset \mathbb{R}^N$ is the space of all vectors tangent to M at x. The standard inner product on $\mathbb{R}^N$ induces an inner product $\langle \cdot\, , \cdot \rangle_x$ on each T_xM. The collection of inner products is called a *Riemannian metric*, and a manifold M together with a Riemannian metric is called a *Riemannian manifold*. The (intrinsic) distance d between two points in M is the infimum of the lengths of differentiable curves in M connecting the two points.

A one-to-one differentiable mapping with a differentiable inverse is called a *diffeomorphism*.

A discrete-time differentiable dynamical system on a differentiable manifold M is a differentiable map $f: M \to M$. The derivative df_x is a linear map from T_xM to $T_{f(x)}M$. In local coordinates df_x is given by the matrix of partial derivatives of f. A continuous-time differentiable dynamical system on M is a *differentiable flow*, i.e. a one-parameter group $\{f^t\}, t \in \mathbb{R}$, of differentiable maps $f^t: M \to M$ that depend differentiably on t. Since $f^{-t} \circ f^t = \mathrm{Id}$, each map f^t is a diffeomorphism. The derivative

$$v(\cdot) = \frac{d}{dt} f^t(\cdot)\bigg|_{t=0}$$

is a *differentiable vector field* tangent to M, and the flow $\{f^t\}$ is the one-parameter group of time-t maps of the differential equation $\dot{x} = v(x)$.

Differentiability, and even subtle differences in the degree of differentiability, have important and sometimes surprising consequences. See, for example, Exercise 2.5.7 and §7.2.

5.1 Expanding endomorphisms revisited

To illustrate and motivate some of the main ideas of this chapter we consider again expanding endomorphisms of the circle $E_m x = mx \bmod 1, x \in [0, 1)$, $m > 1$, introduced in §1.3.

Fix $\epsilon < 1/2$. A finite or infinite sequence of points (x_i) in the circle is called an ϵ-orbit of E_m if $d(x_{i+1}, E_m x_i) < \epsilon$ for all i. The point x_i has m preimages under E_m that are uniformly spread on the circle. Exactly one of them y_i^{i-1} is closer than ϵ/m to x_{i-1}. Similarly y_i^{i-1} has m preimages under E_m; exactly one of them y_i^{i-2} is closer than ϵ/m to x_{i-2}. Continuing in this manner we obtain a point y_i^0 with the property that $d(E_m^j y_i^0, x_j) < \epsilon$ for $0 \leqslant j \leqslant i$. In other words, any finite ϵ-orbit of E_m can be approximated or *shadowed* by a real orbit. If $(x_i)_{i=0}^{\infty}$ is an infinite ϵ-orbit, then the limit $y = \lim_{i\to\infty} y_i^0$ exists (Exercise 5.1.1), and $d(E_m^i y, x_i) \leqslant \epsilon$ for $i \geqslant 0$. Since two different orbits of E_m diverge exponentially, there can be only one shadowing orbit for a given infinite ϵ-orbit. By construction, y depends continuously on (x_i) in the product topology (Exercise 5.1.2).

The above discussion of the ϵ-orbits of E_m is based solely on the uniform forward expansion of E_m. Similar arguments show that if f is C^1-close to E_m, then each infinite ϵ-orbit (x_i) of f is shadowed by a unique real orbit of f that depends continuously on (x_i) (Exercise 5.1.3).

Consider now f that is C^1-close enough to E_m. View each orbit $(f^i(x))$ as an ϵ-orbit of E_m. Let $y = \phi(x)$ be the unique point whose orbit $(E_m^i y)$ shadows $(f^i(x))$. By the above discussion, the map ϕ is a homeomorphism and

$E_m\phi(x) = \phi(f(x))$ for each x (Exercise 5.1.4). This means that any differentiable map that is C^1-close enough to E_m is topologically conjugate to E_m. In other words E_m is *structurally stable*, see §§5.5 and 5.11.

Hyperbolicity is characterized by local expansion and contraction, in complementary directions. This property, which causes local instability of orbits, surprisingly leads to the global stability of the topological pattern of the collection of all orbits.

Exercise 5.1.1. Prove that $\lim_{i\to\infty} y_i^0$ exists.

Exercise 5.1.2. Prove that $\lim_{i\to\infty} y_i^0$ depends continuously on (x_i) in the product topology.

Exercise 5.1.3. Prove that if f is C^1-close to E_m, then each infinite ϵ-orbit (x_i) of f is approximated by a unique real orbit of f that depends continuously on (x_i).

Exercise 5.1.4. Prove that ϕ is a homeomorphism conjugating f and E_m.

5.2 Hyperbolic sets

In this section M is a C^1 Riemannian manifold, $U \subset M$ a non-empty open subset, and $f: U \to f(U) \subset M$ a C^1 diffeomorphism. A compact, f-invariant subset $\Lambda \subset U$ is called *hyperbolic* if there are $\lambda \in (0,1), C > 0$, and families of subspaces $E^s(x) \subset T_xM$ and $E^u(x) \subset T_xM$, $x \in \Lambda$, such that for every $x \in \Lambda$,

1. $T_xM = E^s(x) \oplus E^u(x)$;
2. $\|df_x^n v^s\| \leqslant C\lambda^n \|v^s\|$ for every $v^s \in E^s(x)$ and $n \geqslant 0$;
3. $\|df_x^{-n} v^u\| \leqslant C\lambda^n \|v^u\|$ for every $v^u \in E^u(x)$ and $n \geqslant 0$;
4. $df_x E^s(x) = E^s(f(x))$ and $df_x E^u(x) = E^u(f(x))$.

The subspace $E^s(x)$ (respectively, $E^u(x)$) is called the *stable (unstable) subspace* at x, and the family $\{E^s(x)\}_{x\in\Lambda}$ ($\{E^u(x)\}_{x\in\Lambda}$) is called the *stable (unstable) distribution* of $f|_\Lambda$. The definition allows two extreme cases $E^s(x) = \{0\}$ or $E^u(x) = \{0\}$.

The horseshoe (§1.8) and the solenoid (§1.9) are examples of hyperbolic sets. If $\Lambda = M$, then f is called an *Anosov diffeomorphism*. Hyperbolic toral automorphisms (§1.7) are examples of Anosov diffeomorphisms. Any closed invariant subset of a hyperbolic set is a hyperbolic set.

PROPOSITION 5.2.1. *Let Λ be a hyperbolic set of f. Then the subspaces $E^s(x)$ and $E^u(x)$ depend continuously on $x \in \Lambda$.*

Proof. Let x_i be a sequence of points in Λ converging to $x_0 \in \Lambda$. Passing to a subsequence, we may assume that $\dim E^s(x_i)$ is constant. Let $w_{1,i}, \dots, w_{k,i}$ be an orthonormal basis in $E^s(x_i)$. The restriction of the unit tangent bundle T^1M to Λ is compact. Hence, by passing to a subsequence, $w_{j,i}$ converges to $w_{j,0} \in T^1_{x_0}M$ for each $j = 1, \dots, k$. Since condition (2) of the definition of a hyperbolic set is a closed condition, each vector from the orthonormal frame $w_{1,0}, \dots, w_{k,0}$ satisfies condition (2) and, by the invariance (condition (4)), lies in $E^s(x_0)$. It follows that $\dim E^s(x_0) \geqslant k = \dim E^s(x_i)$. A similar argument shows that $\dim E^u(x_0) \geqslant \dim E^u(x_i)$. Hence, by (1), $\dim E^s(x_0) = \dim E^s(x_i)$ and $\dim E^u(x_0) = \dim E^u(x_i)$ and continuity follows. □

Any two Riemannian metrics on M are equivalent on a compact set in the sense that the ratios of the lengths of non-zero vectors are bounded above and away from zero. Thus the notion of a hyperbolic set does not depend on the choice of the Riemannian metric on M. The constant C depends on the metric, but λ does not (Exercise 5.2.2). However, as the next proposition shows, we can choose a particularly nice metric and $C = 1$ by using a slightly larger λ.

PROPOSITION 5.2.2. *If Λ is a hyperbolic set of f with constants C and λ, then for every $\epsilon > 0$ there is a C^1 Riemannian metric $\langle \cdot, \cdot \rangle'$ in a neighborhood of Λ, called the* Lyapunov, *or* adapted metric *(to f), with respect to which f satisfies the conditions of hyperbolicity with constants $C' = 1$ and $\lambda' = \lambda + \epsilon$, and the subspaces $E^s(x)$, $E^u(x)$ are ϵ-orthogonal, i.e. $\langle v^s, v^u \rangle' < \epsilon$ for all unit vectors $v^s \in E^s(x)$, $v^u \in E^u(x)$, $x \in \Lambda$.*

Proof. For $x \in \Lambda$, $v^s \in E^s(x)$ and $v^u \in E^u(x)$, set

$$\|v^s\|'^2 = \sum_{n=0}^{\infty} (\lambda + \epsilon)^{-2n} \|df_x^n v^s\|^2, \qquad \|v^u\|'^2 = \sum_{n=0}^{\infty} (\lambda + \epsilon)^{-2n} \|df_x^{-2n} v^u\|^2. \tag{5.1}$$

Both series converge uniformly for $\|v^s\|, \|v^u\| \leqslant 1$ and $x \in \Lambda$. We have

$$\begin{aligned} \|df_x v^s\|'^2 &= \sum_{n=0}^{\infty} (\lambda + \epsilon)^{-2n} \|df_x^{n+1} v^s\|^2 = (\lambda + \epsilon)^2 (\|v^s\|'^2 - \|v^s\|^2) \\ &< (\lambda + \epsilon)^2 \|v^s\|'^2 \end{aligned}$$

and similarly for $\|df_x^{-1} v^u\|'$. For $v = v^s + v^u \in T_xM$, $x \in \Lambda$, define $\|v\|' = \sqrt{(\|v^s\|')^2 + (\|v^u\|')^2}$. The metric is recovered from the norm:

$$\langle v, w \rangle' = \frac{1}{2}(\|v + w\|'^2 - \|v\|'^2 - \|w\|'^2).$$

With respect to this continuous metric, E^s and E^u are orthogonal and f satisfies the conditions of hyperbolicity with constant 1 and $\lambda + \epsilon$. Now, by standard methods of differential topology (Hirsch, 1994), $\langle\cdot\,,\cdot\rangle'$ can be uniformly approximated on Λ by a smooth metric defined in a neighborhood of Λ. □

Observe that to construct an adapted metric it is enough to consider sufficiently long finite sums instead of infinite sums in (5.1).

A fixed point x of a differentiable map f is called *hyperbolic* if no eigenvalue of df_x lies on the unit circle. A periodic point x of f of period k is called *hyperbolic* if no eigenvalue of df_x^k lies on the unit circle.

Exercise 5.2.1. Construct a diffeomorphism of the circle that satisfies the first three conditions of hyperbolicity (with Λ being the whole circle) but not the fourth condition.

Exercise 5.2.2. Prove that if Λ is a hyperbolic set of $f\colon U \to M$ for some Riemannian metric on M, then Λ is a hyperbolic set of f for any other Riemannian metric on M with the same constant λ.

Exercise 5.2.3. Let x be a fixed point of a diffeomorphism f. Prove that $\{x\}$ is a hyperbolic set if and only if x is a hyperbolic fixed point. Identify the constants C and λ. Give an example when df_x has exactly 2 eigenvalues $\mu \in (0, 1)$ and μ^{-1}, but $\lambda \neq \mu$.

Exercise 5.2.4. Prove that the horseshoe (§1.8) is a hyperbolic set.

Exercise 5.2.5. Let Λ_i be a hyperbolic set of $f_i\colon U_i \to M_i$, $i = 1, 2$. Prove that $\Lambda_1 \times \Lambda_2$ is a hyperbolic set of $f_1 \times f_2\colon U_1 \times U_2 \to M_1 \times M_2$.

Exercise 5.2.6. Let M be a fiber bundle over N with projection π. Let U be an open set in M and suppose that $\Lambda \subset U$ is a hyperbolic set of $f\colon U \to M$ and that $g\colon N \to N$ is a factor of f. Prove that $\pi(\Lambda)$ is a hyperbolic set of g.

Exercise 5.2.7. What are necessary and sufficient conditions for a periodic orbit to be a hyperbolic set?

5.3 ϵ-orbits

An ϵ-*orbit* of $f\colon U \to M$ is a finite or infinite sequence $(x_n) \subset U$ such that $d(f(x_n), x_{n+1}) \leqslant \epsilon$ for all n. Sometimes ϵ-orbits are referred to as *pseudo-orbits*. For $r \in \{0, 1\}$, denote by dist_r the distance in the space of C^r functions (see §5.13).

THEOREM 5.3.1. *Let Λ be a hyperbolic set of $f: U \to M$. Then there is an open set $O \subset U$ containing Λ, and positive ϵ_0, δ_0 with the following property: for every $\epsilon > 0$ there is $\delta > 0$ such that for any $g: O \to M$ with* $\mathrm{dist}_1(g, f) < \epsilon_0$, *any homeomorphism $h: X \to X$ of a topological space X and any continuous map $\phi: X \to O$ satisfying* $\mathrm{dist}_0(\phi \circ h, g \circ \phi) < \delta$ *there is a continuous map $\psi: X \to O$ with $\psi \circ h = g \circ \psi$ and* $\mathrm{dist}_0(\phi, \psi) < \epsilon$. *Moreover, ψ is unique in the sense that if $\psi' \circ h = g \circ \psi'$ for some $\psi': X \to O$ with* $\mathrm{dist}_0(\phi, \psi') < \delta_0$, *then $\psi' = \psi$.*

Theorem 5.3.1 implies, in particular, that any collection of bi-infinite pseudo-orbits near a hyperbolic set is close to a unique collection of genuine orbits that *shadow* it (Corollary 5.3.2). Moreover, this property holds not only for f itself but for any diffeomorphism C^1-close to f. In the simplest example, if X is a single point x (and h is the identity), Theorem 5.3.1 implies the existence of a fixed point near $h(x)$ for any diffeomorphism C^1-close to f.

Proof[1]. By the Whitney embedding theorem (Hirsch, 1994), we may assume that the manifold M is an m-dimensional submanifold in $\mathbb{R}^N$ for some large N. For $y \in M$, let $D_\alpha(y)$ be the disk of radius α centered at y in the $(N\text{–}m)$-plane $E^\perp(y) \subset \mathbb{R}^N$ that passes through y and is perpendicular to T_yM. Since Λ is compact, by the tubular neighborhood theorem (Hirsch, 1994), for any relatively compact open neighborhood O of Λ in M there is $\alpha \in (0, 1)$ such that the α-neighborhood O_α of O in $\mathbb{R}^N$ is foliated by the disks $D_\alpha(y)$. For each $z \in O_\alpha$ there is a unique point $\pi(z) \in M$ closest to y and the map π is the projection to M along the disks $D_\alpha(y)$. Each map $g: O \to M$ can be extended to a map $\tilde{g}: O_\alpha \to M$ by

$$\tilde{g}(z) = g(\pi(z)).$$

Let $C(X, O_\alpha)$ be the set of continuous maps from X to O_α with distance dist_0. Note that O_α is bounded and $\phi \in C(X, O_\alpha)$. Let Γ be the Banach space of bounded continuous vector fields $v: X \to \mathbb{R}^N$ with the norm $\|v\| = \sup_{x\in X} \|v(x)\|$. The map $\phi' \mapsto \phi' - \phi$ is an isometry from the ball of radius α centered at ϕ in $C(X, O_\alpha)$ onto the ball B_α of radius α centered at 0 in Γ. Define $\Phi: B_\alpha \to \Gamma$ by

$$(\Phi(v))(x) = \tilde{g}\big(\phi(h^{-1}(x)) + v(h^{-1}(x))\big) - \phi(x), \quad v \in B_\alpha,\ x \in X.$$

If v is a fixed point of Φ and $\psi(x) = \phi(x) + v(x)$, then $\tilde{g}(\psi(h^{-1}(x))) = \psi(x)$. Observe that $\tilde{g}(y) \in M$ and hence $\psi(x) \in M$ for $x \in X$ and $g(\psi(h^{-1}(x))) =$

[1] The main idea of this proof was communicated to us by A. Katok.

$\psi(x)$. Thus to prove the theorem it suffices to show that Φ has a unique fixed point near ϕ, which depends continuously on g.

The map Φ is differentiable as a map of Banach spaces, and the derivative

$$(d\Phi_v w)(x) = d\tilde{g}_{\phi(h^{-1}(x))+v(h^{-1}(x))} w(h^{-1}(x))$$

is continuous in v. To establish the existence and uniqueness of a fixed point v and its continuous dependence on g we study the derivative of Φ. By taking the maximum of appropriate derivatives we obtain that $\|(d\Phi_v w)(x)\| \leqslant L\|w\|$, where L depends on the first derivatives of g and on the embedding but does not depend on X, h and ϕ. For $v = 0$,

$$(d\Phi_0 w)(x) = d\tilde{g}_{\phi(h^{-1}(x))} w(h^{-1}(x)).$$

Since Λ is a hyperbolic set, for some constants $\lambda \in (0,1)$ and $C > 1$, we have for every $y \in \Lambda$ and $n \in \mathbb{N}$

$$\|df_y^n v\| \leqslant C\lambda^n \|v\| \qquad \text{if } v \in E^s(y) \tag{5.2}$$

$$\|df_y^{-n} v\| \leqslant C\lambda^n \|v\| \qquad \text{if } v \in E^u(y). \tag{5.3}$$

For $z \in O_\alpha$, let $\widetilde{T}_z$ denote the m-dimensional plane through z that is orthogonal to the disk $D_\alpha(\pi(z))$. The planes $\widetilde{T}_z$ form a differentiable distribution on O_α. Note that $\widetilde{T}_z = T_z M$ for $z \in O$. Extend the splitting $T_y M = E^s(y) \oplus E^u(y)$ continuously from Λ to O_α (decreasing the neighborhood O and α if necessary) so that $E^s(z) \oplus E^u(z) = \widetilde{T}_z$ and $T_z\mathbb{R}^N = E^s(z) \oplus E^u(z) \oplus E^\perp(\pi(z))$. Denote by P^s, P^u and $P^\perp$ the projections in each tangent space $T_z\mathbb{R}^N$ onto $E^s(z)$, $E^u(z)$ and $E^\perp(\pi(z))$, respectively.

Fix $n \in \mathbb{N}$ so that $C\lambda^n < 1/2$. By (5.2)–(5.3) and continuity, for a small enough $\alpha > 0$ and small enough neighborhood $O \supset \Lambda$, there is $\epsilon_0 > 0$ such that for every g with $\mathrm{dist}_1(f, g) < \epsilon_0$, every $z \in O_\alpha$, and every $v^s \in E^s(z)$, $v^u \in E^u(z)$, $v^\perp \in E^\perp(\pi(z))$

$$\|P^s d\tilde{g}_z^n v^s\| \leqslant \frac{1}{2}\|v^s\|, \qquad \|P^u d\tilde{g}_z^n v^s\| \leqslant \frac{1}{100}\|v^s\|, \tag{5.4}$$

$$\|P^u d\tilde{g}_z^n v^u\| \geqslant 2\|v^u\|, \qquad \|P^s d\tilde{g}_z^n v^u\| \leqslant \frac{1}{100}\|v^u\|, \tag{5.5}$$

$$d\tilde{g}_z^n v^\perp = 0. \tag{5.6}$$

Denote

$$\Gamma^\nu = \{v \in \Gamma \colon v(x) \in E^\nu(\phi(x)) \text{ for all } x \in X\}, \quad \nu = s, u, \perp.$$

The subspaces $\Gamma^s, \Gamma^u, \Gamma^\perp$ are closed and $\Gamma = \Gamma^s \oplus \Gamma^u \oplus \Gamma^\perp$. By construction,

$$d\Phi_0 = \begin{pmatrix} A^{ss} & A^{su} & 0 \\ A^{us} & A^{uu} & 0 \\ 0 & 0 & 0 \end{pmatrix},$$

where $A^{ij}\colon \Gamma^i \to \Gamma^j$, $i, j = s, u$. By (5.4)–(5.6), there are positive ϵ_0 and δ such that if $\operatorname{dist}_1(f, g) < \epsilon_0$ and $\operatorname{dist}_0(\phi \circ h, g \circ \phi) < \delta$, then the spectrum of $d\Phi_0$ is separated from the unit circle. Therefore the operator $d\Phi_0 - \mathrm{Id}$ is invertible and

$$\|(d\Phi_0 - \mathrm{Id})^{-1}\| < K,$$

where K depends only on f and ϕ.

As for maps of finite dimensional linear spaces, $\Phi(v) = \Phi(0) + d\Phi_0 v + H(v)$, where $\|H(v_1) - H(v_2)\| \leqslant o(\max\{\|v_1\|, \|v_2\|\}) \cdot \|v_1 - v_2\|$ for small $\|v_1\|$, $\|v_2\|$. A fixed point v of Φ satisfies

$$F(v) = -(d\Phi_0 - \mathrm{Id})^{-1}(\Phi(0) + H(v)) = v.$$

If $\zeta > 0$ is small enough, then for any $v_1, v_2 \in \Gamma$ with $\|v_1\|, \|v_2\| < \zeta$,

$$\|F(v_1) - F(v_2)\| < \frac{1}{2}\|v_1 - v_2\|.$$

Thus for an appropriate choice of constants and neighborhoods in the construction, $F\colon \Gamma \to \Gamma$ is a contraction, and therefore has a unique fixed point, which depends continuously on g. □

Theorem 5.3.1 implies that an ϵ-orbit lying in a small enough neighborhood of a hyperbolic set can be globally (i.e. for all times) approximated by a real f-orbit in the hyperbolic set. This property is called *shadowing* (the real orbit shadows the pseudo-orbit). A continuous map f of a topological space X has the *shadowing property* if for every $\epsilon > 0$ there is $\delta > 0$ such that every δ-orbit is ϵ-approximated by a real orbit.

For $\epsilon > 0$, denote by Λ_ϵ the open ϵ-neighborhood of Λ.

COROLLARY 5.3.2 (Anosov's Shadowing Theorem). *Let Λ be a hyperbolic set of $f\colon U \to M$. Then for every $\epsilon > 0$ there is $\delta > 0$ such that if (x_k) is a finite or infinite δ-orbit of f and* $\operatorname{dist}(x_k, \Lambda) < \delta$ *for all k, then there is $x \in \Lambda_\epsilon$ with* $\operatorname{dist}(f^k(x), x_k) < \epsilon$.

Proof. Choose a neighborhood O satisfying the conclusion of Theorem 5.3.1, and choose $\delta > 0$ such that $\Lambda_\delta \subset O$. If (x_k) is finite or semi-infinite, add to (x_k) the preimages of some $y_0 \in \Lambda$, whose distance to the first point of (x_k) is $<\delta$, and/or the images of some $y_m \in \Lambda$, whose distance to the last point of (x_k) is $<\delta$, to obtain a doubly infinite δ-orbit lying in the δ-neighborhood of Λ. Let $X = (x_k)$ with discrete topology, $g = f$, h be the shift $x_k \mapsto x_{k+1}$, and $\phi\colon X \to U$ be the inclusion, $\phi(x_k) = x_k$. Since (x_k) is a δ-orbit, $\operatorname{dist}(\phi(h(x_k)), f(\phi(x_k)) < \delta$. Theorem 5.3.1 applies and the corollary follows. □

As in Chapter 2 denote by $\mathrm{NW}(f)$ the set of non-wandering points and by $\mathrm{Per}(f)$ the set of periodic points of f. If Λ is f-invariant, denote by $\mathrm{NW}(f|_\Lambda)$ the set of non-wandering points of f restricted to Λ. In general, $\mathrm{NW}(f|_\Lambda) \neq \mathrm{NW}(f) \cap \Lambda$.

PROPOSITION 5.3.3. *Let Λ be a hyperbolic set of $f\colon U \to M$. Then* $\overline{\mathrm{Per}(f|_\Lambda)} = \mathrm{NW}(f|_\Lambda)$.

Proof. Fix $\epsilon > 0$ and let $x \in \mathrm{NW}(f|_\Lambda)$. Choose δ from Theorem 5.3.1 and let $V = B(x, \delta/2) \cap \Lambda$. Since $x \in \mathrm{NW}(f|_\Lambda)$, there is $n \in \mathbb{N}$ such that $f^n(V) \cap V \neq \emptyset$. Let $z \in f^{-n}(f^n(V) \cap V) = V \cap f^{-n}(V)$. Then $\{z, f(z), \dots, f^{n-1}(z)\}$ is a δ-orbit, so by Theorem 5.3.1, there is a periodic point of period n within 2ϵ of z. □

COROLLARY 5.3.4. *If $f\colon M \to M$ is Anosov, then* $\overline{\mathrm{Per}(f)} = \mathrm{NW}(f)$.

Exercise 5.3.1. Interpret Theorem 5.3.1 for $X = \mathbb{Z}_m$ and $h(z) = z + 1 \bmod m$.

Exercise 5.3.2. Let Λ be a hyperbolic set of $f\colon U \to M$. Prove that the restriction $f|_\Lambda$ is expansive.

Exercise 5.3.3. Let $T\colon [0, 1] \to [0, 1]$ be the *tent map*: $T(x) = 2x$ for $0 \leqslant x \leqslant 1/2$ and $T(x) = 2(1 - x)$ for $1/2 \leqslant x \leqslant 1$. Does T have the shadowing property?

Exercise 5.3.4. Prove that a circle rotation does not have the shadowing property. Prove that no isometry of a manifold has the shadowing property.

Exercise 5.3.5. Show that every minimal hyperbolic set consists of exactly one periodic orbit.

5.4 Invariant cones

Although hyperbolic sets are defined in terms of invariant families of linear spaces, it is often convenient, and in more general settings even necessary, to work with invariant families of linear cones instead of subspaces. In this section we characterize hyperbolicity in terms of families of invariant cones.

Let Λ be a hyperbolic set of $f\colon U \to M$. Since the distributions E^s and E^u are continuous (Proposition 5.2.1), we extend them to continuous distributions $\widetilde{E}^s$ and $\widetilde{E}^u$ defined in a neighborhood $U(\Lambda) \supset \Lambda$. If $x \in U(\Lambda)$ and $v \in T_xM$, let $v = v^s + v^u$ with $v^s \in \widetilde{E}^s(x)$ and $v^u \in \widetilde{E}^u(x)$. Assume that the metric is adapted with constant λ. For $\alpha > 0$, define the *stable* and *unstable cones* of size α by

$$K^s_\alpha(x) = \{v \in T_xM\colon \|v^u\| \leqslant \alpha\|v^s\|\},$$

$$K^u_\alpha(x) = \{v \in T_xM\colon \|v^s\| \leqslant \alpha\|v^u\|\}.$$

For a cone K, let $\overset{\circ}{K} = \operatorname{int}(K) \cup \{0\}$. Let $\Lambda_\epsilon = \{x \in U\colon \operatorname{dist}(x, \Lambda) < \epsilon\}$.

PROPOSITION 5.4.1. *For every $\alpha > 0$ there is $\epsilon = \epsilon(\alpha) > 0$, such that $f^i(\Lambda_\epsilon) \subset U(\Lambda)$, $i = -1, 0, 1$, and for every $x \in \Lambda_\epsilon$*

$$df_x K^u_\alpha(x) \subset \overset{\circ}{K}{}^u_\alpha(f(x)) \text{ and } df^{-1}_{f(x)} K^s_\alpha(f(x)) \subset \overset{\circ}{K}{}^s_\alpha(x).$$

Proof. The inclusions hold for $x \in \Lambda$. The statement follows by continuity. □

PROPOSITION 5.4.2. *For every $\delta > 0$ there are $\alpha > 0$ and $\epsilon > 0$ such that $f^i(\Lambda_\epsilon) \subset U(\Lambda)$, $i = -1, 0, 1$, and for every $x \in \Lambda_\epsilon$*

$$\|df_x^{-1} v\| \leqslant (\lambda + \delta)\|v\| \text{ if } v \in K^u_\alpha(x)$$

and

$$\|df_x v\| \leqslant (\lambda + \delta)\|v\| \text{ if } v \in K^s_\alpha(x).$$

Proof. The statement follows by continuity for a small enough α and $\epsilon = \epsilon(\alpha)$ from Proposition 5.4.1. □

The following proposition is the converse of Propositions 5.4.1 and 5.4.2.

PROPOSITION 5.4.3. *Let Λ be a compact invariant set of $f\colon U \to M$. Suppose that there is $\alpha > 0$ and for every $x \in \Lambda$ there are continuous subspaces $\widetilde{E}^s(x)$ and $\widetilde{E}^u(x)$ such that $\widetilde{E}^s(x) \oplus \widetilde{E}^u(x) = T_xM$ and the α-cones $K^s_\alpha(x)$ and $K^u_\alpha(x)$ determined by the subspaces satisfy*

1. *$df_x K^u_\alpha(x) \subset K^u_\alpha(f(x))$ and $df^{-1}_{f(x)} K^s_\alpha(f(x)) \subset K^s_\alpha(x)$; and*
2. *$\|df_x v\| < \|v\|$ for non-zero $v \in K^s_\alpha(x)$ and $\|df_x^{-1} v\| < \|v\|$ for non-zero $v \in K^u_\alpha(x)$.*

Then Λ is a hyperbolic set of f.

Proof. By compactness of Λ and of the unit tangent bundle of M, there is a constant $\lambda \in (0,1)$ such that

$$\|df_x v\| \leqslant \lambda \|v\| \text{ for } v \in K^s_\alpha(x) \text{ and } \|df_x^{-1} v\| \leqslant \lambda \|v\| \text{ for } v \in K^u_\alpha(x).$$

For $x \in \Lambda$, the subspaces

$$E^s(x) = \bigcap_{n \geqslant 0} df^{-n}_{f^n(x)} K^s(f^n(x)) \text{ and } E^u(x) = \bigcap_{n \geqslant 0} df^n_{f^{-n}(x)} K^u(f^{-n}(x))$$

satisfy the definition of hyperbolicity with constants λ and $C = 1$. □

Let

$$\Lambda^s_\epsilon = \{x \in U : \operatorname{dist}(f^n(x), \Lambda) < \epsilon \text{ for all } n \in \mathbb{N}_0\}$$

$$\Lambda^u_\epsilon = \{x \in U : \operatorname{dist}(f^{-n}(x), \Lambda) < \epsilon \text{ for all } n \in \mathbb{N}_0\}.$$

Note that both sets are contained in Λ_ϵ and $f(\Lambda^s_\epsilon) \subset \Lambda^s_\epsilon$, $f^{-1}(\Lambda^u_\epsilon) \subset \Lambda^u_\epsilon$.

PROPOSITION 5.4.4. *Let Λ be a hyperbolic set of f with adapted metric. Then for every $\delta > 0$ there is $\epsilon > 0$ such that the distributions E^s and E^u can be extended to Λ_ϵ so that*

1. *E^s is continuous on Λ^s_ϵ and E^u is continuous on Λ^u_ϵ;*
2. *if $x \in \Lambda_\epsilon \cap f(\Lambda_\epsilon)$ then $df_x E^s(x) = E^s(f(x))$ and $df_x E^u(x) = E^u(f(x))$;*
3. *$\|df_x v\| < (\lambda + \delta)\|v\|$ for every $x \in \Lambda_\epsilon$ and $v \in E^s(x)$;*
4. *$\|df_x^{-1} v\| < (\lambda + \delta)\|v\|$ for every $x \in \Lambda_\epsilon$ and $v \in E^u(x)$.*

Proof. Choose $\epsilon > 0$ small enough that $\Lambda_\epsilon \subset U(\Lambda)$. For $x \in \Lambda^s_\epsilon$, let $E^s(x) = \lim_{n\to\infty} df^{-n}_{f^n(x)}(\widetilde{E}^s(f^n(x)))$. By Proposition 5.4.2, the limit exists if δ, α and ϵ are small enough. If $x \in \Lambda_\epsilon \setminus \Lambda^s_\epsilon$, let $n(x) \in \mathbb{N}$ be such that $f^n(x) \in \Lambda_\epsilon$ for $n = 0, 1, \dots, n(x)$, and $f^{n(x)+1}(x) \notin \Lambda_\epsilon$ and let $E^s(x) = df^{-n(x)}_{f^{n(x)}}(\widetilde{E}^s(f^{n(x)}(x)))$. The continuity of E^s on Λ^s_ϵ and the required properties follow from Proposition 5.4.2. A similar construction with f replaced by f^{-1} gives an extension of E^u. □

Exercise 5.4.1. Prove that the solenoid (§1.9) is a hyperbolic set.

Exercise 5.4.2. Let Λ be a hyperbolic set of f. Prove that there is an open set $O \supset \Lambda$ and $\epsilon > 0$ such that for every g with $\operatorname{dist}_1(f, g) < \epsilon$, the invariant set $\Lambda_g = \bigcap_{n=-\infty}^{\infty} g^n(\bar{O})$ is a hyperbolic set of g.

Exercise 5.4.3. Prove that the topological entropy of an Anosov diffeomorphism is positive.

Exercise 5.4.4. Let Λ be a hyperbolic set of f. Prove that if $\dim E^u(x) > 0$ for each $x \in \Lambda$, then f has sensitive dependence on initial conditions on Λ (see §1.12).

5.5 Stability of hyperbolic sets

In this section we use pseudo-orbits and invariant cones to obtain key properties of hyperbolic sets. The next two propositions imply that hyperbolicity is "persistent".

PROPOSITION 5.5.1. *Let Λ be a hyperbolic set of $f: U \to M$. There is an open set $U(\Lambda) \supset \Lambda$ and $\epsilon_0 > 0$ such that if $K \subset U(\Lambda)$ is a compact invariant subset of a diffeomorphism $g: U \to M$ with $\text{dist}_1(g, f) < \epsilon_0$, then K is a hyperbolic set of g.*

Proof. Assume that the metric is adapted to f and extend the distributions E^s_f and E^u_f to continuous distributions $\widetilde{E}^s_f$ and $\widetilde{E}^u_f$ defined in an open neighborhood $U(\Lambda)$ of Λ. For an appropriate choice of $U(\Lambda)$, ϵ_0 and α, the stable and unstable α-cones determined by $\widetilde{E}^s_f$ and $\widetilde{E}^u_f$ satisfy the assumptions of Proposition 5.4.3 for the map g. □

Denote by $\text{Diff}^1(M)$ the space of C^1 diffeomorphisms of M with the C^1 topology.

COROLLARY 5.5.2. *The set of Anosov diffeomorphisms of a given compact manifold is open in $\text{Diff}^1(M)$.*

PROPOSITION 5.5.3. *Let Λ be a hyperbolic set of $f: U \to M$. For every open set $V \subset U$ containing Λ and every $\epsilon > 0$, there is $\delta > 0$ such that for every $g: V \to M$ with $\text{dist}_1(g, f) < \delta$ there is a hyperbolic set $K \subset V$ of g and a homeomorphism $\chi: K \to \Lambda$ such that $\chi \circ g|_K = f|_\Lambda \circ \chi$ and $\text{dist}_0(\chi, \text{Id}) < \epsilon$.*

Proof. Let $X = \Lambda$, $h = f|_\Lambda$ and let $\phi: \Lambda \hookrightarrow U$ be the inclusion. By Theorem 5.3.1, there is a continuous map $\psi: \Lambda \to U$ such that $\psi \circ f|_\Lambda = g \circ \psi$. Set $K = \psi(\Lambda)$. Now apply Theorem 5.3.1 to $X = K$, $h = g|_K$ and the inclusion $\phi: K \hookrightarrow M$ to get $\psi': K \to U$ with $\psi' \circ g|_K = f|_\Lambda \circ \psi'$. By uniqueness, $\psi^{-1} = \psi'$. For a small enough δ, the map $\chi = \psi'$ is close to the identity and, by Proposition 5.5.1, K is hyperbolic. □

A C^1 diffeomorphism f of a C^1 manifold M is called *structurally stable* if for every $\epsilon > 0$ there is $\delta > 0$ such that if $g \in \text{Diff}^1(M)$ and $\text{dist}_1(g, f) < \delta$, then there is a homeomorphism $h: M \to M$ for which $f \circ h = h \circ g$ and

$\text{dist}_0(h, \text{Id}) < \epsilon$. If one demands that the conjugacy h be C^1, the definition becomes vacuous. For example, if f has a hyperbolic fixed point x, then any small enough perturbation g has a fixed point y nearby; if the conjugation is differentiable, then the matrices dg_y and df_x are similar. This condition restricts g to lie in a proper submanifold of $\text{Diff}^1(M)$.

COROLLARY 5.5.4. *Anosov diffeomorphisms are structurally stable.*

Exercise 5.5.1. Interpret Proposition 5.5.3 when Λ is a hyperbolic periodic point of f.

5.6 Stable and unstable manifolds

Hyperbolicity is defined in terms of infinitesimal objects: a family of linear subspaces invariant by the differential of a map. In this section we construct the corresponding integral objects, the stable and unstable manifolds.

For $\delta > 0$, let $B_\delta = B(0, \delta) \subset \mathbb{R}^m$ be the ball of radius δ at 0.

PROPOSITION 5.6.1 (Hadamard–Perron). *Let $f = (f_n)_{n\in\mathbb{N}_0}$, $f_n \colon B_\delta \to \mathbb{R}^m$, be a sequence of C^1 diffeomorphisms onto their images such that $f_n(0) = 0$. Suppose that for each n there is a splitting $\mathbb{R}^m = E^s(n) \oplus E^u(n)$ and $\lambda \in (0, 1)$ such that*

1. *$df_n(0)E^s(n) = E^s(n+1)$ and $df_n(0)E^u(n) = E^u(n+1)$;*
2. *$\|df_n(0)v^s\| < \lambda\|v^s\|$ for every $v^s \in E^s(n)$;*
3. *$\|df_n(0)v^u\| > \lambda^{-1}\|v^u\|$ for every $v^u \in E^u(n)$;*
4. *the angles between $E^s(n)$ and $E^u(n)$ are uniformly bounded away from 0;*
5. *$\{df_n(\cdot)\}_{n\in\mathbb{N}_0}$ is an equicontinuous family of functions from B_δ to $GL(m, \mathbb{R})$.*

Then there are $\epsilon > 0$ and a sequence $\phi = (\phi_n)_{n\in\mathbb{N}_0}$ of uniformly Lipschitz continuous maps $\phi_n \colon B^s_\epsilon = \{v \in E^s(n) \colon \|v\| < \epsilon\} \to E^u(n)$ such that

1. $\text{graph}(\phi_n) \cap B_\epsilon = W^s_\epsilon(n) := \{x \in B_\epsilon \colon \|f_{n+k-1} \circ \cdots \circ f_{n+1} \circ f_n(x)\| \underset{k\to\infty}{\longrightarrow} 0\}$;
2. $f_n(\text{graph}(\phi_n)) \subset \text{graph}(\phi_{n+1})$;
3. *if $x \in \text{graph}(\phi_n)$, then $\|f_n(x)\| \leqslant \lambda\|x\|$, so by (2), $f_n^k(x) \to 0$ exponentially as $k \to \infty$;*
4. *for $x \in B_\epsilon \setminus \text{graph}(\phi_n)$,*

$$\|P^u_{n+1} f_n(x) - \phi_{n+1}(P^s_{n+1} f_n(x))\| > \lambda^{-1}\|P^u_n x - \phi_n(P^s_n x)\|,$$

where P_n^s (P_n^u) denotes the projection onto $E^s(n)$ ($E^u(n)$) parallel to $E^u(n)$ ($E^s(n)$);

5. *ϕ_n is differentiable at 0 and $d\phi_n(0) = 0$, i.e. the tangent plane to $\text{graph}(\phi_n)$ is $E^s(n)$;*
6. *ϕ depends continuously on f in the topologies induced by the following distance functions*

$$d_0(\phi, \psi) = \sup_{n\in\mathbb{N}_0,\, x\in B_\epsilon} 2^{-n}|\phi_n(x) - \psi_n(x)|$$

$$d(f, g) = \sup_{n\in\mathbb{N}_0} 2^{-n} \operatorname{dist}_1(f_n, g_n),$$

where dist_1 is the C^1 distance.

Proof. For positive constants L and ϵ, let $\Phi(L, \epsilon)$ be the space of sequences $\phi = (\phi_n)_{n\in\mathbb{N}_0}$, where $\phi_n \colon B_\epsilon^s \to E^u(n)$ is a Lipschitz-continuous map with Lipschitz constant L and $\phi_n(0) = 0$. Define distance on $\Phi(L, \epsilon)$ by $d(\phi, \psi) = \sup_{n\in\mathbb{N}_0,\, x\in B_\epsilon} |\phi_n(x) - \psi_n(x)|$. This metric is complete.

We now define an operator $F \colon \Phi(L, \epsilon) \to \Phi(L, \epsilon)$ called the *graph transform*. Suppose $\phi = (\phi_n) \in \Phi$. We prove in the next lemma that for a small enough ϵ, the projection of the set $f_n^{-1}(\text{graph}(\phi_{n+1}))$ onto $E^s(n)$ covers $E_\epsilon^s(n)$, and $f_n^{-1}(\text{graph}(\phi_{n+1}))$ contains the graph of a continuous function $\psi_n \colon B_\epsilon^s \to E_\epsilon^u(n)$ with Lipschitz constant L. We set $F(\phi)_n = \psi_n$.

Note that a map $h \colon \mathbb{R}^k \to \mathbb{R}^l$ is Lipschitz continuous at 0 with Lipschitz constant L if and only if the graph of h lies in the L-cone about $\mathbb{R}^k$ and is Lipschitz continuous at $x \in \mathbb{R}^k$ if and only if its graph lies in the L-cone about the translate of $\mathbb{R}^k$ by $(x, h(x))$.

LEMMA 5.6.2. *For any $L > 0$, there exists $\epsilon > 0$ such that the graph transform F is a well-defined operator on $\Phi(L, \epsilon)$.*

Proof. For $L > 0$ and $x \in B_\epsilon$, let $K_L^s(n)$ denote the stable cone

$$K_L^s(n) = \{v \in \mathbb{R}^m \colon v = v^s + v^u,\ v^s \in E^s(n),\ v^u \in E^u(n),\ |v^u| \leqslant L|v^s|\}.$$

Note that $df_n^{-1}(0)K_L^s(n+1) \subset K_L^s(n)$ for any $L > 0$. Therefore, by the uniform continuity of df_n, for any $L > 0$ there is $\epsilon > 0$ such that $df_n^{-1}(x)K_L^s(n+1) \subset K_L^s(n)$ for any $n \in \mathbb{N}_0$ and $x \in B_\epsilon$. Hence the preimage under f_n of the graph of a Lipschitz continuous function is the graph of a Lipschitz continuous function. For $\phi \in \Phi(L, \epsilon)$, consider the composition $\beta = P^s(n) \circ f_n^{-1} \circ \phi_n$, where $P^s(n)$ is the projection onto $E^s(n)$ parallel to $E^u(n)$. If ϵ is small enough, then β is an expanding map and its image covers $B_\epsilon^s(n)$ (Exercise 5.6.1). Hence $F(\phi) \in \Phi(L, \epsilon)$. □

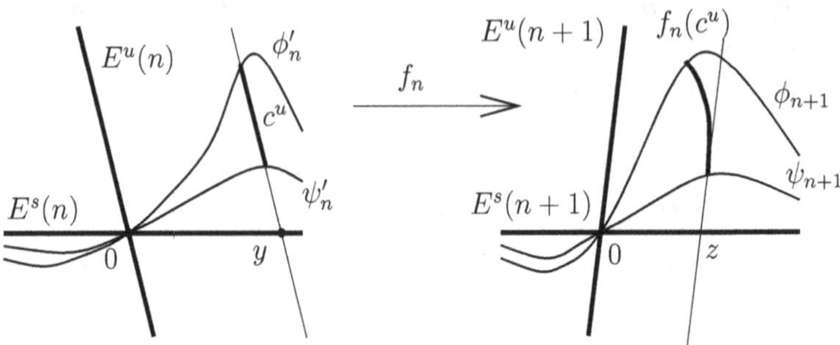

Figure 5.1. Graph transform applied to ϕ and ψ.

The next lemma shows that F is a contracting operator for an appropriate choice of ϵ and L.

LEMMA 5.6.3. *There are $\epsilon > 0$ and $L > 0$ such that F is a contracting operator on $\Phi(L, \epsilon)$.*

Proof. For $L \in (0, 0.1)$, let $K_L^u(n)$ denote the unstable cone

$$K_L^u(n) = \{v \in \mathbb{R}^m : v = v^s + v^u,\ v^s \in E^s(n),\ v^u \in E^u(n),\ |v^u| \geqslant L^{-1}|v^s|\}$$

and note that $df_n(0)K_L^u(n) \subset K_L^u(n+1)$. As in Lemma 5.6.2, by the uniform continuity of df_n, for any $L > 0$ there is $\epsilon > 0$ such that $df_n K_L^u(n) \subset K_L^u(n+1)$ for every $n \in \mathbb{N}_0$ and $x \in B_\epsilon$.

Let $\phi, \psi \in \Phi(L, \epsilon), \phi' = F(\phi), \psi' = F(\psi)$ (see Figure 5.1). For any $\eta > 0$ there are $n \in \mathbb{N}_0$ and $y \in B_\epsilon^s$ such that $|\phi'_n(y) - \psi'_n(y)| > d(\phi', \psi') - \eta$. Let c^u be the straight line segment from $(y, \phi'_n(y))$ to $(y, \phi'_n(y))$. Since c^u is parallel to $E^u(n)$, we have that $\text{length}(f_n(c^u)) > \lambda^{-1}\,\text{length}(c^u)$. Let $f_n(y, \psi'_n(y)) = (z, \psi_{n+1}(z))$ and consider the curvilinear triangle formed by the straight line segment from $(z, \phi_{n+1}(z))$ to $(z, \psi_{n+1}(z))$, $f_n(c^u)$ and the shortest curve on the graph of ψ_{n+1} connecting the ends of these curves. For a small enough $\epsilon > 0$ the tangent vectors to the image $f_n(c^u)$ lie in $K_L^u(n+1)$ and the tangent vectors to the graph of ϕ_{n+1} lie in $K_L^s(n+1)$. Therefore,

$$\begin{aligned}|\phi_{n+1}(z) - \psi_{n+1}(z)| &\geqslant \frac{\text{length}(f_n(c^u))}{1+2L} - L(1+L)\cdot\text{length}(f_n(c^u)) \\ &\geqslant (1-4L)\,\text{length}(f_n(c^u))\end{aligned}$$

and

$$\begin{aligned}d(\phi, \psi) &\geqslant |\phi_{n+1}(z) - \psi_{n+1}(z)| \geqslant (1-4L)\,\text{length}(f_n(c^u)) \\ &> (1-4L)\lambda^{-1}\,\text{length}(c^u) = (1-4L)\lambda^{-1}(d(\phi', \psi') - \eta).\end{aligned}$$

Since η is arbitrary, F is contracting for small enough L and ϵ. □

We can now finish the proof of Proposition 5.6.1. Since F is contracting (Lemma 5.6.3) and depends continuously on f, it has a unique fixed point $\phi \in \Phi(L, \epsilon)$, which depends continuously on f (property (6)) and automatically satisfies (2). The invariance of the stable and unstable cones (with a small enough ϵ) implies that ϕ satisfies (3) and (4). Property (1) follows immediately from (3) and (4). Since property (1) gives a geometric characterization of graph(ϕ_n), the fixed point of F for a smaller ϵ is a restriction of the fixed point of F for a larger ϵ to a smaller domain. As $\epsilon \to 0$ and $L \to 0$, the stable cone $K^s_L(0, n)$ (which contains graph(ϕ_n)) tends to $E^s(n)$. Therefore $E^s(n)$ is the tangent plane to graph(ϕ_n) at 0 (property (5)). □

The following theorem establishes the existence of *local stable manifolds* for points in a hyperbolic set Λ and in Λ^s_δ; and of *local unstable manifolds* for points in Λ and in Λ^u_δ (see §5.4); recall that $\Lambda^s_\delta \supset \Lambda$ and $\Lambda^u_\delta \supset \Lambda$.

THEOREM 5.6.4. *Let $f\colon M \to M$ be a C^1 diffeomorphism of a differentiable manifold and let $\Lambda \subset M$ be a hyperbolic set of f with constant λ (the metric is adapted).*

Then there are $\epsilon, \delta > 0$ such that for every $x^s \in \Lambda^s_\delta$ and every $x^u \in \Lambda^u_\delta$ (see §5.4):

1. *the sets*

$$W^s_\epsilon(x^s) = \{y \in M\colon \operatorname{dist}(f^n(x^s), f^n(y)) < \epsilon \text{ for all } n \in \mathbb{N}_0\}, \text{ and}$$
$$W^u_\epsilon(x^u) = \{y \in M\colon \operatorname{dist}(f^{-n}(x^u), f^{-n}(y)) < \epsilon \text{ for all } n \in \mathbb{N}_0\}$$

   called the local stable manifold of x^s *and the* local unstable manifold of x^u, *are C^1 embedded discs;*
2. *$T_{y^s}W^s_\epsilon(x^s) = E^s(y^s)$ for every $y^s \in W^s_\epsilon(x^s)$ and $T_{y^u}W^u_\epsilon(x^u) = E^u(y^u)$ for every $y^u \in W^u_\epsilon(x^u)$ (see Proposition 5.4.4);*
3. *$f(W^s_\epsilon(x^s)) \subset W^s_\epsilon(f(x^s))$ and $f^{-1}(W^u_\epsilon(f(x^u))) \subset W^u_\epsilon(x^u)$;*
4. *if $y^s, z^s \in W^s_\epsilon(x^s)$; then $d^s(f(y^s), f(z^s)) < \lambda d^s(y^s, z^s)$, where d^s is the distance along $W^s_\epsilon(x^s)$;*
   if $y^u, z^u \in W^u_\epsilon(x^u)$, then $d^u(f^{-1}(y^u), f^{-1}(z^u)) < \lambda d^u(y^u, z^u)$, where d^u is the distance along $W^u_\epsilon(x^u)$;
5. *if $0 < \operatorname{dist}(x^s, y) < \epsilon$ and $\exp^{-1}_{x^s}(y)$ lies in the δ-cone $K^u_\delta(x^s)$, then $\operatorname{dist}(f(x^s), f(y)) > \lambda^{-1}\operatorname{dist}(x^s, y)$;*
   if $0 < \operatorname{dist}(x^u, y) < \epsilon$ and $\exp^{-1}_{x^u}(y)$ lies in the δ-cone $K^s_\delta(x^u)$, then $\operatorname{dist}(f(x^u), f(y)) < \lambda \operatorname{dist}(x^s, y)$;
6. *if $y^s \in W^s_\epsilon(x^s)$, then $W^s_\alpha(y^s) \subset W^s_\epsilon(x^s)$ for some $\alpha > 0$;*
   if $y^u \in W^u_\epsilon(x^u)$, then $W^u_\beta(y^u) \subset W^u_\epsilon(x^u)$ for some $\beta > 0$.

Proof. Since $\Lambda^s_\delta \supset \Lambda$ is compact, for a small enough δ, there is a collection $\mathcal{U}$ of coordinate charts (U_x, ψ_x), $x \in \Lambda^s_\delta$, such that U_x covers the

δ-neighborhood of x and the changes of coordinates $\psi_x \circ \psi_y^{-1}$ between the charts have equicontinuous first derivatives. For $x^s \in \Lambda_\delta^s$, let $f_n = \psi_{f^n(x^s)} \circ f \circ \psi_{f^{n-1}(x^s)}^{-1}$, $E^s(n) = d\psi_{f^n(x^s)}(x^s)E^s(f^n(x^s))$, $E^u(n) = d\psi_{f^n(x)}(x)E^u(f^n(x))$, apply Proposition 5.6.1, and set $W_\epsilon^s(x) = W_0^s(\epsilon)$. Similarly, apply Proposition 5.6.1 to f^{-1} to construct the local unstable manifolds. Properties (1)–(6) follow immediately from Proposition 5.6.1. □

Let Λ be a hyperbolic set of $f\colon U \to M$ and $x \in \Lambda$. The *(global) stable and unstable manifolds* of x are defined by:

$$W^s(x) = \{y \in M\colon d(f^n(x), f^n(y)) \to 0 \text{ as } n \to \infty\},$$

$$W^u(x) = \{y \in M\colon d(f^{-n}(x), f^{-n}(y)) \to 0 \text{ as } n \to \infty\}.$$

PROPOSITION 5.6.5. *There is $\epsilon_0 > 0$ such that for every $\epsilon \in (0, \epsilon_0)$, for every $x \in \Lambda$*

$$W^s(x) = \bigcup_{n=0}^{\infty} f^{-n}(W_\epsilon^s(f^n(x)) \qquad W^u(x) = \bigcup_{n=0}^{\infty} f^n(W_\epsilon^u(f^{-n}(x)).$$

Proof. Exercise 5.6.2. □

COROLLARY 5.6.6. *The global stable and unstable manifolds are embedded C^1 submanifolds of M homeomorphic to the unit balls in corresponding dimensions.*

Proof. Exercise 5.6.3. □

Exercise 5.6.1. Suppose $f\colon \mathbb{R}^m \to \mathbb{R}^m$ is a continuous map such that $|f(x) - f(y)| \geqslant a|x - y|$ for some $a > 1$, for all $x, y \in \mathbb{R}^m$. If $f(0) = 0$ show that the image of a ball of radius $r > 0$ centered at 0 contains the ball of radius ar centered at 0.

Exercise 5.6.2. Prove Proposition 5.6.5.

Exercise 5.6.3. Prove Corollary 5.6.6.

5.7 Inclination Lemma

Let M be a differentiable manifold. Recall that two submanifolds $N_1, N_2 \subset M$ of complementary dimensions *intersect transversely* (or are *transverse*) at a point $p \in N_1 \cap N_2$ if $T_pM = T_pN_1 \oplus T_pN_2$. We write $N_1 \pitchfork N_2$ if every point of intersection of N_1 and N_2 is a point of transverse intersection.

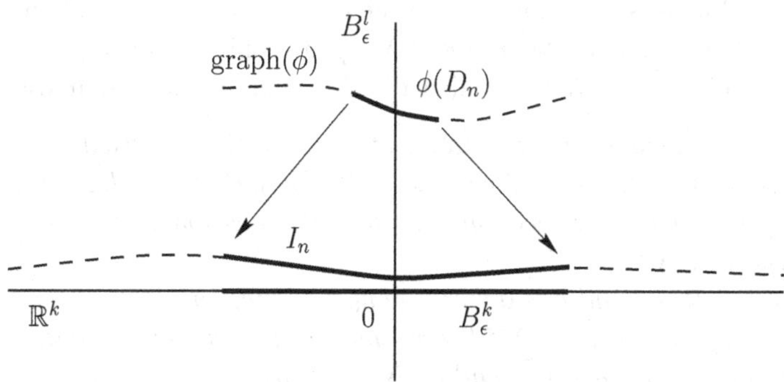

Figure 5.2. The image of the graph of ϕ under f^n.

Denote by B^i_ϵ the open ball of radius ϵ centered at 0 in $\mathbb{R}^i$. For $v \in \mathbb{R}^m = \mathbb{R}^k \times \mathbb{R}^l$, denote by $v^u \in \mathbb{R}^k$ and $v^s \in \mathbb{R}^l$ the components of $v = v^u + v^s$ and by $\pi^u \colon \mathbb{R}^m \to \mathbb{R}^k$ the projection to $\mathbb{R}^k$. For $\delta > 0$, let $K^u_\delta = \{v \in \mathbb{R}^m \colon \|v^s\| \leqslant \delta\|v^u\|\}$ and $K^s_\delta = \{v \in \mathbb{R}^m \colon \|v^u\| \leqslant \delta\|v^s\|\}$.

LEMMA 5.7.1. *Let $\lambda \in (0,1)$, $\epsilon, \delta \in (0, 0.1)$. Suppose $f \colon B^k_\epsilon \times B^l_\epsilon \to \mathbb{R}^m$ and $\phi \colon B^k_\epsilon \to B^l_\epsilon$ are C^1 maps such that:*

1. *0 is a hyperbolic fixed point of f;*
2. *$W^u_\epsilon(0) = B^k_\epsilon \times \{0\}$ and $W^s_\epsilon(0) = \{0\} \times B^l_\epsilon$;*
3. *$\|df_x(v)\| \geqslant \lambda^{-1}\|v\|$ for every $v \in K^u_\delta$ whenever both $x, f(x) \in B^k_\epsilon \times B^l_\epsilon$;*
4. *$\|df_x(v)\| \leqslant \lambda\|v\|$ for every $v \in K^s_\delta$ whenever both $x, f(x) \in B^k_\epsilon \times B^l_\epsilon$;*
5. *$df_x(K^u_\delta) \subset K^u_\delta$ whenever both $x, f(x) \in B^k_\epsilon \times B^l_\epsilon$;*
6. *$d(f^{-1})_x(K^s_\delta) \subset K^s_\delta$ whenever both $x, f^{-1}(x) \in B^k_\epsilon \times B^l_\epsilon$;*
7. *$T_{(y,\phi(y))}\,\mathrm{graph}(\phi) \subset K^u_\delta$ for every $y \in B^k_\epsilon$.*

Then for every $n \in \mathbb{N}$ there is a subset $D_n \subset B^k_\epsilon$ diffeomorphic to B^k and such that the image I_n under f^n of the graph of the restriction $\phi|_{D_n}$ has the properties $\pi^u(I_n) \supset B^k_{\epsilon/2}$ and $T_xI_n \subset K^u_{\delta\lambda^{2n}}$ for each $x \in I_n$.

Proof. The lemma follows from the invariance of the cones (Exercise 5.7.2). □

The meaning of the lemma is that the tangent planes to the image of the graph of ϕ under f^n are exponentially (in n) close to the "horizontal" space $\mathbb{R}^k$, and the image spreads over B^k_ϵ in the horizontal direction (see Figure 5.2).

The following theorem, which is also sometimes called the Lambda Lemma, implies that if f is C^r with $r \geqslant 1$, and D is any C^1 disc that intersects transversely the stable manifold $W^s(x)$ of a hyperbolic fixed point x, then the forward images of D converge in the C^r topology to the unstable manifold $W^u(x)$ (Palis and de Melo, 1982). We prove only C^1 convergence. Let B^u_R be the ball of radius R centered at x in $W^u(x)$ in the induced metric.

THEOREM 5.7.2 (Inclination Lemma). *Let x be a hyperbolic fixed point of a diffeomorphism $f: U \to M$, $\dim W^u(x) = k$ and $\dim W^s(x) = l$. Let $y \in W^s(x)$ and suppose that $D \ni y$ is a C^1 submanifold of dimension k intersecting $W^s(x)$ transversely at y.*

Then for every $R > 0$ and $\beta > 0$ there are $n_0 \in \mathbb{N}$ and, for each $n \geqslant n_0$, a subset $\widetilde{D} = \widetilde{D}(R, \beta, n)$, $y \in \widetilde{D} \subset D$, diffeomorphic to an open k-disc and such that the C^1 distance between $f^n(\widetilde{D})$ and B^u_R is less than β.

Proof. We will show that for some $n_1 \in \mathbb{N}$, an appropriate subset $D_1 \subset f^{n_1}(D)$ satisfies the assumptions of Lemma 5.7.1. Since $\{x\}$ is a hyperbolic set of f, for any $\delta > 0$ there is $\epsilon > 0$ such that $E^s(x)$ and $E^u(x)$ can be extended to invariant distributions $\widetilde{E}^s$ and $\widetilde{E}^u$ in the ϵ-neighborhood B_ϵ of x and the hyperbolicity constant is at most $(\lambda + \delta)$ (Proposition 5.4.4). Since $f^n(y) \to x$, there is $n_2 \in \mathbb{N}$ such that $z = f^{n_2}(y) \in B_\epsilon$. Since D intersects $W^s(x)$ transversely, so does $f^{n_2}(D)$. Therefore there is $\eta > 0$ such that if $v \in T_z f^{n_2}(D)$, $\|v\| = 1$, $v = v^s + v^u$, $v^s \in \widetilde{E}^s(z)$, $v^u \in \widetilde{E}^u(z)$, and $v^u \neq 0$, then $\|v^u\| \geqslant \eta \|v^s\|$. By Proposition 5.4.4, for a small enough $\delta > 0$, the norm $\|df^n v^s\|$ decays exponentially and $\|df^n v^u\|$ grows exponentially. Therefore for an arbitrarily small cone size, there exists $n_2 \in \mathbb{N}$ such that $T_{f^{n_2}(y)} f^{n_2}(D)$ lies inside the unstable cone at $f^{n_2}(y)$. □

Let p be a hyperbolic periodic point of a diffeomorphism $f: U \to M$. A point $q \in U$ is called *homoclinic* (for p) if $q \neq p$ and $q \in W^s(p) \cap W^u(p)$; it is called *transverse homoclinic* (for p) if in addition $W^s(p)$ and $W^u(p)$ intersect transversely at q.

Exercise 5.7.1. Prove that if x is a homoclinic point, then x is non-wandering but not recurrent.

Exercise 5.7.2. Prove Lemma 5.7.1.

Exercise 5.7.3. Let p be a hyperbolic fixed point of f and $q \in W^s(p) \cap W^u(p)$ a transverse homoclinic point. Prove that the union of p with the orbit of q is a hyperbolic set of Λ.

5.8 Horseshoes and transverse homoclinic points

Let $\mathbb{R}^m = \mathbb{R}^k \times \mathbb{R}^l$. We will refer to $\mathbb{R}^k$ and $\mathbb{R}^l$ as the unstable and stable subspaces, respectively, and denote by π^u and π^s the projections to those

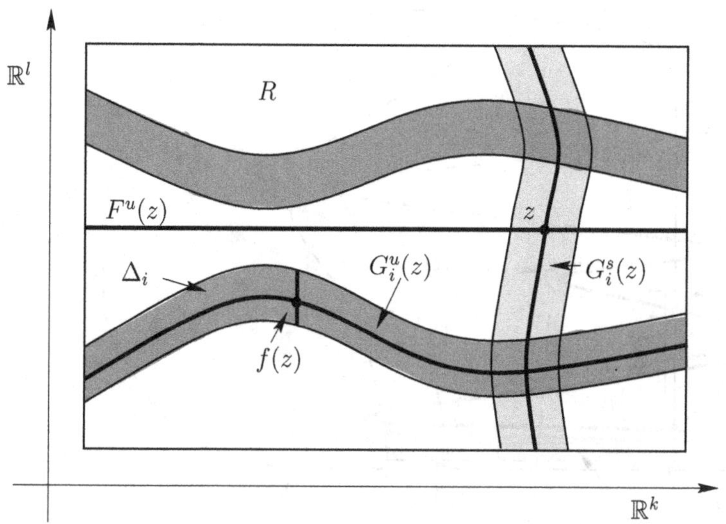

Figure 5.3. A non-linear horseshoe.

subspaces. For $v \in \mathbb{R}^m$, denote $v^u = \pi^u(v) \in \mathbb{R}^k$ and $v^s = \pi^s(v) \in \mathbb{R}^l$. For $\alpha \in (0, 1)$, call the sets $K_\alpha^u = \{v \in \mathbb{R}^m\colon |v^s| \leqslant \alpha|v^u|\}$ and $K_\alpha^s = \{v \in \mathbb{R}^m\colon |v^u| \leqslant \alpha|v^s|\}$ the unstable and stable cones, respectively. Let $R^u = \{x \in \mathbb{R}^k\colon |x| \leqslant 1\}$, $R^s = \{x \in \mathbb{R}^l\colon |x| \leqslant 1\}$ and $R = \mathbb{R}^u \times \mathbb{R}^s$. For $z = (x, y)$, $x \in \mathbb{R}^k$, $y \in \mathbb{R}^l$, the sets $F^s(z) = \{x\} \times R^s$ and $F^u(z) = R^u \times \{y\}$ will be referred to as unstable and stable fibers, respectively. We say that a C^1 map $f\colon R \to \mathbb{R}^m$ has a *horseshoe* if there are $\lambda, \alpha \in (0, 1)$ such that:

1. f is one-to-one on R;
2. $f(R) \cap R$ has at least two components $\Delta_0, \dots, \Delta_{q-1}$;
3. if $z \in R$ and $f(z) \in \Delta_i$ with $0 \leqslant i < q$, then the sets $G_i^u(z) = f(F^u(z)) \cap \Delta_i$ and $G_i^s(z) = f^{-1}(F^s(f(z)) \cap \Delta_i)$ are connected and the restrictions of π^u to $G_i^u(z)$ and of π^s to $G_i^s(z)$ are onto and one-to-one;
4. if $z, f(z) \in R$ then the derivative df_z preserves the unstable cone K_α^u and $\lambda|df_z v| \geqslant |v|$ for every $v \in K_\alpha^u$, the inverse $df_{f(z)}^{-1}$ preserves the stable cone K_α^s, and $\lambda|df_{f(z)}^{-1} v| \geqslant |v|$ for every $v \in K_\alpha^s$.

The intersection $\Lambda = \bigcap_{n\in\mathbb{Z}} f^n(R)$ is called a *horseshoe* (see Figure 5.3).

THEOREM 5.8.1. *The horseshoe $\Lambda = \bigcap_{n\in\mathbb{Z}} f^n(R)$ is a hyperbolic set of f. If $f(R) \cap R$ has q components, then the restriction of f to Λ is topologically conjugate to the full two-sided shift σ in the space Σ_q of bi-infinite sequences in the alphabet $\{0, 1, \dots, q-1\}$.*

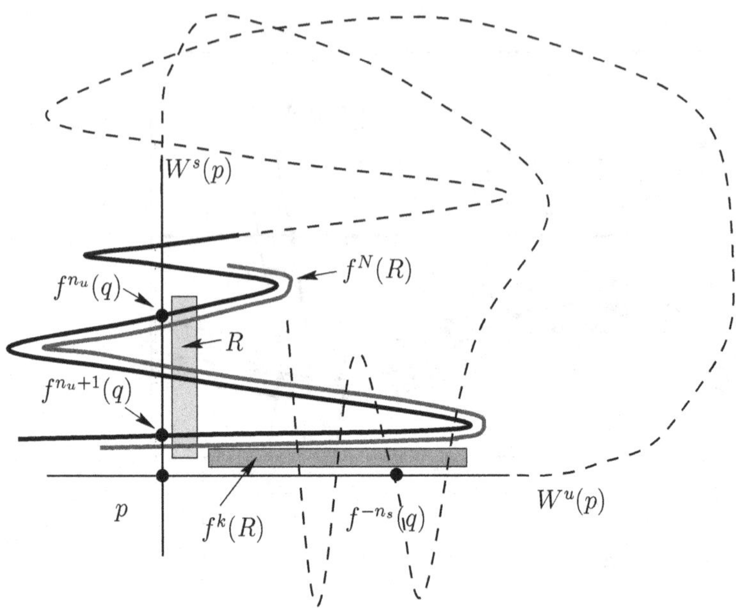

Figure 5.4. A horseshoe at a homoclinic point.

Proof. The hyperbolicity of Λ follows from the invariance of the cones and the stretching of vectors inside the cones. The topological conjugacy of $f|_\Lambda$ to the two-sided shift is left as an exercise (Exercise 5.8.2). □

COROLLARY 5.8.2. *If a diffeomorphism f has a horseshoe, then the topological entropy of f is positive.*

The next theorem shows that horseshoes, and hence hyperbolic sets in general, are rather common.

THEOREM 5.8.3. *Let p be a hyperbolic periodic point of a diffeomorphism $f: U \to M$ and let q be a transverse homoclinic point of p. Then for every $\epsilon > 0$ the union of the ϵ-neighborhoods of the orbits of p and q contains a horseshoe of f.*

Proof. We consider only the two-dimensional case; the argument for higher dimensions is a routine generalization of the proof below. We assume without loss of generality that $f(p) = p$ and f preserves orientation. There is a C^1 coordinate system in a neighborhood $V = V^u \times V^s$ of p such that p is the origin and the stable and unstable manifolds of p coincide locally with the coordinate axes (Figure 5.4). For a point $x \in V$ and a vector $v \in \mathbb{R}^2$, we write

$x = (x^u, x^s)$ and $v = (v^u, v^s)$, where s and u indicate the stable (vertical) and unstable (horizontal) components, respectively. We also assume that there is $\lambda \in (0, 1)$ such that $|df_p v^s| < \lambda |v^s|$ and $|df_p^{-1} v^u| < \lambda |v^u|$ for every $v \neq 0$. Fix $\delta > 0$ and let $K^s_{\delta/2}$ and $K^u_{\delta/2}$ be the stable and unstable $\delta/2$-cones. Choose V small enough so that for every $x \in V$

$$df_x(K^u_{\delta/2}) \subset K^u_{\delta/2}, \quad |df_x^{-1} v| < \lambda |v| \text{ if } v \in K^u_{\delta/2},$$

$$df_x^{-1} K^s_{\delta/2} \subset K^s_{\delta/2}, \quad |df_x v| < \lambda |v| \text{ if } v \in K^s_{\delta/2}.$$

Since $q \in W^s(p) \cap W^u(p)$, we have that $f^n(q) \in V$ and $f^{-n}(q) \in V$ for all sufficiently large n. By invariance, $W^s(p)$ and $W^u(p)$ pass through all images $f^n(q)$. Since $W^u(p)$ intersects $W^s(p)$ transversely at q, by Theorem 5.7.2, there is n_u such that $f^n(q) \in V$ for $n \geqslant n_u$ and an appropriate neighborhood D^u of $f^n(q)$ in $W^u(p)$ is a C^1 submanifold that "stretches across" V and whose tangent planes lie in $K^u_{\delta/2}$, i.e. D^u is the graph of a C^1 function $\phi^u \colon V^u \to V^s$ with $\|d\phi^u\| < \delta/2$. Similarly since $q \in W^u(p)$, there is $n_s \in \mathbb{N}$ such that $f^{-n}(q) \in U$ for $n \geqslant n_s$ and a small neighborhood D^s of $f^{-n}(q)$ in $W^s(p)$ is the graph of a C^1 function $\phi^s \colon V^s \to V^u$ with $\|d\phi^s\| < \delta/2$. Note that since f preserves orientation, the point $f^{n_u+1}(q)$ is not the next intersection of $W^u(p)$ with $W^s(p)$ after $f^{n_u}(q)$; in Figure 5.4 it is shown as the second intersection after $f^{n_u}(q)$ along $W^s(p)$.

Consider a narrow "box" R shown in Figure 5.4 and let $N = k + n_u + n_s + 1$. We will show that for an appropriate choice of the size and position of R and of $k \in \mathbb{N}$, the map $\tilde{f} = f^N$, box R, and its image $\tilde{f}(R)$ satisfy the definition of a horseshoe. The smaller the width of the box and the closer it lies to $W^s(p)$, the larger is k for which $f^k(R)$ reaches the vicinity of $f^{-n_s}(q)$. The number $\bar{n} = n_u + n_s + 1$ is fixed. If v^u is a horizontal vector at $f^{-n_s}(q)$, its image $w = df^{\bar{n}}_{f^{-n_s}(q)} v^u$ is tangent to $W^u(p)$ at $f^{n_u+1}(q)$ and therefore lies in $K^u_{\delta/2}$. Moreover, $|w| \geqslant 2\beta |v^u|$ for some $\beta > 0$. For any sufficiently close vector v at a close enough base point, the image will lie in K^u_δ and $|df^{\bar{n}} v| \geqslant \beta |v|$. The same holds for "almost horizontal" vectors at points close to $f^{-n_s-1}(q)$.

On the other hand, $df_x(K^u_\alpha) \subset K^u_{\lambda\alpha}$ for every small enough $\alpha > 0$ and every $x \in V$. Therefore if $x \in R, f(x), \dots, f^k(x) \in V$ and $v \in K^u_\delta$ is a tangent vector at x, then $df_x^k v \in K^u_{\delta\lambda^k}$ and $|df_x^k v| > \lambda^{-k} |v|$. Suppose now that $x \in R$ is such that $f^k(x)$ is close to either $f^{-n_s}(q)$ or $f^{-n_s-1}(q)$. Let k be large enough so that $\beta/\lambda^k > 10$. There is $\lambda' \in (0, 1)$ such that if $x \in R$ and $f^N(x)$ is close to either $f^{n_u}(q)$ or $f^{n_u+1}(q)$ (i.e. $f^k(x)$ is close to $f^{-n_s}(q)$ or $f^{-n_s-1}(q)$), then K^u_δ is invariant under df_x^N and $\lambda' |df_x^N v| \geqslant |v|$ for every $v \in K^u_\delta$. Similarly, for an appropriate choice of R and k, the stable δ-cones are invariant under df^{-N} and vectors from K^s_δ are stretched by df^{-N} by a factor at least $(\lambda')^{-1}$.

To guarantee the correct intersection of $f^N(R)$ with R we must choose R carefully. Choose the horizontal boundary segments of R to be straight line segments and let R stretch vertically so that it crosses $W^u(p)$ near $f^{n_u}(q)$ and $f^{n_u+1}(q)$. By Theorem 5.7.2, the images of these horizontal segments under f^k are almost horizontal line segments. To construct the vertical boundary segments of R, take two vertical segments s_1 and s_2 to the left of $f^{-n_s-1}(q)$ and to the right of $f^{-n_s}(q)$ and truncate their preimages $f^{-k}(s_i)$ by the horizontal boundary segments. By Theorem 5.7.2, the preimages are almost vertical line segments. This choice of R satisfies the definition of a horseshoe. □

Exercise 5.8.1. Let $f: U \to M$ be a diffeomorphism, p a periodic point of f, and q a (non-transverse) homoclinic point (for p). Prove that every arbitrarily small C^1 neighborhood of f contains a diffeomorphism g such that p is a periodic point of g and q is a transverse homoclinic point (for p).

Exercise 5.8.2. Prove that if $f(R) \cap R$ in Theorem 5.8.1 has q connected components, then the restriction of f to Λ is topologically conjugate to the full two-sided shift in the space Σ_q of bi-infinite sequences in the alphabet $\{1, \dots, q\}$.

Exercise 5.8.3. Let $p_1, \dots, p_k$ be periodic points (of possibly different periods) of $f: U \to M$. Suppose $W^u(p_i)$ intersects $W^s(p_{i+1})$ transversely at q_i, $i = 1, \dots, k$, $p_{k+1} = p_1$ (in particular, $\dim W^s(p_i) = \dim W^s(p_1)$ and $\dim W^u(p_i) = \dim W^u(p_1), i = 2, \dots, k$). The points q_i are called *transverse heteroclinic points*. Prove the following generalization of Theorem 5.8.3: any neighborhood of the union of the orbits of p_i's and q_i's contains a horseshoe.

5.9 Local product structure and locally maximal hyperbolic sets

A hyperbolic set Λ of $f: U \to M$ is called *locally maximal* if there is an open set V such that $\Lambda \subset V \subset U$ and $\Lambda = \bigcap_{n=-\infty}^{\infty} f^n(V)$. The horseshoe (§5.8) and the solenoid(§1.9) are examples of locally maximal hyperbolic sets (Exercise 5.9.1).

Since every closed invariant subset of a hyperbolic set is also a hyperbolic set, the geometric structure of a hyperbolic set may be very complicated and difficult to describe. However, due to their special properties, locally maximal hyperbolic sets allow a geometric characterization.

Since $E^s(x) \cap E^u(x) = \{0\}$, the local stable and unstable manifolds of x intersect at x transversely. By continuity, this transversality extends to a neighborhood of the diagonal in $\Lambda \times \Lambda$.

PROPOSITION 5.9.1. *Let Λ be a hyperbolic set of f. For every small enough $\epsilon > 0$ there is $\delta > 0$ such that if $x, y \in \Lambda$ and $d(x, y) < \delta$, then the intersection $W^s_\epsilon(x) \cap W^u_\epsilon(y)$ is transverse and consists of exactly one point $[x, y]$, which depends continuously on x and y. Furthermore, there is $C = C(\delta) > 0$ such that if $x, y \in \Lambda$ and $d(x, y) < \delta$, then $d^s(x, [x, y]) \leqslant Cd(x, y)$ and $d^u(x, [x, y]) \leqslant Cd(x, y)$, where d^s and d^u denote distances along the stable and unstable manifolds.*

Proof. The proposition follows immediately from the uniform transversality of E^s and E^u and Lemma 5.9.2. □

Let $\epsilon > 0, k, l \in \mathbb{N}$ and let $B^k_\epsilon \subset \mathbb{R}^k, B^l_\epsilon \subset \mathbb{R}^l$ be the ϵ-balls centered at the origin.

LEMMA 5.9.2. *For every $\epsilon > 0$ there is $\delta > 0$ such that if $\phi\colon B^k_\epsilon \to \mathbb{R}^l$ and $\psi\colon B^l_\epsilon \to \mathbb{R}^k$ are differentiable maps and $|\phi(x)|, \|d\phi(x)\|, |\psi(y)|, \|d\phi(y)\| < \delta$ for all $x \in B^k_\epsilon$ and $y \in B^l_\epsilon$, then the intersection* $\text{graph}(\phi) \cap \text{graph}(\psi) \subset \mathbb{R}^{k+l}$ *is transverse and consists of exactly one point, which depends continuously on ϕ and ψ in the C^1 topology.*

Proof. Exercise 5.9.3. □

The following property of hyperbolic sets plays a major role in their geometric description and is equivalent to local maximality. A hyperbolic set Λ has a *local product structure* if there are (small enough) $\epsilon > 0$ and $\delta > 0$ such that (i) for all $x, y \in \Lambda$ the intersection $W^s_\epsilon(x) \cap W^u_\epsilon(y)$ consists of at most one point, which belongs to Λ; and (ii) for $x, y \in \Lambda$ with $d(x, y) < \delta$, the intersection consists of exactly one point of Λ denoted $[x, y] = W^s_\epsilon(x) \cap W^u_\epsilon(y)$, and the intersection is transverse (Proposition 5.9.1). If a hyperbolic set Λ has a local product structure, then for every $x \in \Lambda$ there is a neighborhood $U(x)$ such that

$$U(x) \cap \Lambda = \{[y, z]\colon y \in U(x) \cap W^s_\epsilon(x),\ z \in U(x) \cap W^u_\epsilon(x)\}.$$

PROPOSITION 5.9.3. *A hyperbolic set Λ is locally maximal if and only if it has a local product structure.*

Proof. Suppose Λ is locally maximal. If $x, y \in \Lambda$ and $\text{dist}(x, y)$ is small enough, then by Proposition 5.9.1, $W^s_\epsilon(x) \cap W^u_\epsilon(y) = [x, y] := z$ exists and, by Theorem 5.6.4(4), the forward and backward semiorbits of z stay close to Λ. Since Λ is locally maximal, $z \in \Lambda$.

Conversely, assume that Λ has a local product structure with constants ϵ, δ and C from Proposition 5.9.1. We must show that if the whole orbit of a point q lies close to Λ, then the point lies in Λ. Fix $\alpha \in (0, \delta C/3)$ such that $f(p) \in W^u_{\delta/3}(f(x))$ for each $x \in \Lambda$ and $p \in W^u_\alpha(x)$. Assume first that $q \in W^u_\alpha(x_0)$ for some $x_0 \in \Lambda$ and that there are $y_n \in \Lambda$ such that $d(f^n(q), y_n) < \alpha/C$ for all $n > 0$. Since $f(x_0)$, $y_1 \in \Lambda$ and $d(f(x_0), y_1) < d(f(x_0), f(q)) + d(f(q), y_1) < \delta/3 + \alpha/C < \delta$, we have that $x_1 = [y_1, f(x_0)] \in \Lambda$ and, by Proposition 5.9.1, $f(q) \in W^u_\alpha(x_1)$. Similarly, $x_2 = [y_2, f(x_1)] \in \Lambda$ and $f^2(q) \in W^u_\alpha(x_2)$. By repeating this argument we construct points $x_n = [y_n, f^n(q)] \in \Lambda$ with $f^n(q) \in W^u_\alpha(x_n)$. Observe that $q_n = f^{-n}(x_n) \to q$ as $n \to \infty$. Since Λ is closed, $q \in \Lambda$. Similarly, if $q \in W^s_\alpha(x_0)$ for some $x_0 \in \Lambda$ and $f^n(q)$ stays close enough to Λ for all $n < 0$, then $q \in \Lambda$.

Assume now that $f^n(y)$ is close enough to $x_n \in \Lambda$ for all $n \in \mathbb{Z}$. Then $y \in \Lambda^s_\epsilon$ and $y \in \Lambda^u_\epsilon$. Hence, by Propositions 5.4.4 and 5.4.3, the union $\Lambda \cup y$ is a hyperbolic set (with close constants) and the local stable and unstable manifolds of y are well-defined. Observe that the forward semiorbit of $p = [y, x_0]$ and the backward semiorbit of $q = [x_0, y]$ stay close to Λ. Therefore, by the above argument, $p, q \in \Lambda$ and (by the local product structure) $y = [p, q] \in \Lambda$. □

Exercise 5.9.1. Prove that horseshoes (§5.8) and the solenoid (§1.9) are locally maximal hyperbolic sets.

Exercise 5.9.2. Let p be a hyperbolic fixed point of f and $q \in W^s(p) \cap W^u(p)$ a transverse homoclinic point. By Exercise 5.7.3, the union of p with the orbit of q is a hyperbolic set of f. Is it locally maximal?

Exercise 5.9.3. Prove Lemma 5.9.2.

5.10 Anosov diffeomorphisms

Recall that a C^1 diffeomorphism f of a connected differentiable manifold M is called *Anosov* if M is a hyperbolic set for f; it follows directly from the definition that M is locally maximal and compact.

The simplest example of an Anosov diffeomorphism is the automorphism of $\mathbb{T}^2$ given by the matrix $\left(\begin{smallmatrix}2 & 1\\ 1 & 1\end{smallmatrix}\right)$. More generally, any linear hyperbolic automorphism of the n-torus $\mathbb{T}^n$ is Anosov. Such an automorphism is given by an $n \times n$ integer matrix with determinant ± 1 and with no eigenvalues of modulus 1.

Toral automorphisms can be generalized as follows. Let N be a simply-connected nilpotent Lie group and Γ a uniform discrete subgroup of N. The quotient $M = N/\Gamma$ is a compact *nilmanifold*. Let $\overline{f}$ be an automorphism of N that preserves Γ and whose derivative at the identity is hyperbolic. The induced diffeomorphism f of M is Anosov. For specific examples of this type see Smale (1967). Up to finite coverings, all known Anosov diffeomorphisms are topologically conjugate to automorphisms of nilmanifolds.

The families of stable and unstable manifolds of an Anosov diffeomorphism form two foliations (see §5.13) called the *stable foliation* W^s and *unstable foliation* W^u (Exercise 5.10.1). These foliations are in general not C^1, or even Lipschitz (Anosov, 1967), but they are Hölder continuous (Theorem 6.1.3). In spite of the lack of Lipschitz continuity, the stable and unstable foliations possess a uniqueness property similar to the uniqueness theorem for ordinary differential equations (Exercise 5.10.2).

Proposition 5.10.1 states basic properties of the stable and unstable distributions E^s and E^u, and the stable and unstable foliations W^s and W^u, of an Anosov diffeomorphism f. These properties follow immediately from the previous sections of this chapter. We assume that the metric is adapted to f and denote by d^s and d^u the distances along the stable and unstable leaves.

PROPOSITION 5.10.1. *Let $f\colon M \to M$ be an Anosov diffeomorphism. Then there are $\lambda \in (0,1)$, $C > 0$, $\epsilon > 0$, $\delta > 0$ and, for every $x \in M$, a splitting $T_xM = E^s(x) \oplus E^u(x)$ such that:*

1. *$df_x(E^s(x)) = E^s(f(x))$ and $df_x(E^u(x)) = E^u(f(x))$;*
2. *$\|df_x v^s\| \leqslant \lambda \|v^s\|$ and $\|df_x^{-1} v^u\| \leqslant \lambda \|v^u\|$ for all $v^s \in E^s(x)$, $v^u \in E^u(x)$;*
3. *$W^s(x) = \{y \in M\colon d(f^n(x), f^n(y)) \to 0$ as $n \to \infty\}$ and $d^s(f(x), f(y)) \leqslant \lambda d^s(x,y)$ for every $y \in W^s(x)$;*
4. *$W^u(x) = \{y \in M\colon d(f^{-n}(x), f^{-n}(y)) \to 0$ as $n \to \infty\}$ and $d^u(f^{-1}(x), f^{-1}(y)) \leqslant \lambda^n d^u(x,y)$ for every $y \in W^u(x)$;*
5. *$f(W^s(x)) = W^s(f(x))$ and $f(W^u(x)) = W^u(f(x))$;*
6. *$T_xW^s(x) = E^s(x)$ and $T_xW^u(x) = E^u(x)$;*
7. *if $d(x,y) < \delta$, then the intersection $W^s_\epsilon(x) \cap W^u_\epsilon(y)$ is exactly one point $[x,y]$, which depends continuously on x and y and $d^s([x,y],x) \leqslant Cd(x,y)$, $d^u([x,y],y) \leqslant Cd(x,y)$.*

For convenience we restate several properties of Anosov diffeomorphisms. Recall that a diffeomorphism $f\colon M \to M$ is *structurally stable* if for every $\epsilon > 0$ there is a neighborhood $\mathcal{U} \subset \mathrm{Diff}^1(M)$ of f such that for every $g \in \mathcal{U}$ there is a homeomorphism $h\colon M \to M$ with $h \circ f = g \circ h$ and $\mathrm{dist}_0(h, \mathrm{Id}) < \epsilon$.

PROPOSITION 5.10.2

1. *Anosov diffeomorphisms form an open (possibly empty) subset in the C^1 topology (Corollary 5.5.2).*
2. *Anosov diffeomorphisms are structurally stable (Corollary 5.5.4).*

3. *The set of periodic points of an Anosov diffeomorphism is dense in the set of non-wandering points (Corollary 5.3.4).*

Here is a more direct proof of the density of periodic points. Let ϵ and δ satisfy Proposition 5.10.1. If $x \in M$ is non-wandering, then there is $n \in \mathbb{N}$ and $y \in M$ such that $\operatorname{dist}(x, y)$, $\operatorname{dist}(f^n(y), y) < \delta/(2C)$. Assume that $\lambda^n < 1/(2C)$. Then the map $z \mapsto [y, f^n(z)]$ is well-defined for $z \in W^s_\delta(y)$. It maps $W^s_\delta(y)$ into itself and, by the Brouwer Fixed Point Theorem, has a fixed point y_1 such that $d^s(y_1, y) < \delta$, $f^n(y_1) \in W^u(y_1)$ and $d^u(y_1, f^n(y_1)) < \delta$. The map f^{-n} sends $W^u_\delta(f^n(y_1))$ to itself and therefore has a fixed point.

THEOREM 5.10.3. *Let $f\colon M \to M$ be an Anosov diffeomorphism. Then the following are equivalent:*

1. $\mathrm{NW}(f) = M$;
2. *every unstable manifold is dense in M;*
3. *every stable manifold is dense in M;*
4. *f is topologically transitive;*
5. *f is topologically mixing.*

Proof. We say that a set A is *ϵ-dense* in a metric space (X, d) if $d(x, A) < \epsilon$ for every $x \in X$.

(1)$\Rightarrow$(2). We will show that every unstable manifold is ϵ-dense in M for an arbitrary $\epsilon > 0$. By Proposition 5.10.2(3), the periodic points are dense. Assume that $\epsilon > 0$ satisfies Proposition 5.10.1(7) and that periodic points x_i, $i = 1, \dots, N$, form an $\epsilon/4$-net in M. Let P be the product of the periods of the x_i's and set $g = f^P$. Note that the stable and unstable manifolds of g are the same as those of f.

LEMMA 5.10.4. *There is $q \in \mathbb{N}$ such that if* $\operatorname{dist}(W^u(y), x_i) < \epsilon/2$ *and* $\operatorname{dist}(x_i, x_j) < \epsilon/2$ *for some $y \in M$, i, j, then* $\operatorname{dist}(g^{nq}(W^u(y)), x_i) < \epsilon/2$ *and* $\operatorname{dist}(g^{nq}(W^u(y)), x_j) < \epsilon/2$ *for every $n \in \mathbb{N}$.*

Proof. By Proposition 5.10.2(3), there is $z \in W^u(y) \cap W^s_{C\epsilon}(x_i)$. Therefore $\operatorname{dist}(g^t(z), x_i) < \epsilon/2$ for any $t \geqslant t_0$, where t_0 depends on ϵ but not on z. Since $\operatorname{dist}(g^t(z), x_j) < \epsilon$, by Proposition 5.10.2(3), there is $w \in W^u(g^t(z)) \cap W^s_{C\epsilon}(x_j)$. Hence $\operatorname{dist}(g^\tau(w), x_j) < \epsilon/2$ for any $\tau \geqslant \tau_0$ which depends only on ϵ but not on w. The lemma follows with $q = s_0 + t_0$. □

Since M is compact and connected, any x_i can be connected to any x_j by a chain of not more than N periodic points x_k with distance $<\epsilon/2$ between any 2 consecutive points. By Lemma 5.10.4, $g^{Nq}(W^u(y))$ is ϵ-dense in M for any $y \in M$. Hence $W^u(x)$ is ϵ-dense for any $x = g^{-Nq}(y) \in M$. Therefore $W^u(x)$ is dense for each x. Reversing the time gives (1)$\Rightarrow$(3).

LEMMA 5.10.5. *If every (un)stable manifold is dense in M, then for every $\epsilon > 0$ there is $R = R(\epsilon) > 0$ such that every ball of radius R in every (un)stable manifold is ϵ-dense in M.*

Proof. Let $x \in M$. Since $W^u(x) = \bigcup_R W^u_R(x)$ is dense, there is $R(x)$ such that $W^u_{R(x)}(x)$ is $\epsilon/2$ dense. Since W^u is a continuous foliation, there is $\delta(x) > 0$ such that $W^u_{R(x)}(y)$ is ϵ-dense for any $y \in B(x, \delta(x))$. By the compactness of M, a finite collection $\mathcal{B}$ of the $\delta(x)$-balls covers M. The maximal $R(x)$ for the balls from $\mathcal{B}$ satisfies the lemma. □

(2) ⇒ (5). Let $U, V \subset M$ be non-empty open sets. Let $x, y \in M$ and $\delta > 0$ be such that $W^u_\delta(x) \subset U$ and $B(y, \delta) \subset V$ and let $R = R(\delta)$ (see Lemma 5.10.5). Since f expands unstable manifolds exponentially and uniformly, there is $N \in \mathbb{N}$ such that $f^n(W^u_\delta(x)) \supset W^u_R(f^n(x))$ for $n \geqslant N$. By Lemma 5.10.5, $f(U) \cap V \neq \emptyset$ and hence f is topologically mixing. Similarly (3) ⇒ (5).

(1) ⇒ (3) follows by reversing the time. Obviously (5) ⇒ (4) and (4) ⇒ (1). □

Exercise 5.10.1. Prove that the stable and unstable manifolds of an Anosov diffeomorphism form foliations (see §5.13).

Exercise 5.10.2. Although the stable and unstable distributions of an Anosov diffeomorphism, in general, are not Lipschitz continuous, the following uniqueness property holds true. Let $\gamma(\cdot)$ be a differentiable curve such that $\dot{\gamma}(t) \in E^s(\gamma(t))$ for every t. Prove that γ lies in one stable manifold.

5.11 Axiom A and structural stability

Some of the results of §5.10 extend to a natural wider class of hyperbolic dynamical systems. Throughout this section we assume that f is a diffeomorphism of a compact manifold M. Recall that the set of non-wandering points $\mathrm{NW}(f)$ is closed and f-invariant, and that $\overline{\mathrm{Per}(f)} \subset \mathrm{NW}(f)$.

A diffeomorphism f satisfies Smale's *Axiom A* if the set $\mathrm{NW}(f)$ is hyperbolic and $\overline{\mathrm{Per}(f)} = \mathrm{NW}(f)$. The second condition does not follow from the first. By Proposition 5.3.3, the set $\mathrm{Per}(f)$ is dense in the set $\mathrm{NW}(f|_{\mathrm{NW}(f)})$ of non-wandering points of the restriction of f to $\mathrm{NW}(f)$. However, in general $\mathrm{NW}(f|_{\mathrm{NW}(f)}) \neq \mathrm{NW}(f)$ (Exercises 5.11.1 and 5.11.2).

For a hyperbolic periodic point p of f, denote by $W^s(O(p))$ and $W^u(O(p))$ the unions of the stable and unstable manifolds of p and its images, respectively. If p and q are hyperbolic periodic points, we write $p \leqslant q$ when $W^s(O(p))$ and $W^u(O(q))$ have a point of transverse intersection. The relation $\leqslant$ is reflexive. It follows from Theorem 5.7.2 that $\leqslant$ is transitive (Exercise 5.11.3). If $p \leqslant q$ and $q \leqslant p$, we write $p \sim q$ and say that p and q are *heteroclinically related*. The relation $\sim$ is an equivalence relation.

The following theorem can be proved in a straightforward manner by applying the results and techniques of this chapter.

THEOREM 5.11.1 (Smale's Spectral Decomposition (Smale, 1967)). *If f satisfies Axiom A, then there is a unique representation of* $\mathrm{NW}(f)$

$$\mathrm{NW}(f) = \Lambda_1 \cup \Lambda_2 \cup \cdots \cup \Lambda_k$$

as a disjoint union of closed f-invariant sets (called basic sets*) such that*

1. *each Λ_i is a locally maximal hyperbolic set of f;*
2. *f is topologically transitive on each Λ_i; and*
3. *each Λ_i is a disjoint union of closed sets Λ_i^j, $1 \leqslant j \leqslant m_i$, the diffeomorphism f cyclically permutes the sets Λ_i^j and f^{m_i} is topologically mixing on each Λ_i^j.*

The basic sets are precisely the closures of the equivalence classes of $\sim$. For two basic sets, we write $\Lambda_i \leqslant \Lambda_j$ if there are periodic points $p \in \Lambda_i$ and $q \in \Lambda_j$ such that $p \leqslant q$.

Let f satisfy Axiom A. We say that f satisfies the *strong transversality condition* if $W^s(x)$ intersects $W^u(y)$ transversely (at all common points) for all $x, y \in \mathrm{NW}(f)$.

THEOREM 5.11.2 (Structural Stability Theorem). *A C^1 diffeomorphism is structurally stable if and only if it satisfies Axiom A and the strong transversality condition.*

Robbin (1971) showed that a C^2 diffeomorphism satisfying Axiom A and the strong transversality condition is structurally stable. Robinson (1976) weakened C^2 to C^1. Mañé (1988) proved that a structurally stable C^1 diffeomorphism satisfies Axiom A and the strong transversality condition.

Exercise 5.11.1. Give an example of a diffeomorphism f with $\mathrm{NW}(f|_{\mathrm{NW}(f)}) \neq \mathrm{NW}(f)$.

Exercise 5.11.2. Give an example of a diffeomorphism f for which $\mathrm{NW}(f)$ is hyperbolic and $\mathrm{NW}(f|_{\mathrm{NW}(f)}) \neq \mathrm{NW}(f)$.

Exercise 5.11.3. Prove that $\leqslant$ is a transitive relation.

Exercise 5.11.4. Suppose that f satisfies Axiom A. Prove that $NW(f)$ is a locally maximal hyperbolic set.

5.12 Markov partitions

Recall (Chapters 1, 3) that a partition of the "phase" space of a dynamical system induces a coding of the orbits and hence a semiconjugacy with a subshift. For hyperbolic dynamical systems, there is a special class of partitions – *Markov partitions* – for which the target subshift is a subshift of finite type. A Markov partition $\mathcal{P}$ for an invariant subset Λ of a diffeomorphism f of a compact manifold M is a collection of sets R_i called *rectangles* such that for all i, j, k:

1. each R_i is the closure of its interior;
2. $\operatorname{int} R_i \cap \operatorname{int} R_j = \emptyset$ if $i \neq j$;
3. $\Lambda \subset \bigcup_i R_i$;
4. if $f^m(\operatorname{int} R_i) \cap \operatorname{int} R_j \cap \Lambda \neq \emptyset$ for some $m \in \mathbb{Z}$ and $f^n(\operatorname{int} R_j) \cap \operatorname{int} R_k \cap \Lambda \neq \emptyset$ for some $n \in \mathbb{Z}$, then $f^{m+n}(\operatorname{int} R_i) \cap \operatorname{int} R_k \cap \Lambda \neq \emptyset$.

The last condition guarantees the Markov property of the subshift corresponding to $\mathcal{P}$, i.e. the independence of the future from the past. For hyperbolic dynamical systems, each rectangle is closed under the local product structure "commutator" $[x, y]$, i.e. if $x, y \in R_i$, then $[x, y] \in R_i$. For $x \in R_i$ let $W^s(x, R_i) = \bigcup_{y \in R_i} [x, y]$ and $W^u(x, R_i) = \bigcup_{y \in R_i} [y, x]$. The last condition implies that if $x \in \operatorname{int} R_i$ and $f(x) \in \operatorname{int} R_j$, then $W^u(f(x), R_j) \subset f(W^u(x, R_i))$ and $W^s(x, R_i) \subset f^{-1}(W^s(f(x), R_j))$.

The partition of the unit interval $[0, 1]$ into m intervals $[k/m, (k+1)/m]$ is a Markov partition for the expanding endomorphism E_m. The target subshift in this case is the full shift on m symbols.

We now describe a Markov partition for the hyperbolic toral automorphism $f = f_M$ given by the matrix $M = \left(\begin{smallmatrix} 2 & 1 \\ 1 & 1 \end{smallmatrix}\right)$ which was constructed by Adler and Weiss (1967). The eigenvalues are $(3 \pm \sqrt{5})/2$. We begin by partitioning the unit square representing the torus $\mathbb{T}^2$ in Figure 5.5 into two "rectangles": A, consisting of three parts A_1, A_2, A_3; and B, consisting of two parts B_1, B_2. The longer sides of the rectangles are parallel to the eigendirection of the larger eigenvalue $(3 + \sqrt{5})/2$ and the shorter sides are parallel to the eigendirection of the smaller eigenvalue $(3 - \sqrt{5})/2$. In Figure 5.5, the identified points and regions are marked by the same symbols. The images of A and B are shown in Figure 5.6. We subdivide A and B into five subrectangles $\Delta_1, \Delta_2, \Delta_3, \Delta_4, \Delta_5$ that are the connected components of the intersections of A and B with $f(A)$ and $f(B)$. This guarantees that all future non-empty intersections

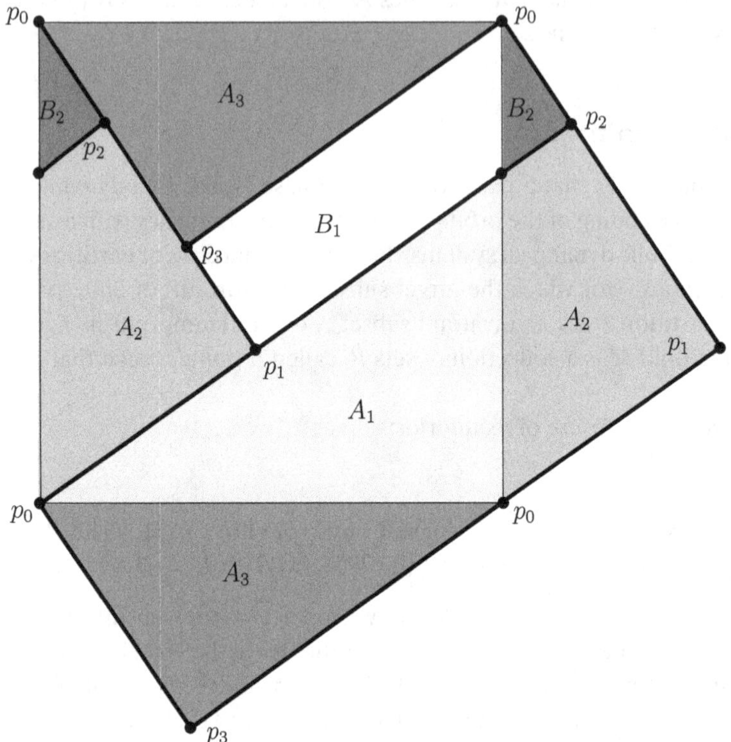

Figure 5.5. Markov partition for the toral automorphism f_M.

will have exactly one connected component. The image of A consists of Δ_1, Δ_3' and Δ_4'; the image of B consists of Δ_2' and Δ_5'. The part of the boundary of the Δ_i's that is parallel to the eigendirection of the larger eigenvalue is called unstable; the part that is parallel to the eigendirection of the smaller eigenvalue is called stable. By construction, the partition Δ of $\mathbb{T}^2$ into five rectangles Δ_i has the property that the image of the stable boundary is contained in the stable boundary and the preimage of the unstable boundary is contained in the unstable boundary (Exercise 5.12.1). In other words, for each i, j, the intersection $\Delta_{ij} = \Delta_i \cap f(\Delta_j)$ consists of one or two rectangles that stretch "all the way" through Δ_i and the stable boundary of Δ_{ij} is contained in the stable boundary of Δ_i; similarly, the intersection $\Delta_{ij}^{-1} = \Delta_i \cap f^{-1}(\Delta_j)$ consists of one or two rectangles that stretch "all the way" through Δ_i and the unstable boundary of Δ_{ij}^{-1} is contained in the unstable boundary of Δ_i. Let $a_{ij} = 1$ if the interior of $f(\Delta_i) \cap \Delta_j$ is not empty and $a_{ij} = 0$ otherwise,

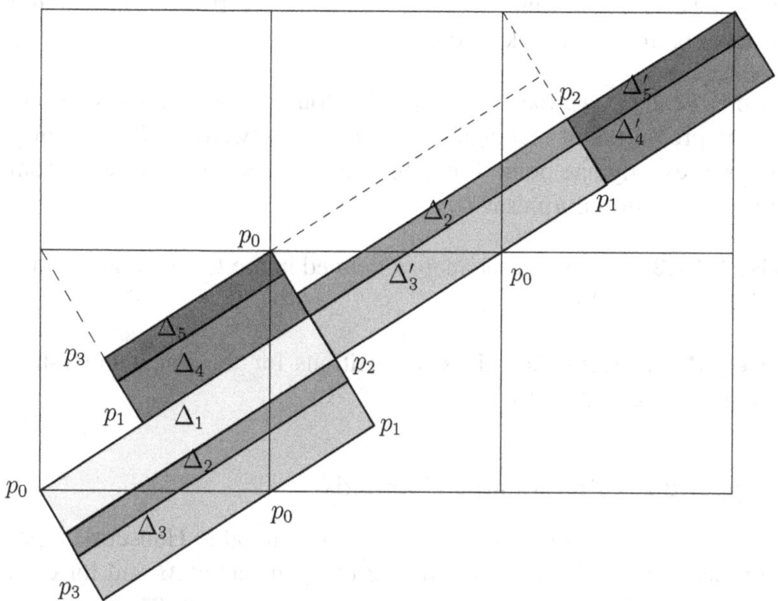

Figure 5.6. The image of the Markov partition under f_M.

$i, j = 1, \dots, 5$. This defines the adjacency matrix

$$A = \begin{pmatrix} 1 & 0 & 1 & 1 & 0 \\ 1 & 0 & 1 & 1 & 0 \\ 1 & 0 & 1 & 1 & 0 \\ 0 & 1 & 0 & 1 & 1 \\ 0 & 1 & 0 & 1 & 1 \end{pmatrix}.$$

If $\omega = (\dots, \omega_{-1}, \omega_0, \omega_1, \dots)$ is an allowed infinite sequence for this adjacency matrix, then the intersection $\bigcap_{i=-\infty}^{\infty} f^{-i}(\Delta_{\omega_i})$ consists of exactly one point $\phi(\omega)$; it follows that there is a continuous semiconjugacy $\phi \colon \Sigma_A \to \mathbb{T}^2$, i.e. $f \circ \phi = \phi \circ \sigma$, where σ is the shift in Σ_A (Exercise 5.12.2). Conversely, let B_0 be the union of the boundaries of the Δ_i's and let $B = \bigcup_{i=-\infty}^{\infty} f^i(B_0)$. For $x \in \mathbb{T}^2 \setminus B$, set $\psi_i(x) = j$ if $f^i(x) \in \Delta_j$. The itinerary sequence $(\psi_i(x))_{i=-\infty}^{\infty}$ is an element of Σ_A and $\phi \circ \psi = \mathrm{Id}$ (Exercise 5.12.3).

In higher dimensions, this direct geometric construction does not work. Even for a hyperbolic toral automorphism, the boundary is nowhere differentiable. Nevertheless, as Bowen (1970) showed, any locally maximal hyperbolic set Λ has a Markov partition which provides a semiconjugacy from a subshift of finite type to Λ.

Exercise 5.12.1. Prove that the stable boundary is forward invariant and the unstable boundary is backward invariant under f_M.

Exercise 5.12.2. Prove that for the toral automorphism f_M, the intersection of the preimages of rectangles Δ_i along an allowed infinite sequence ω consists of exactly one point. Prove that there is a semiconjugacy ϕ from $\sigma|_{\Sigma_A}$ to the toral automorphism f_M.

Exercise 5.12.3. Prove that the map ψ defined in the text satisfies $\psi(x) \in \Sigma_A$ and that $\phi \circ \psi = \text{Id}$.

Exercise 5.12.4. Construct Markov partitions for the linear horseshoe (§1.8) and the solenoid (§1.9).

5.13 Appendix: Differentiable manifolds

An *m-dimensional C^k manifold M* is a second-countable Hausdorff topological space together with a collection $\mathcal{U}$ of open sets in M and for each $U \in \mathcal{U}$ a homeomorphism ϕ_U from U onto the unit ball $B^m \subset \mathbb{R}^m$ such that:

1. $\mathcal{U}$ is a cover of M; and
2. for $U, V \in \mathcal{U}$, if $U \cap V \neq \emptyset$, the map $\phi_U \circ \phi_V^{-1}\colon \phi_V(U \cap V) \to \phi_U(U \cap V)$ is C^k.

We may take $k \in \mathbb{N} \cup \{\infty, \omega\}$, where C^ω denotes the class of real analytic functions.

We write M^m to indicate that M has dimension m. If $x \in M$ and $U \in \mathcal{U}$ contains x, then the pair (U, ϕ_U), $U \in \mathcal{U}$, is called a *coordinate chart at x*, and the n component functions $x_1, x_2, \ldots, x_m$ of ϕ_U are called *coordinates* on U. The collection of coordinate charts $\{(U, \phi_U)\}_{U \in \mathcal{U}}$ is called an *atlas* on M. Note that any open subset of $\mathbb{R}^m$ is a C^k manifold, for any $k \in \mathbb{N} \cup \{\infty, \omega\}$.

If M^m and N^n are C^k manifolds, then a continuous map $f\colon M \to N$ is C^k if for any coordinate chart (U, ϕ_U) on M, and any coordinate chart (V, ψ_V) on N, the map $\psi_V \circ f \circ \phi_U^{-1}\colon \phi_U(U \cap f^{-1}(V)) \to \mathbb{R}^n$ is a C^k map. For $k \geqslant 0$, the set of C^k maps from M to N is denoted $C^k(M, N)$. We say that a sequence of functions $f_n \in C^k(M, N)$ converges if the functions and all their derivatives up to order k converge uniformly on compact sets. This defines a topology on $C^k(M, N)$ called the *C^k topology*.

We set $C^k(M) = C^k(M, \mathbb{R})$. The subset of $C^k(M, M)$ consisting of diffeomorphisms of M is denoted $\text{Diff}^k(M)$.

A C^k *curve* in M^m is a C^k map $\alpha\colon (-\epsilon, \epsilon) \to M$. The *tangent vector* to α at $\alpha(0) = p$ is the linear map $v\colon C^1(M) \to \mathbb{R}$ defined by

$$v(f) = \left.\frac{d}{dt}\right|_{t=0} (f(\alpha(t)),$$

for $f \in C^1(M)$. The *tangent space* at p is the linear space T_pM of all tangent vectors at p.

Suppose (U, ϕ) is a coordinate chart, with coordinate functions $x_1, \dots, x_m$, and let $p \in U$. For $i = 1, \dots, m$, consider the curves

$$\alpha_i^p(t) = \phi^{-1}(x_1(p), \dots, x_{i-1}(p), x_i(p) + t, x_{i+1}(p), \dots, x_m(p)).$$

Define $(\frac{\partial}{\partial x_i})_p$ to be the tangent vector to α_i^p at p, i.e. for $g \in C^1(M)$,

$$\left(\frac{\partial}{\partial x_i}\right)_p (g) = \left.\frac{d}{dt}\right|_{t=0} g\left(\alpha_i^p(t)\right) = \left(\frac{\partial}{\partial x_i}(g \circ \phi)\right)_{\phi(p)}.$$

The vectors $\frac{\partial}{\partial x_i}$, $i = 1, \dots, m$, are linearly independent at p, and span T_pM. In particular, T_pM is a vector space of dimension m.

Let $f\colon M \to N$ be a C^k map, $k \geqslant 1$. For $p \in M$, the *tangent map* $df_p\colon T_pM \to T_{f(p)}N$ is defined by $df_p(v)(g) = v(g \circ f)$, for $g \in C^1(N)$. In terms of curves, if v is tangent to α at $p = \alpha(0)$, then $df_p(v)$ is tangent to $f \circ \alpha$ at $f(p)$.

The *tangent bundle* $TM = \bigcup_{x \in M} T_xM$ of M is a C^{k-1} manifold of twice the dimension of M with coordinate charts defined as follows. Let (U, ϕ_U) be a coordinate chart on M, $\phi_U = (x_1, \dots, x_m)\colon U \to \mathbb{R}^m$. For each i, the derivative dx_i is a function from $TU = \bigcup_{p \in U} T_pM$ to $\mathbb{R}$, defined by $dx_i(v) = v(x_i)$, for $v \in TU$. The function $(x_1, \dots, x_m, dx_1, \dots, dx_m)\colon TU \to \mathbb{R}^{2m}$ is a coordinate chart on TU, which we denote $d\phi_U$. Note that if $y, w \in \mathbb{R}^m$, then

$$d\phi_U \circ d\phi_V^{-1}(y, w) = (\phi_U \circ \phi_V^{-1}(y), d(\phi_U \circ \phi_V^{-1})_y(w)).$$

Let $\pi\colon TM \to M$ be the *projection map* that sends a vector $v \in T_pM$ to its base point p. A C^r *vector field* X on M is a C^r map $X\colon M \to TM$ such that $\pi \circ X$ is the identity on M. We write $X_p = X(p)$.

Let M^m and N^n be C^k manifolds. We say that M is a C^k *submanifold* of N if M is a subset of N and the inclusion map $i\colon M \to N$ is C^k and has rank m for each $x \in M$. If the topology of M coincides with the subspace topology, then M is an *embedded submanifold*. For each $x \in M$, the tangent space T_xM is naturally identified with a subspace of T_xN. Two submanifolds $M_1, M_2 \subset N$ of complementary dimensions *intersect transversely* (or are *transverse*) at a point $p \in M_1 \cap M_2$ if $T_pN = T_pM_1 \oplus T_pM_2$.

A *distribution* E on a differentiable manifold M is a family of k-dimensional subspaces $E(x) \subset T_x M$, $x \in M$. The distribution is C^l, $l \geqslant 0$, if locally it is spanned by k C^l vector fields.

Suppose W is a partition of a differentiable manifold M into C^1 submanifolds of dimension k. For $x \in M$, let $W(x)$ be the submanifold containing x. We say that W is a *k-dimensional continuous foliation with C^1 leaves* (or simply a *foliation*) if every $x \in M$ has a neighborhood U and a homeomorphism $h\colon B^k \times B^{m-k} \to U$ such that:

1. for each $z \in B^{m-k}$, the set $h(B^k \times \{z\})$ is the connected component of $W(h(0,z)) \cap U$ containing $h(0,z)$; and
2. $h(\cdot, z)$ is C^1 and depends continuously on z in the C^1 topology.

The pair (U, h) is called a *foliation coordinate chart*. The sets $h(B^k \times \{z\})$ are called *local leaves* (or *plaques*), and the sets $h(\{y\} \times B^{m-k})$ are called *local transversals*. For $x \in U$, we denote by $W_U(x)$ the local leaf containing x. More generally, a differentiable submanifold $L^{m-k} \subset M$ is a *transversal* if L is transverse to the leaves of the foliation. Each submanifold $W(x)$ of the foliation is called a *leaf* of W.

A continuous foliation W is a C^k foliation, $k \geqslant 1$, if the maps h can be chosen to be C^k. For example, lines of constant slope on $\mathbb{T}^2$ form a C^∞ foliation.

A foliation W defines a distribution $E = TW$ consisting of the tangent spaces to the leaves. A distribution E is *integrable* if it is tangent to a foliation.

A C^k *Riemannian metric* on a C^{k+1} manifold M consists of a positive definite symmetric bilinear form $\langle\, ,\rangle_p$ in each tangent space T_pM such that for any C^k vector fields X and Y, the function $p \mapsto \langle X_p, Y_p\rangle_p$ is C^k. For $v \in T_pM$, we write $\|v\| = (\langle v, v\rangle_p)^{1/2}$. If $\alpha\colon [a,b] \to M$ is a differentiable curve, we define the length of α to be $\int_a^b \|\dot\alpha(s)\|\, ds$. The (intrinsic) distance d between two points in M is defined to be the infimum of the lengths of differentiable curves in M connecting the two points.

A C^k *Riemannian manifold* is a C^{k+1} manifold with a C^k Riemannian metric. We denote by T^1M the set of tangent vectors of length 1 in a Riemannian manifold M.

A Riemannian manifold carries a natural measure called the *Riemannian volume*. Roughly speaking, the Riemannian metric allows one to compute the Jacobian of a differentiable map, and therefore allows one to define integration in a coordinate-free way.

If X is a topological space and (Y, d) is a metric space with metric, define a metric $dist_0$ on $C(X, Y)$ by

$$dist_0(f, g) = \min\left\{1, \sup_{x\in X} \max\{d(f(x), g(x))\}\right\}.$$

If X is compact, then this metric induces the topology of uniform convergence on compact sets. If X is not compact, this metric induces a finer topology. For example, the sequence of functions $f_n(x) = x^n$ in $C((0,1),\mathbb{R})$ converges to 0 in the topology of uniform convergence on compact sets, but not in the metric $dist_0$. The topology of uniform convergence on compact sets is metrizable even for non-compact sets, but we will not need this metric.

If M^m and N^n are C^1 Riemannian manifolds we define a distance function $dist_1$ on $C^1(M,N)$ as follows: the Riemannian metric on N induces a metric (distance function) on the tangent bundle TN, making TN a metric space. For $f \in C^1(M,N)$, the differential of f gives a map $df\colon T^1M \to TN$ on the unit tangent bundle of M. We set $dist_1(f,g) = dist_0(df,dg)$. If M is compact, the topology induced by this metric is the C^1 topology.

A differentiable manifold M is a (differentiable) *fiber bundle* over a differentiable manifold N with *fiber* F and (differentiable) *projection* $\pi : M \to N$ if for every $x \in N$ there is a neighborhood $V \ni x$ such that $\pi^{-1}(V)$ is diffeomorphic to $V \times F$ and $\pi^{-1}(y) \cong y \times F$. A diffeomorphism $f\colon M \to M$ is an *extension of* or a *skew product over* a diffeomorphism $g\colon N \to N$ if $\pi \circ f = g \circ \pi$; in this case g is called a *factor* of f.

CHAPTER SIX

Ergodicity of Anosov diffeomorphisms

The purpose of this chapter is to establish the ergodicity of volume-preserving Anosov diffeomorphisms (Theorem 6.3.1). This result, which was first obtained by Anosov (1969) (see also Anosov and Sinai (1967)), shows that hyperbolicity has strong implications for the ergodic properties of a dynamical system. Moreover, since a small perturbation of an Anosov diffeomorphism is also Anosov (Proposition 5.10.2), this gives an open set of ergodic diffeomorphisms.

Our proof is an improvement of the arguments in Anosov (1969) and Anosov and Sinai (1967). It is based on the classical approach called *Hopf's argument*. The first observation is that any f-invariant function is constant mod 0 on the stable and unstable manifolds, Lemma 6.3.2. Since these manifolds have complementary dimensions, one would expect the Fubini Theorem to imply that the function is constant mod 0, and ergodicity would follow. The major difficulty is that, although the stable and unstable manifolds are differentiable, they need not depend differentiably on the point they pass through, even if f is real-analytic. Thus the local product structure defined by the stable and unstable foliations does not yield a differentiable coordinate system, and we cannot apply the usual Fubini Theorem. So we establish a property of the stable and unstable foliations called *absolute continuity* that implies the Fubini Theorem.

The stable and unstable manifolds do not vary differentiably because they depend on the infinite future and past, respectively.

6.1 Hölder continuity of the stable and unstable distributions

For a subspace $A \subset \mathbb{R}^N$ and a vector $v \in \mathbb{R}^N$, set

$$\operatorname{dist}(v, A) = \min_{w \in A} \|v - w\|.$$

For subspaces A, B in $\mathbb{R}^N$, define

$$\operatorname{dist}(A, B) = \max\left(\max_{v\in A, \|v\|=1} \operatorname{dist}(v, B), \max_{w\in B, \|w\|=1} \operatorname{dist}(w, A)\right).$$

The following lemmas can be used to prove the Hölder continuity of invariant distributions for a variety of dynamical systems. Our objective is the Hölder continuity of the stable and unstable distributions of an Anosov diffeomorphism which was first established by Anosov (1967). We consider only the stable distribution; Hölder continuity of the unstable distribution follows by reversing the time.

LEMMA 6.1.1. *Let $L_n^i\colon \mathbb{R}^N \to \mathbb{R}^N$, $i = 1, 2$, $n \in \mathbb{N}$, be two sequences of linear maps. Assume that for some $b > 0$ and $\delta \in (0, 1)$,*

$$\|L_n^1 - L_n^2\| \leqslant \delta b^n$$

for each positive integer n.

Suppose that there are two subspaces $E^1, E^2 \subset \mathbb{R}^N$ and positive constants $C > 1$ and $\lambda < \mu$ with $\lambda < b$ such that

$$\begin{aligned} \|L_n^i v\| &\leqslant C\lambda^n \|v\| && \text{if } v \in E^i,\\ \|L_n^i w\| &\geqslant C^{-1}\mu^n \|w\| && \text{if } w \perp E^i. \end{aligned}$$

Then

$$\operatorname{dist}(E^1, E^2) \leqslant 3C^2 \frac{\mu}{\lambda} \delta^{\frac{\log\mu - \log\lambda}{\log b - \log\lambda}}.$$

Proof. Set $K_n^1 = \{v \in \mathbb{R}^N\colon \|L_n^1 v\| \leqslant 2C\lambda^n \|v\|\}$. Let $v \in K_n^1$. Write $v = v^1 + v_\perp^1$, where $v^1 \in E^1$ and $v_\perp^1 \perp E^1$. Then

$$\|L_n^1 v\| = \|L_n^1(v^1 + v_\perp^1)\| \geqslant \|L_n^1 v_\perp^1\| - \|L_n^1 v^1\| \geqslant C^{-1}\mu^n \|v_\perp^1\| - C\lambda^n \|v^1\|,$$

and hence

$$\|v_\perp^1\| \leqslant C\mu^{-n}(\|L_n^1 v\| + C\lambda^n \|v^1\|) \leqslant 3C^2 \left(\frac{\lambda}{\mu}\right)^n \|v\|.$$

It follows that

$$\operatorname{dist}(v, E^1) \leqslant 3C^2 \left(\frac{\lambda}{\mu}\right)^n \|v\|. \tag{6.1}$$

Set $\gamma = \lambda/b < 1$. There is a unique non-negative integer k such that $\gamma^{k+1} < \delta \leqslant \gamma^k$. Let $v^2 \in E^2$. Then

$$\begin{aligned} \|L_k^1 v^2\| &\leqslant \|L_k^2 v^2\| + \|L_k^1 - L_k^2\| \cdot \|v^2\| \\ &\leqslant C\lambda^k \|v^2\| + b^k \delta \|v^2\| \\ &\leqslant (C\lambda^k + (b\gamma)^k)\|v^2\| \leqslant 2C\lambda^k \|v^2\|. \end{aligned}$$

It follows that $v^2 \in K_k^1$ and hence $E^2 \subset K_k^1$. By symmetry, $E^1 \subset K_k^2$. By (6.1) and by the choice of k,

$$\operatorname{dist}(E^1, E^2) \leqslant 3C^2 \left(\frac{\lambda}{\mu}\right)^k \leqslant 3C^2 \frac{\mu}{\lambda} \delta^{\frac{\log \mu - \log \lambda}{\log b - \log \lambda}}. \qquad \square$$

LEMMA 6.1.2. *Let f be a C^2 diffeomorphism of a compact C^2 submanifold $M \subset \mathbb{R}^N$. Then for each $n \in \mathbb{N}$ and all $x, y \in M$*

$$\left\| df_x^n - df_y^n \right\| \leqslant b^n \cdot \|x - y\|,$$

where $b = b_1(b_1 + b_2)$, with $b_1 = \max_{z \in M} \|df_z\| \geqslant 1$ and $b_2 = \max_{z \in M} \|d_z^2 f\|$.

Proof. Observe that $\|f^n(x) - f^n(y)\| \leqslant (b_1)^n \|x - y\|$ for all $x, y \in M$. The lemma obviously holds for $n = 1$. For the inductive step we have

$$\begin{aligned} \|df_x^{n+1} - df_y^{n+1}\| &\leqslant \|df_{f^n(x)}\| \cdot \|df_x^n - df_y^n\| + \|df_{f^n(x)} - df_{f^n(y)}\| \cdot \|df_y^n\| \\ &\leqslant b_1 b^n \|x - y\| + b_2 b_1^n \|x - y\| b_1 \leqslant b^{n+1} \|x - y\|. \qquad \square \end{aligned}$$

Let M be a manifold embedded in $\mathbb{R}^N$, and suppose E is a distribution on M. We say that E is *Hölder continuous* with *Hölder exponent* $\alpha \in (0, 1]$ and *Hölder constant* L if

$$\operatorname{dist}(E(x), E(y)) \leqslant L \cdot \|x - y\|^\alpha$$

for all $x, y \in M$ with $\|x - y\| \leqslant 1$.

One can define Hölder continuity for a distribution on an abstract Riemannian manifold by using parallel transport along geodesics to identify tangent spaces at nearby points. However, for a compact manifold M it suffices to consider Hölder continuity for some embedding of M in $\mathbb{R}^N$. This is so because on a compact manifold M, the ratio of any two Riemannian metrics is bounded above and below. So is the ratio between the intrinsic distance function on M and the extrinsic distance on M obtained by restricting the distance in $\mathbb{R}^N$ to M. Thus the Hölder exponent is independent of both the Riemannian metric and the embedding, but the Hölder constant does change. So, without loss of generality, and to simplify the arguments in this section and the next one, we will deal only with manifolds embedded in $\mathbb{R}^N$.

THEOREM 6.1.3. *Let M be a compact C^2 manifold and $f \colon M \to M$ a C^2 Anosov diffeomorphism. Suppose that $0 < \lambda < 1 < \mu$ and $C > 0$ are such that $\|df_x^n v^s\| \leqslant C\lambda^n \|v^s\|$ and $\|df_x^n v^u\| \geqslant C\mu^n \|v^u\|$ for all $x \in M$, $v^s \in E^s(x)$, $v^u \in E^u(x)$, and $n \in \mathbb{N}$. Set $b_1 = \max_{z \in M} \|df_z\|$, $b_2 = \max_{z \in M} \|d_z^2 f\|$, $b = b_1(b_1 + b_2)$. Then the stable distribution E^s is Hölder continuous with* *exponent*

$$\alpha = \frac{\log \mu - \log \lambda}{\log b - \log \lambda}.$$

Proof. As indicated above, we may assume that M is embedded in $\mathbb{R}^N$. For $x \in M$, let $E^{\perp}(x)$ denote the orthogonal complement to the tangent plane T_xM in $\mathbb{R}^N$. Since $E^{\perp}$ is a smooth distribution, it is sufficient to prove the Hölder continuity of $E^s \oplus E^{\perp}$ on M.

Since M is compact, there is a constant $\overline{C} > 1$ such that for any $x \in M$, if $v \in T_xM$ is perpendicular to E^s, then $\|df_x^n v\| \geqslant \overline{C}^{-1} \mu^n \|v\|$.

For $x \in M$, extend df_x to a linear map $L(x)\colon \mathbb{R}^N \to \mathbb{R}^N$ by setting $L(x)|E^{\perp}(x) = 0$, and set $L_n(x) = L(f^{n-1}(x)) \circ \cdots \circ L(f(x)) \circ L(x)$. Note that $L_n(x)|_{T_xM} = df_x^n$.

Fix $x_1, x_2 \in M$ with $\|x_1 - x_2\| < 1$. By Lemma 6.1.2, the conditions of Lemma 6.1.1 are satisfied with $L_n^i = L_n(x_i)$ and $E^i = E^s(x_i), i = 1, 2$, and the theorem follows. □

Exercise 6.1.1. Let $\beta \in (0, 1]$ and M be a compact $C^{1+\beta}$ manifold, i.e. the first derivatives of the coordinate functions are Hölder continuous with exponent β. Let $f\colon M \to M$ be a $C^{1+\beta}$ Anosov diffeomorphism. Prove that the stable and unstable distributions of f are Hölder continuous.

6.2 Absolute continuity of the stable and unstable foliations

Let M be a smooth n-dimensional manifold. Recall (§5.13) that a continuous k-dimensional foliation W with C^1 leaves is a partition of M into C^1 submanifolds $W(x) \ni x$ which locally depend continuously in the C^1 topology on $x \in M$. Denote by m the Riemannian volume in M and by m_N the induced Riemannian volume in a C^1 submanifold N. Note that every leaf $W(x)$ and every transversal carry an induced Riemannian volume.

Let (U, h) be a foliation coordinate chart on M (§5.13), and let $L = h(\{y\} \times B^{n-k})$ be a C^1 local transversal. The foliation W is called *absolutely continuous* if for any such L and U there is a measurable family of positive measurable functions $\delta_x\colon W_U(x) \to \mathbb{R}$ (called the *conditional densities*) such that for any measurable subset $A \subset U$

$$m(A) = \int_L \int_{W_U(x)} \mathbf{1}_A(x, y)\delta_x(y)\, dm_{W(x)}(y)\, dm_L(x).$$

Note that the conditional densities are automatically integrable.

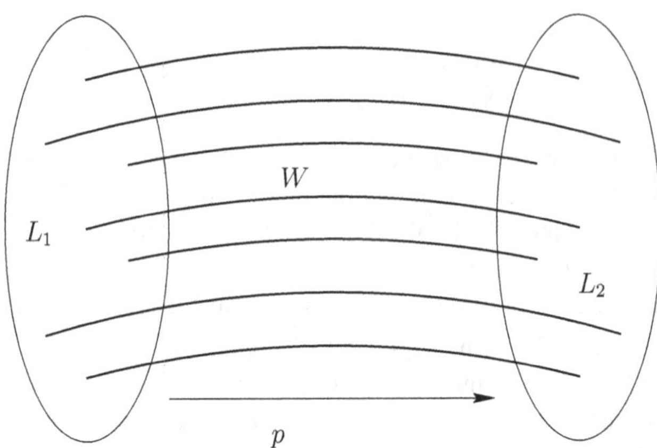

Figure 6.1. Holonomy map p for a foliation W and transversals L_1 and L_2.

PROPOSITION 6.2.1. *Let W be an absolutely continuous foliation of a Riemannian manifold M and let $f\colon M \to \mathbb{R}$ be a measurable function. Suppose there is a set $A \subset M$ of measure 0 such that f is constant on $W(x)\setminus A$ for every leaf $W(x)$.*

Then f is essentially constant on almost every leaf, i.e. for any transversal L, the function f is $m_{W(x)}$-essentially constant for m_L-almost every $x \in L$.

Proof. Absolute continuity implies that $m_{W(x)}(A \cap W(x)) = 0$ for m_L-almost every $x \in L$. □

Absolute continuity of the stable and unstable foliations is the property we need in order to prove the ergodicity of Anosov diffeomorphisms. However, we will prove a stronger property, called transverse absolute continuity, see Proposition 6.2.2.

Let W be a foliation of M, and (U, h) a foliation coordinate chart. Let $L_i = h(\{y_i\} \times B^{m-k})$, for $y_i \in B^k, i = 1, 2$. Define a homeomorphism $p\colon L_1 \to L_2$ by $p(h(y_1, z)) = h(y_2, z)$, for $z \in B^{m-k}$; p is called the *holonomy map* (see Figure 6.1). The foliation W is *transversely absolutely continuous* if the holonomy map p is absolutely continuous for any foliation coordinate chart and any transversals L_i as above, i.e. if there is a positive measurable function $q\colon L_1 \to \mathbb{R}$ (called the *Jacobian* of p) such that for any measurable subset $A \subset L_1$

$$m_{L_2}(p(A)) = \int_{L_1} \mathbf{1}_A q(z) dm_{L_1}(z).$$

If the Jacobian q is bounded on compact subsets of L_1 then W is said to be *transversely absolutely continuous with bounded Jacobians*.

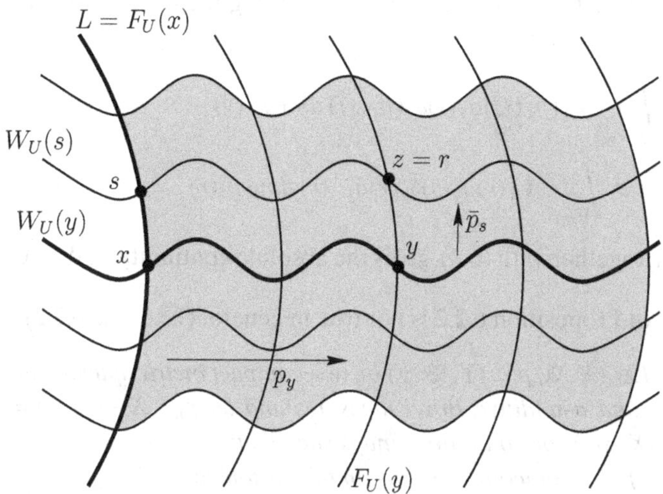

Figure 6.2. Holonomy maps for W and F.

PROPOSITION 6.2.2. *If W is transversely absolutely continuous, then it is absolutely continuous.*

Proof. Let L and U be as in the definition of an absolutely continuous foliation, $x \in L$ and let F be an $(n-k)$-dimensional C^1 foliation such that $F(x) \supset L$, $F_U(x) = L$ and $U = \bigcup_{y \in W_U(x)} F_U(y)$, see Figure 6.2. Obviously F is absolutely continuous and transversely absolutely continuous. Let $\bar{\delta}_y(\cdot)$ denote the conditional densities for F. Since F is a C^1 foliation, $\bar{\delta}$ is continuous, and hence, measurable. For any measurable set $A \subset U$, by the Fubini Theorem,

$$m(A) = \int_{W_U(x)} \int_{F_U(y)} \mathbf{1}_A(y, z)\bar{\delta}_y(z)\, dm_{F(y)}(z)\, dm_{W(x)}(y). \tag{6.2}$$

Let p_y denote the holonomy map along the leaves of W from $F_U(x) = L$ to $F_U(y)$ and let $q_y(\cdot)$ denote the Jacobian of p_y. We have

$$\int_{F_U(y)} \mathbf{1}_A(y, z)\bar{\delta}_y(z)\, dm_{F(y)}(z) = \int_L \mathbf{1}_A(p_y(s))q_y(s)\bar{\delta}_y(p_y(s))\, dm_L(s),$$

and by changing the order of integration in (6.2), which is an integral with respect to the product measure, we get

$$m(A) = \int_L \int_{W_U(x)} \mathbf{1}_A(p_y(s))q_y(s)\bar{\delta}_y(p_y(s))\, dm_{W(x)}(y)\, dm_L(s). \tag{6.3}$$

Similarly, let $\bar{p}_s$ denote the holonomy map along the leaves of F from $W_U(x)$ to $W_U(s)$, $s \in L$, and let $\bar{q}_s$ denote the Jacobian of $\bar{p}_s$. We transform the integral over $W_U(x)$ into an integral over $W_U(s)$ using the change of variables $r = p_y(s)$, $y = \bar{p}_s^{-1}(r)$:

$$\int_{W_U(x)} \mathbf{1}_A(p_y(s))q_y(s)\bar{\delta}_y(p_y(s))\, dm_{W(x)}(y)$$
$$= \int_{W_U(s)} \mathbf{1}_A(r)q_y(s)\bar{\delta}_y(r)\bar{q}_s^{-1}(r)\, dm_{W(s)}(r).$$

The last formula, together with (6.3), gives the absolute continuity of W. □

The converse of Proposition 6.2.2 is not true in general (Exercise 6.2.2).

LEMMA 6.2.3. *Let $(X, \mathfrak{A}, \mu)$, $(Y, \mathfrak{B}, \nu)$ be two compact metric spaces with Borel σ-algebras and σ-additive Borel measures and let $p_n\colon X \to Y$, $n = 1, 2, \ldots$, and $p\colon X \to Y$ be continuous maps such that:*

1. *each p_n and p are homeomorphisms onto their images;*
2. *p_n converges to p uniformly as $n \to \infty$;*
3. *there is a constant J such that $\nu(p_n(A)) \leqslant J\mu(A)$ for every $A \in \mathfrak{A}$.*

Then $\nu(p(A)) \leqslant J\mu(A)$ for every $A \in \mathfrak{A}$.

Proof. It is sufficient to prove the statement for an arbitrary open ball $B_r(x)$ in X. If $\delta < r$ then $p(B_{r-\delta}(x)) \subset p_n(B_r(x))$ for n large enough, and hence, $\nu(p(B_{r-\delta}(x))) \leqslant \nu(p_n(B_r(x))) \leqslant J\mu(B_r(x))$. Observe now that $\nu(p(B_{r-\delta}(x))) \nearrow \nu(p(B_r(x)))$ as $\delta \searrow 0$. □

For subspaces $A, B \subset \mathbb{R}^N$, set

$$\Theta(A, B) = \min\{\|v - w\|\colon\ v \in A, \|v\| = 1;\ w \in B, \|w\| = 1\}.$$

For $\theta \in [0, \sqrt{2}]$, we say that a subspace $A \subset \mathbb{R}^N$ is *θ-transverse*, to a subspace $B \subset \mathbb{R}^N$ if $\Theta(A, B) \geqslant \theta$.

LEMMA 6.2.4. *Let $\hat{E}$ be a smooth k-dimensional distribution on a compact subset of $\mathbb{R}^N$. Then for every $\xi > 0$ and $\epsilon > 0$ there is $\delta > 0$ with the following property. Suppose $Q_1, Q_2 \subset \mathbb{R}^N$ are $(N-k)$-dimensional C^1 submanifolds with a smooth holonomy map $\hat{p}\colon Q_1 \to Q_2$ such that $\hat{p}(x) \in Q_2$, $\hat{p}(x) - x \in \hat{E}(x)$, $\Theta(T_xQ_1, \hat{E}(x)) \geqslant \xi$, $\Theta(T_{\hat{p}(x)}Q_2, \hat{E}(x)) \geqslant \xi$, $\operatorname{dist}(T_xQ_1, T_{\hat{p}(x)}Q_2) \leqslant \delta$ and $\|\hat{p}(x) - x\| \leqslant \delta$ for each $x \in Q_1$. Then the Jacobian of $\hat{p}$ does not exceed $(1 + \epsilon)$.*

Proof. Since only the first derivatives of Q_1 and Q_2 affect the Jacobian of $\hat{p}$ at $x \in Q_1$, it equals the Jacobian at x of the holonomy map $\tilde{p}\colon T_xQ_1 \to T_{\hat{p}(x)}Q_2$ along $\hat{E}$. By applying an appropriate linear transformation L (whose determinant depends only on ξ), switching to new coordinates (u, v) in $\mathbb{R}^N$, and using the same notation for the images of all objects under L, we may assume that (a) $x = (0, 0)$; (b) $T_{(0,0)}Q_1 = \{v = 0\}$; (c) $p(x) = (0, v_0)$, where $\|v_0\| = \|\hat{p}(x) - x\|$; (d) $T_{(0,v_0)}Q_2$ is given by the equation $v = v_0 + Bu$, where B is a $k \times (N-k)$ matrix whose norm depends only on δ; (e) $\hat{E}(0, 0) = \{u = 0\}$, and $\hat{E}(w, 0)$ is given by the equation $u = w + A(w)v$, where $A(w)$ is an $(N-k) \times k$ matrix which is C^1 in w and $A(0) = 0$.

The image of $(w, 0)$ under $\hat{p}$ is the intersection point of the planes $v = v_0 + Bu$ and $u = w + A(w)v$. Since the norm of B is bounded from above in terms of ξ, it suffices to estimate the determinant of the derivative $\partial u/\partial w$ at $w = 0$. We substitute the first equation into the second one

$$u = w + A(w)v_0 + A(w)Bu,$$

differentiate with respect to w

$$\frac{\partial u}{\partial w} = I + \frac{\partial A(w)}{\partial w}v_0 + \frac{\partial A(w)}{\partial w}Bu + A(w)B\frac{\partial u}{\partial w},$$

and obtain for $w = 0$ (using $u(0) = 0$ and $A(0) = 0$)

$$\left.\frac{\partial u}{\partial w}\right|_{w=0} = I + \left.\frac{\partial A(w)}{\partial w}\right|_{w=0} v_0 \qquad \square$$

THEOREM 6.2.5. *The stable and unstable foliations of a C^2 Anosov diffeomorphism are transversely absolutely continuous.*

Proof. Let $f \colon M \to M$ be a C^2 Anosov diffeomorphism with stable and unstable distributions E^s and E^u, and hyperbolicity constants C and $0 < \lambda < 1 < \mu$. We will prove the absolute continuity of the stable foliation W^s. Absolute continuity of the unstable foliation W^u follows by reversing the time. To prove the theorem, we are going to uniformly approximate the holonomy map by continuous maps with uniformly bounded Jacobians.

As in the proof of Theorem 6.1.3, we assume that M is a compact submanifold in $\mathbb{R}^N$ (Hirsch, 1994) and denote by $T_xM^\perp$ the orthogonal complement of T_xM in $\mathbb{R}^N$. Let $\hat{E}^s$ be a smooth distribution that approximates the continuous distribution $\widetilde{E}^s(x) = E^s(x) \oplus T_xM^\perp$.

LEMMA 6.2.6. *For every $\theta > 0$ there is a constant $C_1 > 0$ such that for every $x \in M$, for every subspace $H \subset T_xM$ of the same dimension as $E^u(x)$ and θ-transverse to $E^s(x)$, and for every $k \in \mathbb{N}$,*

1. *$\|df_x^k v\| \geqslant C_1\mu^k\|v\|$, for every $v \in H$;*
2. $\operatorname{dist}(df_x^k H, df_x^k E^u(x)) \leqslant C_1 \left(\frac{\lambda}{\mu}\right)^k \operatorname{dist}(H, E^u(x))$.

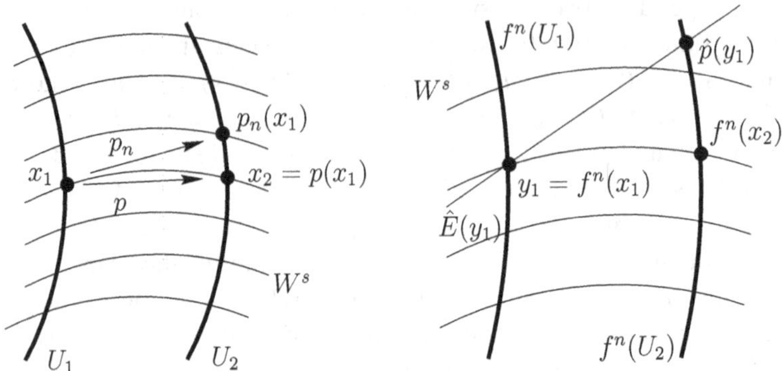

Figure 6.3. Construction of approximating maps p_n.

Proof. Exercise 6.2.3. □

By compactness of M, there is a $\theta_0 > 0$ such that $\Theta(E^s(x), E^u(x)) \geqslant \theta_0$ for every $x \in M$. Also by compactness, there is a covering of M by finitely many foliation coordinate charts $(U_i, h_i), i = 1, \ldots, l$ of the stable foliation W^s. It follows that there are positive constants ϵ and δ such that every $y \in M$ is contained in a coordinate chart U_j with the following property: if L is a compact connected submanifold of U_j such that:

1. L intersects transversely every local stable leaf of U_j;
2. $\Theta(T_z L, E^s) > \theta_0/3$ for all $z \in L$; and
3. $\operatorname{dist}(y, L) < \delta$;

then for any subspace $E \subset \mathbb{R}^n$ with $dist(E, E^s(y) \oplus T_y M^\perp) < \epsilon$, the affine plane $y + E$ intersects L transversely in a unique point z_y, and $\|y - z_y\| < 6\delta/\theta_0$.

Let (U, h) be a foliation coordinate chart, and L_1, L_2 local transversals in U with holonomy map $p\colon L_1 \to L_2$. Define a map $\hat{p}\colon f^n(L_1) \to f^n(L_2)$ as follows: for $x \in L_1$, let $\hat{p}(f^n(x))$ be the unique intersection point of the affine plane $f^n(x) + \hat{E}(f^n(x))$ with $f^n(L_2)$ that is closest to $f^n(p(x))$ along $f^n(L_2)$ (note that there may be several such intersection points). The map $\hat{p}$ is well-defined by Lemma 6.2.6 and the remarks in the preceding paragraph.

For $x \in L_1$, set $p_n(x) = f^{-n}(\hat{p}(f^n(x)))$. Let $x_1 \in L_1, x_2 = p_n(x_1)$ and set $y_i = f^n(x_i)$, see Figure 6.3. Observe that:

$$\operatorname{dist}(f^k(x_1), f^k(p(x_1))) \leqslant C\lambda^k \operatorname{dist}(x_1, p(x_1)), \quad \text{for } k = 0, 1, 2, \ldots. \tag{6.4}$$

Assuming that $\hat{E}^s$ is C^0-close enough to $\widetilde{E}^s$, it is, by Lemma 6.2.6, uniformly transverse to $f^n(L_1)$ and $f^n(L_2)$. Therefore there is $C_2 > 0$ such that

$$\begin{aligned}\operatorname{dist}(\hat{p}(f^n(x_1)), f^n(p(x_1))) &\leqslant C_2 \operatorname{dist}(f^n(x_1), f^n(p(x_1)))\\ &\leqslant C_2 C\lambda^n \operatorname{dist}(x_1, p(x_1)).\end{aligned}$$

Therefore, by (6.4) and Lemma 6.2.6,

$$\operatorname{dist}(p_n(x_1), p(x_1)) \leqslant \frac{C_2 C}{C_1}\left(\frac{\lambda}{\mu}\right)^n \operatorname{dist}(x_1, p(x_1)), \tag{6.5}$$

and hence p_n converges uniformly to p as $n \to \infty$.

Combining (6.4) and (6.5) we get

$$\begin{aligned}&\operatorname{dist}(f^k(x_1), f^k(x_2))\\ &\quad\leqslant \operatorname{dist}(f^k(x_1), f^k(p(x_1))) + \operatorname{dist}(f^k(p(x_1)), f^k(x_2)) \leqslant C_3\lambda^k.\end{aligned} \tag{6.6}$$

Let $J(f^k(x_i))$ be the Jacobian of $\tilde{f}$ in the direction of the tangent plane $T_i^k(x_i) = T_{f^k(x_i)}L_i, i = 1, 2, k = 0, 1, 2, \ldots$. Also denote by Jac_{p_n} the Jacobian of p_n and by $\mathrm{Jac}_{\hat{p}}$ the Jacobian of $\hat{p}\colon\ f^n(L_1) \to f^n(L_2)$, which is uniformly bounded by Lemma 6.2.4. Then,

$$\mathrm{Jac}_{p_n}(x_1) = \prod_{k=0}^{n-1}(J(f^k(x_2)))^{-1} \cdot \mathrm{Jac}_{\hat{p}}(f^n(x_1)) \cdot \prod_{k=0}^{n-1} J(f^k(x_1)).$$

To obtain a uniform bound on Jac_{p_n} we need to estimate from above $P = \prod_{k=0}^{n-1}(J(f^k(x_1))/J(f^k(x_2)))$. By Theorem 6.1.3, Lemma 6.2.6, and (6.6), for some $C_4, C_5, C_6 > 0$ and $\bar{\alpha}$:

$$\begin{aligned}&\operatorname{dist}(T_1^k(x_1), T_2^k(x_2))\\ &\quad\leqslant \operatorname{dist}(T_1^k(x_1), \widetilde{E}^u(f^k(x_1))) + \operatorname{dist}(\widetilde{E}^u(f^k(x_1)), \widetilde{E}^u(f^k(x_2)))\\ &\qquad + \operatorname{dist}(T_2^k(x_2), \widetilde{E}^u(f^k(x_2)))\\ &\quad\leqslant 2C_1\left(\frac{\lambda}{\mu}\right)^k + C_4(\operatorname{dist}(f^k(x_1), f^k(x_2)))^{\alpha}\\ &\quad\leqslant 2C_1\left(\frac{\lambda}{\mu}\right)^k + C_5\lambda^{\alpha k} \leqslant C_6\lambda^{\alpha k}.\end{aligned} \tag{6.7}$$

Since f is a C^2 diffeomorphism, its derivative is Lipschitz continuous and the Jacobians $J(f^k(x_1))$ and $J(f^k(x_2))$ are bounded away from 0 and ∞.

Therefore it follows from (6.7) that

$$\frac{|J(f^k(x_1)) - J(f^k(x_2))|}{|J(f^k(x_2))|} < C_7 \lambda^{\alpha k}.$$

Hence the product P converges and is bounded. □

Exercise 6.2.1. Let W be a k-dimensional foliation of M and let L be an $(n-k)$-dimensional local transversal to W at $x \in M$, i.e. $T_xM = T_xW(x) \oplus T_xL$. Prove that there is a neighborhood $U \ni x$ and a C^1 coordinate chart $w\colon B^k \times B^{n-k} \to U$ such that the connected component of $L \cap U$ containing x is $w(0, B^{n-k})$ and there are C^1 functions $f_y\colon B^k \to B^{n-k}, y \in B^{n-k}$, with the following properties:

(i) f_y depends continuously on y in the C^1 topology;
(ii) $w(\text{graph}(f_y)) = W_U(w(0,y))$.

Exercise 6.2.2. Give an example of an absolutely continuous foliation which is not transversely absolutely continuous.

Exercise 6.2.3. Prove Lemma 6.2.6.

Exercise 6.2.4. Let W_i, $i = 1, 2$, be two transverse foliations of dimensions k_i on a smooth manifold M, i.e. $T_xW_1(x) \cap T_xW_2(x) = \{0\}$ for each $x \in M$. The foliations W_1 and W_2 are called *integrable* if there is a $(k_1 + k_2)$-dimensional foliation W (called the *integral hull* of W_1 and W_2) such that $W(x) = \bigcup_{y \in W_1(x)} W_2(y) = \bigcup_{y \in W_2(x)} W_1(y)$ for every $x \in M$.

Let W_1 be a C^1 foliation and W_2 be an absolutely continuous foliation and assume that W_1 and W_2 are integrable with integral hull W. Prove that W is absolutely continuous.

6.3 Proof of ergodicity

The proof of Theorem 6.3.1 below follows the main ideas of E. Hopf's argument for the ergodicity of the geodesic flow on a compact surface of variable negative curvature.

We say that a measure μ on a differentiable Riemannian manifold M is *smooth* if it has a continuous density q with respect to the Riemannian volume m, i.e. $\mu(A) = \int_A q(x)\, dm(x)$ for each bounded Borel set $A \subset M$.

THEOREM 6.3.1. *A C^2 Anosov diffeomorphism preserving a smooth measure is ergodic.*

Proof. Let $(X, \mathfrak{A}, \mu)$ be a finite measure space such that X is a compact metric space with distance d, μ is a Borel measure and $\mathfrak{A}$ is the μ-completion of the Borel σ-algebra. Let $f\colon X \to X$ be a homeomorphism. For $x \in X$, define the *stable* set $V^s(x)$ and *unstable* set $V^u(x)$ by the formulas

$$V^s(x) = \{y \in X\colon\ d(f^n(x), f^n(y)) \to 0 \text{ as } n \to \infty\},$$
$$V^u(x) = \{y \in X\colon\ d(f^n(x), f^n(y)) \to 0 \text{ as } n \to -\infty\}.$$

LEMMA 6.3.2. *Let $\phi\colon X \to \mathbb{R}$ be an f-invariant measurable function. Then ϕ is constant* mod 0 *on stable and unstable sets, i.e. there is a null set N such that ϕ is constant on $V^s(x) \setminus N$ and on $V^u(x) \setminus N$ for every $x \in X \setminus N$.*

Proof. We will only deal with the stable sets. Without loss of generality assume that ϕ is non-negative. For a real C set $\phi_C(x) = \min(\phi(x), C)$. The function ϕ_C is f-invariant and it suffices to prove the lemma for ϕ_C with arbitrary C. For $k \in \mathbb{N}$, let $\psi_k\colon X \to \mathbb{R}$ be a continuous function such that $\int_X |\phi_C - \psi_k|\, d\mu(x) < \frac{1}{k}$. By the Birkhoff ergodic theorem, the limit

$$\psi_k^+(x) = \lim_{n\to\infty} \frac{1}{n} \sum_{i=0}^{n-1} \psi_k(f^i(x))$$

exists for μ-a.e. x. By the invariance of μ and ϕ_C, for every $j \in \mathbb{Z}$,

$$\begin{aligned}\frac{1}{k} > \int_X |\phi_C(x) - \psi_k(x)|\, d\mu(x) &= \int_X |\phi_C(f^j(y)) - \psi_k(f^j(y))|\, d\mu(y)\\ &= \int_X |\phi_C(y) - \psi_k(f^j(y))|\, d\mu(y)\end{aligned}$$

and hence

$$\begin{aligned}&\int_X \left|\phi_C(y) - \frac{1}{n}\sum_{i=0}^{n-1} \psi_k(f^i(y))\right| d\mu(y)\\ &\qquad\leqslant \frac{1}{n}\sum_{i=0}^{n-1} \int_X |\phi_C(y) - \psi_k(f^i(y))|\, d\mu(y) < \frac{1}{k}.\end{aligned}$$

Since ψ_k is uniformly continuous, $\psi_k^+(y) = \psi_k^+(x)$ whenever $y \in V^s(x)$ and $\psi_k^+(x)$ is defined. Therefore there is a null set N_k such that ψ_k^+ exists and is constant on the stable sets in $X \setminus N_k$. It follows that $\phi_C^+(x) = \lim_{k\to\infty} \psi_k^+(x)$ is constant on the stable sets in $X \setminus \cup N_k$. Clearly $\phi_C(x) = \phi_C^+(x) \bmod 0$. □

Let ϕ be a μ-measurable f-invariant function. By Lemma 6.3.2, there is a μ-null set N_s such that ϕ is constant on the leaves of W^s in $M \setminus N_s$ and another μ-null set N_u such that ϕ is constant on the leaves of W^u in $M \setminus N_u$.

Let $x \in M$ and let $U \ni x$ be a small neighborhood as in the definition of absolute continuity for W^s and W^u. Let $G_s \subset U$ be the set of points $z \in U$ for which $m_{W^s(z)}(N_s \cap W^s(z)) = 0$ and $z \notin N_s$. Let $G_u \subset U$ be the set of points $z \in U$ for which $m_{W^u(z)}(N_u \cap W^u(z)) = 0$ and $z \notin N_u$. By Proposition 6.2.1 and the absolute continuity of W^u and W^s (Theorem 6.2.5), both sets G_s and G_u have full μ-measure in U, and hence so does $G_s \cap G_u$. Again, by the absolute continuity of W^u, there is a full μ-measure subset of points $z \in U$ such that $z \in G_s \cap G_u$ and $m_{W^u(z)}$-a.e. point from $W^u(z)$ also lies in $G_s \cap G_u$. It follows that $\phi(x) = \phi(z)$ for μ-a.e. point $x \in U$. Since M is connected, ϕ is constant mod 0 on M. □

Exercise 6.3.1. Prove that a C^2 Anosov diffeomorphism preserving a smooth measure is weak mixing.

CHAPTER SEVEN

Low dimensional dynamics

As we have seen in the previous chapters, general dynamical systems exhibit a wide variety of behaviors and cannot be completely classified by their invariants. The situation is considerably better in low dimensional dynamics and especially in one-dimensional dynamics. The two crucial tools for studying one-dimensional dynamical systems are the Intermediate Value Theorem (for continuous maps) and conformality (for non-singular differentiable maps). A differentiable map f is *conformal* if the derivative at each point is a non-zero scalar multiple of an orthogonal transformation, i.e. if the derivative expands or contracts distances by the same amount in all directions. In dimension one, any non-singular differentiable map is conformal. The same is true for complex analytic maps, which we study in Chapter 8. But in higher dimensions, differentiable maps are rarely conformal.

7.1 Circle homeomorphisms

The circle $S^1 = [0, 1] \bmod 1$ can be considered as the quotient space $\mathbb{R}/\mathbb{Z}$. The quotient map $\pi : \mathbb{R} \to S^1$ is a *covering map*, i.e. each $x \in S^1$ has a neighborhood U_x such that $\pi^{-1}(U_x)$ is a disjoint union of connected open sets, each of which is mapped homeomorphically onto U_x by π.

Let $f\colon S^1 \to S^1$ be a homeomorphism. We will assume throughout this section that f is orientation-preserving (see Exercise 7.1.3 for the orientation-reversing case). Since π is a covering map, we can lift f to an increasing homeomorphism $F\colon \mathbb{R} \to \mathbb{R}$ such that $\pi \circ F = f \circ \pi$. For each $x_0 \in \pi^{-1}(f(0))$ there is a unique lift F such that $F(0) = x_0$, and any two lifts differ by an integer translation. For any lift F and any $n \in \mathbb{Z}$, $F(x + n) = F(x) + n$, for any $x \in \mathbb{R}$.

THEOREM 7.1.1. *Let $f\colon S^1 \to S^1$ be an orientation-preserving homeomorphism, and $F\colon \mathbb{R} \to \mathbb{R}$ a lift of f. Then for every $x \in \mathbb{R}$, the limit*

$$\rho(F) = \lim_{n\to\infty} \frac{F^n(x) - x}{n}$$

exists, and is independent of the point x. The number $\rho(f) = \pi(\rho(F))$ is independent of the lift F, and is called the rotation number *of f. If f has a periodic point, then $\rho(f)$ is rational.*

Proof. Suppose for the moment that the limit exists for some $x \in [0, 1)$. Since F maps any interval of length 1 to an interval of length 1, it follows that $|F^n(x) - F^n(y)| \leqslant 1$ for any $y \in [0, 1)$. Thus

$$\big|(F^n(x) - x) - (F^n(y) - y)\big| \leqslant |F^n(x) - F^n(y)| + |x - y| \leqslant 2,$$

so $\lim_{n\to\infty} \frac{F^n(x)-x}{n} = \lim_{n\to\infty} \frac{F^n(y)-y}{n}$. Since $F^n(y + k) = F^n(y) + k$, the same holds for any $y \in \mathbb{R}$.

Suppose $F^q(x) = x + p$ for some $x \in [0, 1)$ and some $p, q \in \mathbb{N}$. This is equivalent to asserting that $\pi(x)$ is a periodic point for f with period q. For $n \in \mathbb{N}$, write $n = kq + r$, $0 \leqslant r < q$. Then $F^n(x) = F^r(F^{kq}x)) = F^r(x + kp) = F^r(x) + kp$, and since $|F^r(x) - x|$ is bounded for $0 \leqslant r < q$,

$$\lim_{n\to\infty} \frac{F^n(x) - x}{n} = \frac{p}{q}.$$

Thus the rotation number exists and is rational whenever f has a periodic point.

Suppose now that $F^q(x) \neq x + p$ for all $x \in \mathbb{R}$ and $p, q \in \mathbb{N}$. By continuity, for each pair $p, q \in \mathbb{N}$, either $F^q(x) > x + p$ for all $x \in \mathbb{R}$, or $F^q(x) < x + p$ for all $x \in \mathbb{R}$. For $n \in \mathbb{N}$, choose $p_n \in \mathbb{N}$ so that $p_n - 1 < F^n(x) - x < p_n$ for all $x \in \mathbb{R}$. Then for any $m \in \mathbb{N}$,

$$m(p_n - 1) < F^{mn}(x) - x = \sum_{k=0}^{m-1} F^n(F^{kn}(x)) - F^{kn}(x) < mp_n,$$

which implies that

$$\frac{p_n}{n} - \frac{1}{n} < \frac{F^{mn}(x) - x}{mn} < \frac{p_n}{n}.$$

Interchanging the roles of m and n, we also have

$$\frac{p_m}{m} - \frac{1}{m} < \frac{F^{mn}(x) - x}{mn} < \frac{p_m}{m}.$$

Thus $|p_m/m - p_n/n| < |1/m + 1/n|$, so $\{p_n/n\}$ is a Cauchy sequence. It follows that $(F^n(x) - x)/n$ converges as $n \to \infty$.

If $G = F + k$ is another lift of f, then $\rho(G) = \rho(F) + k$, so $\rho(f)$ is independent of the lift F. Moreover, there is a unique lift F such that $\rho(F) = \rho(f)$ (Exercise 7.1.1). □

Since $S^1 = [0, 1] \bmod 1$, we will often abuse notation by writing $\rho(f) = x$ for some $x \in [0, 1]$.

PROPOSITION 7.1.2. *The rotation number depends continuously on the map in the C^0 topology.*

Proof. Let f be an orientation-preserving circle homeomorphism, and choose $p, q, p', q' \in \mathbb{N}$ such that $p/q < \rho(f) < p'/q'$. Let F be the lift of f such that $p < F^q(x) - x < p + q$. Then for all $x \in \mathbb{R}, p < F^q(x) - x < p + q$, since otherwise we would have $\rho(f) = p/q$. If g is another circle homeomorphism close to F, then there is a lift G close to F, and for g sufficiently close to f, the same inequality $p < G^q(x) - x < p + q$ holds for all $x \in \mathbb{R}$. Thus $p/q < \rho(g)$. A similar argument involving p' and q' completes the proof. □

PROPOSITION 7.1.3. *Rotation number is an invariant of topological conjugacy.*

Proof. Let f and h be orientation-preserving homeomorphisms of S^1, and let F and H be lifts of f and h. Then $H \circ F \circ H^{-1}$ is a lift of $h \circ f \circ h^{-1}$ and for $x \in \mathbb{R}$,

$$\begin{aligned}\frac{(HFH^{-1})^n(x) - x}{n} &= \frac{(HF^nH^{-1})(x) - x}{n} \\ &= \frac{H(F^nH^{-1}(x)) - F^nH^{-1}(x)}{n} + \frac{F^nH^{-1}(x) - H^{-1}(x)}{n} \\ &\quad + \frac{H^{-1}(x) - x}{n}.\end{aligned}$$

Since the numerators in the first and third terms of the last expression are bounded independent of n, we conclude that

$$\rho(hfh^{-1}) = \lim_{n\to\infty} \frac{(HFH^{-1})^n(x) - x}{n} = \lim_{n\to\infty} \frac{F^n(x) - x}{n} = \rho(f). \qquad \square$$

PROPOSITION 7.1.4. *If $f\colon S^1 \to S^1$ is a homeomorphism, then $\rho(f)$ is rational if and only if f has a periodic point. Moreover, if $\rho(f) = p/q$ where p and q are relatively prime non-negative integers, then every periodic point of f has minimal period q, and if $x \in \mathbb{R}$ projects to a periodic point of f, then $F^q(x) = x + p$ for the unique lift F with $\rho(F) = p/q$.*

Proof. The "if" part of the first assertion is contained in Theorem 7.1.1.

Suppose $\rho(f) = p/q$, where $p, q \in \mathbb{N}$. If F and $\tilde{F} = F + l$ are two lifts of f, then $\tilde{F}^q = F^q + lq$. Thus we may choose F to be the unique lift with $p \leqslant F^q(0) < p + q$. To show the existence of a periodic point of f, it suffices to show the existence of a point $x \in [0, 1]$ such that $F^q(x) = x + k$, for some $k \in \mathbb{N}$. We may assume that $x + p < F^q(x) < x + p + q$ for all $x \in [0, 1]$, since otherwise we have $F^q(x) = x + l$ for $k = p$ or $k = p + q$, and we are done. Choose $\epsilon > 0$ such that for any $x \in [0, 1]$, $x + p + \epsilon < F^q(x) < x + p + q - \epsilon$. The same inequality then holds for all $x \in \mathbb{R}$ since $F^q(x + k) = F^q(x) + k$ for all $k \in \mathbb{N}$. Thus

$$\frac{p+\epsilon}{q} = \frac{k(p+\epsilon)}{kq} < \frac{F^{kq}(x) - x}{kq} < \frac{k(p+q-\epsilon)}{kq} = \frac{p+1-\epsilon}{q}$$

for all $k \in \mathbb{N}$, contradicting $\rho(f) = p/q$. We conclude that $F^q(x) = x + p$ or $F^q(x) = x + p + q$ for some x, and x is periodic with period q.

Now assume $\rho(f) = p/q$, with p and q relatively prime, and suppose $x \in [0, 1)$ is a periodic point of f. Then there are integers $p', q' \in \mathbb{N}$ such that $F^{q'}(x) = x + p'$. By the proof of Theorem 7.1.1, $\rho(f) = p'/q'$, so if d is the greatest common divisor of p' and q', then $q' = qd$ and $p' = pd$. We claim that $F^q(x) = x + p$. If not, then either $F^q(x) > x + p$ or $F^q(x) < x + p$. Suppose the former holds (the other case is similar). Then by monotonicity,

$$F^{dq}(x) > F^{(d-1)q}(x) + p > \cdots > x + dp,$$

contradicting the fact that $F^{q'}(x) = x + p'$. Thus x is periodic with period q. □

Suppose f is a homeomorphism of S^1. Given any subset $A \subset S^1$ and a distinguished point $x \in A$, we define an ordering on A by lifting A to the interval $[\tilde{x}, \tilde{x} + 1) \subset \mathbb{R}$, where $\tilde{x} \in \pi^{-1}(x)$, and using the natural ordering on $\mathbb{R}$. In particular, if $x \in S^1$, then the orbit $\{x, f(x), f^2(x), \dots\}$ has a natural order (using x as the distinguished point).

THEOREM 7.1.5. *Let $f\colon S^1 \to S^1$ be an orientation-preserving homeomorphism with rational rotation number $\rho = p/q$, where p and q are relatively prime. Then for any periodic point $x \in S^1$, the ordering of the orbit $\{x, f(x), f^2(x), \dots, f^{q-1}(x)\}$ is the same as the ordering of the set $\{0, p/q, 2p/q, \dots, (q-1)p/q\}$, which is the orbit of 0 under the rotation R_ρ.*

Proof. Let x be a periodic point of f and let $i \in \{0, \dots, q-1\}$ be the unique number such that $f^i(x)$ is the first point to the right of x in the orbit of x. Then $f^{2i}(x)$ must be the first point to the right of $f^i(x)$, since if $f^l(x) \in$

$(f^i(x), f^{2i}(x))$ then $l > i$ and $f^{l-i}(x) \in (x, f^i(x))$, contradicting the choice of i. Thus the points of the orbit are ordered as $x, f^i(x), f^{2i}(x), \ldots, f^{(q-1)i}(x)$.

Let $\tilde{x}$ be a lift of x. Since f^i carries each interval $[f^{ki}(x), f^{(k+1)i}(x)]$ to its successor, and there are q of these intervals, there is a lift $\bar{F}$ of f^i such that $\bar{F}^q\tilde{x} = \tilde{x}+1$. Let F be the lift of f with $F^q(x) = x+p$. Then F^i is a lift of f^i, so $F^i = \bar{F}+k$ for some k. We have

$$x + ip = F^{qi}(x) = (\bar{F}+k)^q(x) = \bar{F}^q(x) + qk = x + 1 + qk.$$

Thus $ip = 1+qk$, so i is the unique number between 0 and q such that $ip = 1 \bmod q$. Since the points of the set $\{0, p/q, 2p/q, \ldots, (q-1)p/q\}$ are ordered as $0, ip/q, \ldots, (q-1)ip/q$, the theorem follows. □

Now we turn to the study of orientation-preserving homeomorphisms with irrational rotation number. If x and y are two points in S^1, then we define the interval $[x, y] \subset S^1$ to be $\pi([\tilde{x}, \tilde{y}])$, where $\tilde{x} \in \pi^{-1}(x)$ and $\tilde{y} = \pi^{-1}(y) \cap [\tilde{x}, \tilde{x}+1)$. Open and half-open intervals are defined in a similar way.

LEMMA 7.1.6. *Suppose $\rho(f)$ is irrational. Then for any $x \in S^1$, and any distinct integers m and n, every forward orbit of f intersects the interval $I = [f^m(x), f^n(x)]$.*

Proof. We assume that $m > n$; the case $m < n$ is similar. It suffices to show that $S^1 = \bigcup_{k=0}^{\infty} f^{-k}I$. Suppose not. Then

$$S^1 \not\subset \bigcup_{k=1}^{\infty} f^{-k(m-n)}I = \bigcup_{k=1}^{\infty}[f^{-(k-1)m+kn}(x), f^{-km+(k+1)n}(x)].$$

Since the intervals $f^{-k(m-n)}I$ abut at the endpoints, we conclude that $f^{-k(m-n)}f^n(x)$ converges monotonically to a point $z \in S^1$, which is a fixed point for f^{m-n}, contradicting the irrationality of $\rho(f)$. □

PROPOSITION 7.1.7. *If $\rho(f)$ is irrational, then $\omega(x) = \omega(y)$ for any $x, y \in S^1$, and either $\omega(x) = S^1$ or $\omega(x)$ is perfect and nowhere dense.*

Proof. Fix $x, y \in S^1$. Suppose $f^{a_n}(x) \to x_0 \in \omega(x)$ for some sequence $a_n \nearrow \infty$. By Lemma 7.1.6, for each $n \in \mathbb{N}$, we can choose b_n such that $f^{b_n}(y) \in [f^{a_{n-1}}(x), f^{a_n}(x)]$. Then $f^{b_n}(y) \to x_0$, so $\omega(x) \subset \omega(y)$. By symmetry, $\omega(x) = \omega(y)$.

To show that $\omega(x)$ is perfect, we fix $z \in \omega(x)$. Since $\omega(x)$ is invariant, we have that $z \in \omega(z)$ is a limit point of $\{f^n(z)\} \subset \omega(x)$, so $\omega(x)$ is perfect.

To prove the last claim, we suppose that $\omega(x) \neq S^1$. Then $\partial\omega(x)$ is a non-empty closed invariant set. If $z \in \partial\omega(x)$, then $\omega(z) = \omega(x)$. Hence $\omega(x) \subset \partial\omega(x)$ and $\omega(x)$ is nowhere dense. □

LEMMA 7.1.8. *Suppose $\rho(f)$ is irrational. Let F be a lift of f, and $\rho = \rho(F)$. Then for any $x \in \mathbb{R}$, $n_1\rho + m_1 < n_2\rho + m_2$ if and only if $F^{n_1}(x) + m_1 < F^{n_2}(x) + m_2$, for any $m_1, m_2, n_1, n_2 \in \mathbb{Z}$.*

Proof. Suppose $F^{n_1}(x) + m_1 < F^{n_2}(x) + m_2$ or, equivalently,

$$F^{(n_1-n_2)}(x) < x + m_2 - m_1.$$

This inequality holds for all x, since otherwise the rotation number would be rational. In particular, for $x = 0$ we have $F^{(n_1-n_2)}(0) < m_2 - m_1$. By an inductive argument, $F^{k(n_1-n_2)}(0) < k(m_2 - m_1)$. If $n_1 - n_2 > 0$, it follows that

$$\frac{F^{k(n_1-n_2)}(0) - 0}{k(n_1 - n_2)} < \frac{m_2 - m_1}{n_1 - n_2},$$

so

$$\rho = \lim_{k\to\infty} \frac{F^{k(n_1-n_2)}(0)}{k(n_1 - n_2)} \leqslant \frac{m_2 - m_1}{n_1 - n_2}.$$

Irrationality of ρ implies strict inequality, so $n_1\rho + m_1 < n_2\rho + m_2$. The same result holds in the case $n_1 - n_2 < 0$ by a similar argument. The converse follows by reversing the inequality. □

THEOREM 7.1.9 (Poincaré Classification). *Let $f\colon S^1 \to S^1$ be an orientation-preserving homeomorphism with irrational rotation number ρ.*

1. *If f is topologically transitive, then f is topologically conjugate to the rotation R_ρ.*
2. *If f is not topologically transitive, then R_ρ is a factor of f, and the factor map $h\colon S^1 \to S^1$ can be chosen to be monotone.*

Proof. Let F be a lift of f and fix $x \in \mathbb{R}$. Let $A = \{F^n(x) + m\colon n, m \in \mathbb{Z}\}$ and $B = \{n\rho + m\colon n, m \in \mathbb{Z}\}$. Then B is dense in $\mathbb{R}$ (§1.2). Define $H\colon A \to B$ by $H(F^n(x) + m) = n\rho + m$. By the preceding lemma, H preserves order and is bijective. Extend H to a map $H\colon \mathbb{R} \to \mathbb{R}$ by defining

$$H(y) = \sup\{n\rho + m\colon F^n(x) + m < y\}.$$

Then $H(y) = \inf\{n\rho + m\colon F^n(x) + m > y\}$, since otherwise $\mathbb{R}\backslash B$ would contain an interval.

We claim that $H\colon \mathbb{R} \to \mathbb{R}$ is continuous. If $y \in \bar{A}$, then $H(y) = \sup\{H(z)\colon z \in A, z < y\}$ and $H(y) = \inf\{H(z)\colon z \in A, z > y\}$ implies that H

is continuous on $\bar{A}$. If I is an interval in $\mathbb{R}\backslash\bar{A}$, then H is constant on I and the constant agrees with the values at the endpoints. Thus $H\colon \mathbb{R} \to \mathbb{R}$ is a continuous extension of $H\colon A \to B$.

Note that H is surjective, non-decreasing, and

$$\begin{aligned} H(y+1) &= \sup\{n\rho + m\colon F^n(x) + m < y+1\} \\ &= \sup\{n\rho + m\colon F^n(x) + (m-1) < y\} = H(y) + 1. \end{aligned}$$

Moreover,

$$\begin{aligned} H\big(F(y)\big) &= \sup\{n\rho + m\colon F^n(x) + m < F(y)\} \\ &= \sup\{n\rho + m\colon F^{n-1}(x) + m < y\} \\ &= \rho + H(y). \end{aligned}$$

We conclude that H descends to a map $h\colon S^1 \to S^1$ and $h \circ f = R_\rho \circ h$.

Finally, note that f is transitive if and only if $\{F^n(x) + m\colon n, m \in \mathbb{Z}\}$ is dense in $\mathbb{R}$. Since H is constant on any interval in $\mathbb{R}\backslash\bar{A}$, we conclude that h is injective if and only if f is transitive. (Note that by Proposition 7.1.7, either every orbit is dense or no orbit is dense.) □

Exercise 7.1.1. Show if F and $G = F + k$ are two lifts of f, then $\rho(F) = \rho(G) + k$, so $\rho(f)$ is independent of the choice of lift used in its definition. Show that there is a unique lift F of f such that $\rho(F) = \rho(f)$.

Exercise 7.1.2. Show that $\rho(f^m) = m\rho(f)$.

Exercise 7.1.3. Show if f is an orientation-reversing homeomorphism of S^1, then $\rho(f^2) = 0$.

Exercise 7.1.4. Suppose f has rational rotation number. Show that:

(a) if f has exactly one periodic orbit, then every non-periodic point is both forwards and backwards asymptotic to the periodic orbit; and

(b) if f has more than one periodic orbit, then every non-periodic orbit is forward asymptotic to some periodic orbit and backwards asymptotic to a different periodic orbit.

Exercise 7.1.5. Show that Theorems 7.1.1 and 7.1.5 hold under the weaker hypothesis that $f\colon S^1 \to S^1$ is a continuous map such that any (and thus every) lift F of f is non-decreasing.

7.2 Circle diffeomorphisms

The total variation of a function $f\colon S^1 \to \mathbb{R}$ is

$$\mathrm{Var}(f) = \sup \sum_{k=1}^{n} |f(x_k) - f(x_{k+1})|,$$

where the supremum is taken over all partitions $0 \leqslant x_1 < \cdots < x_n \leqslant 1$, for all $n \in \mathbb{N}$. We say that g has *bounded variation* if $\mathrm{Var}(g)$ is finite. Note that any Lipshitz function has bounded variation. In particular, any C^1 function has bounded variation.

THEOREM 7.2.1 (Denjoy). *Let f be an orientation-preserving C^1 diffeomorphism of the circle with irrational rotation number $\rho = \rho(f)$. If f' has bounded variation, then f is topologically conjugate to the rigid rotation R_ρ.*

Proof. We know from Theorem 7.1.9 that if f is transitive, it is conjugate to R_ρ. Thus we assume that f is not transitive, and argue to obtain a contradiction. By Proposition 7.1.7, we may assume that $\omega(0)$ is a perfect, nowhere dense set. Then $S^1\backslash\omega(0)$ is a disjoint union of open intervals. Let $I = (a, b)$ be one of these intervals. Then the intervals $\{f^n(I)\}_{n\in\mathbb{Z}}$ are pairwise disjoint, since otherwise f would have a periodic point. Thus $\sum_{n\in\mathbb{Z}} \mathrm{l}((f^n(I)) \leqslant 1$, where $\mathrm{l}(f^n(I)) = \int_a^b (f^n)'(t)\,dt$ is the length of $f^n(I)$.

LEMMA 7.2.2. *Let J be an interval in S^1 and suppose the interiors of the intervals $J, f(J), \ldots, f^{n-1}(J)$ are pairwise disjoint. Let $g = \log f'$, and fix $x, y \in J$. Then for any $n \in \mathbb{Z}$,*

$$\mathrm{Var}(g) \geqslant |\log(f^n)'(x) - \log(f^n)'(y)|.$$

Proof. Using the fact that the intervals $J, f(J), \ldots, f^n(J)$ are disjoint, we get

$$\mathrm{Var}(g) \geqslant \sum_{k=0}^{n-1} |g(f^k(y)) - g(f^k(x))| \geqslant \left|\sum_{k=0}^{n-1} g(f^k(y)) - g(f^k(x))\right|$$

$$= \left|\log\prod_{k=0}^{n-1} f'(f^k(y)) - \log\prod_{k=0}^{n-1} f'(f^k(x))\right| = |\log(f^n)'(y) - \log(f^n)'(x)|.$$

□

Fix $x \in S^1$. We claim that there are infinitely many $n \in \mathbb{N}$ such that the intervals $(x, f^{-n}(x)), (f(x), f^{1-n}(x)), \ldots, (f^n(x), x)$ are pairwise disjoint. It suffices to show that there are infinitely many n such that $f^k(x)$ is not in the interval $(x, f^n(x))$ for $0 \leqslant |k| \leqslant n$. Lemma 7.1.8 implies that the orbit of x is ordered in the same way as the orbit of a point under the irrational rotation

R_ρ. Since the orbit of a point under an irrational rotation is dense, the claim follows.

Choose n as in the preceding paragraph. Then by applying Lemma 7.2.2 with $y = f^{-n}(x)$, we obtain

$$\operatorname{Var}(g) \geqslant \left|\log \frac{(f^n)'(x)}{(f^n)'(y)}\right| = |\log((f^n)'(x)(f^{-n})'(x))|.$$

Thus for infinitely many $n \in \mathbb{N}$, we have

$$\begin{aligned} \mathrm{l}(f^n(I)) + \mathrm{l}(f^{-n}(I)) &= \int_I (f^n)'(x)\,dx + \int_I (f^{-n})'(x)\,dx \\ &= \int_I [(f^n)'(x) + (f^{-n})'(x)]\,dx \\ &\geqslant \int_I \sqrt{(f^n)'(x)(f^{-n})'(x)}\,dx \\ &\geqslant \int_I \sqrt{\exp(-\operatorname{Var}(g))}\,dx = \exp\left(-\frac{1}{2}\operatorname{Var}(g)\right)\mathrm{l}(I). \end{aligned}$$

This contradicts the fact that $\sum_{n\in\mathbb{Z}} \mathrm{l}(f^n(I)) \leqslant \infty$, so we conclude that f is transitive, and therefore conjugate to R_ρ. □

THEOREM 7.2.3 (Denjoy Example). *For any irrational number $\rho \in (0, 1)$, there is a non-transitive C^1 orientation-preserving diffeomorphism $f\colon S^1 \to S^1$ with rotation number ρ.*

Proof. We know from Lemma 7.1.8 that if $\rho(f) = \rho$, then for any $x \in S^1$, the orbit of x is ordered the same way as any orbit of R_ρ, i.e. $f^k(x) < f^l(x) < f^m(x)$ if and only if $R_\rho^k(x) < R_\rho^l(x) < R_\rho^m(x)$. Thus in constructing f, we have no choice about the order of the orbit of any point. We do, however, have a choice about the spacing between points in the orbit.

Let $\{l_n\}_{n\in\mathbb{Z}}$ be a sequence of positive real numbers such that $\sum_{n\in\mathbb{Z}} l_n = 1$ and l_n is decreasing as $n \to \pm\infty$ (we will impose additional constraints later). Fix $x_0 \in S^1$ and define

$$a_n = \sum_{\{k\in\mathbb{Z}:\ R_\rho^k(x_0)\in[x_0, R_\rho^n(x_0))\}} l_k, \qquad b_n = a_n + l_n.$$

The intervals $[a_n, b_n]$ are pairwise disjoint. Since $\sum_{n\in\mathbb{Z}} l_n = 1$, the union of these intervals covers a set of measure 1 in $[0, 1]$, and is therefore dense.

To define a C^1 homeomorphism $f\colon S^1 \to S^1$ it suffices to define a continuous, positive function g on S^1 with total integral 1. Then f will be defined to be the integral of g. The function g should satisfy

1. $\int_{a_n}^{b_n} g(t)\,dt = l_{n+1}$.

To construct such a g it suffices to define g on each interval $[a_n, b_n]$ so that it also satisfies

2. $g(a_n) = g(b_n) = 1$.
3. For any sequence $\{x_k\} \subset \bigcup_{n\in\mathbb{Z}}[a_n, b_n]$, if $\lim x_k \notin \bigcup_{n\in\mathbb{Z}}[a_n, b_n]$, then $g(x_k) \to 1$.

We then define g to be 1 on $S^1 \setminus \bigcup_{n\in\mathbb{Z}}[a_n, b_n]$.

There are many such possibilities for $g|[a_n, b_n]$. We use the quadratic polynomial

$$g(x) = 1 + \frac{6(l_{n+1} - l_n)}{l_n^3}(b_n - x)(x - a_n),$$

which clearly satisfies condition (1). For $n \geqslant 0$, we have $l_{n+1} - l_n < 0$, so

$$1 \geqslant g(x) \geqslant 1 - \frac{6(l_n - l_{n+1})}{l_n^3}\left(\frac{l_n}{2}\right)^2 = \frac{3l_{n+1} - l_n}{2l_n}.$$

For $n < 0$, we have $l_{n+1} - l_n > 0$, so

$$1 \leqslant g(x) \leqslant \frac{3l_{n+1} - l_n}{2l_n}.$$

Thus if we choose l_n such that $(3l_{n+1} - l_n)/(2l_n) \to 1$ as $n \to \pm\infty$, then (3) is satisfied. For example, we could choose $l_n = \alpha(|n| + 2)^{-1}(|n| + 3)^{-1}$, where $\alpha = 1/\sum_{n\in\mathbb{Z}}((|n| + 2)^{-1}(|n| + 3)^{-1})$.

Now define $f(x) = a_1 + \int_0^x g(t)\,dt$. Using the results above, it follows that $f\colon S^1 \to S^1$ is a C^1 homeomorphism of S^1 with rotation number ρ (Exercise 7.2.1). Moreover, $f^n(0) = a_n$, and $\omega(0) = S^1 \setminus \bigcup_{n\in\mathbb{Z}}(a_n, b_n)$ is a closed, perfect, invariant set of measure zero. □

Exercise 7.2.1. Verify the statements in the last paragraph of the proof of Theorem 7.2.3.

Exercise 7.2.2. Show directly that the example constructed in the proof of Theorem 7.2.3 is not C^2.

7.3 The Sharkovsky Theorem

We consider the set $\mathbb{N}_{\mathrm{Sh}} = \mathbb{N} \cup \{2^\infty\}$ obtained by adding the formal symbol 2^∞ to the set of natural numbers. The Sharkovsky ordering of this set is

$$\begin{gathered} 1 \prec 2 \prec \cdots \prec 2^n \prec \cdots \prec 2^\infty \prec \\ \prec \cdots 2^m \cdot (2n+1) \prec \cdots \prec 2^m \cdot 7 \prec 2^m \cdot 5 \prec 2^m \cdot 3 \prec \\ \prec \cdots 2(2n+1) \prec \cdots \prec 14 \prec 10 \prec 6 \prec \\ \prec \cdots 2n+1 \prec \cdots \prec 7 \prec 5 \prec 3. \end{gathered}$$

The symbol 2^∞ is added so that $\mathbb{N}_{\mathrm{Sh}}$ has the least upper bound property, i.e. every subset of $\mathbb{N}_{\mathrm{Sh}}$ has a supremum. The Sharkovsky ordering is preserved by multiplication by 2^k, for any $k \geqslant 0$ (where $2^k \cdot 2^\infty = 2^\infty$, by definition).

For $\alpha \in \mathbb{N}_{\mathrm{Sh}}$, let $S(\alpha) = \{k \in \mathbb{N} \colon k \preceq \alpha\}$ (note that $S(\alpha)$ is defined to be a subset of $\mathbb{N}$, not $\mathbb{N}_{\mathrm{Sh}}$). For a map $f\colon [0,1] \to [0,1]$, we denote by $\mathrm{MinPer}(f)$ the set of minimal periods of periodic points of f.

THEOREM 7.3.1 (Sharkovsky, 1964). *For every continuous map $f\colon [0,1] \to [0,1]$, there is $\alpha \in \mathbb{N}_{\mathrm{Sh}}$ such that $\mathrm{MinPer}(f) = S(\alpha)$. Conversely, for every $\alpha \in \mathbb{N}_{\mathrm{Sh}}$, there is a continuous map $f\colon [0,1] \to [0,1]$ with $\mathrm{MinPer}(f) = S(\alpha)$.*

The proof of the first assertion of the Sharkovsky Theorem proceeds as follows. We assume that f has a periodic point x of minimal period $n > 1$, since otherwise there is nothing to show. The orbit of x partitions the interval $[0, 1]$ into a finite collection of subintervals whose endpoints are elements of the orbit. The endpoints of these intervals are permuted by f. By examining the combinatorial possibilities for the permutations of pairs of endpoints, and using the Intermediate Value Theorem, one establishes the existence of periodic points of the desired periods.

The second assertion of the Sharkovsky Theorem is proved as Lemma 7.3.9.

If I and J are intervals in $[0, 1]$ and $f(I) \supset J$, we say that I *f-covers* J, and we write $I \to J$. If $a, b \in [0, 1]$, then we will use $[a, b]$ to represent the closed interval between a and b, regardless of whether $a \geqslant b$ or $a \leqslant b$.

LEMMA 7.3.2

1. *If $f(I) \supset I$, then the closure of I contains a fixed point of f.*
2. *If $\{I_k\}_{k\in\mathbb{N}}$ is a sequence of non-empty closed intervals in $[0, 1]$ such that $f(I_k) \supset I_{k+1}$ for all k, then there is a point $x \in I_1$ such that $f^k(x) \in I_{k+1}$, for all $k \in \mathbb{N}$. Moreover, if $I_n = I_1$ for some $n > 0$, then I_1 contains a periodic point x of period n such that $f^k(x) \in I_{k+1}$ for $k = 1, \ldots, n-1$.*

Proof. The proof of part (1) is a simple application of the Intermediate Value Theorem.

To prove part (2), note that since $f(I_1) \supset I_2$, there are points $a_0, b_0 \in I_1$ that map to the endpoints of I_2. Let J_1 be the subinterval of I_1 with endpoints a_0, b_0. Then $f(J_1) = I_2$. Suppose we have defined subintervals $J_1 \supset J_2 \supset \cdots \supset J_n$ in I_1 such that $f^k(J_k) = I_{k+1}$. Then $f^{n+1}(J_n) = f(I_{n+1}) \supset I_{n+2}$, so there is an interval $J_{n+1} \subset J_n$ such that $f^{n+1}(J_{n+1}) = I_{n+2}$. Thus we obtain a nested sequence $\{J_n\}$ of non-empty closed intervals. The intersection $\bigcap_{i=1}^{m-1} J_i$ is non-empty, and for any x in the intersection, $f^k(x) \in I_{k+1}$ for $1 \leqslant k < m-1$.

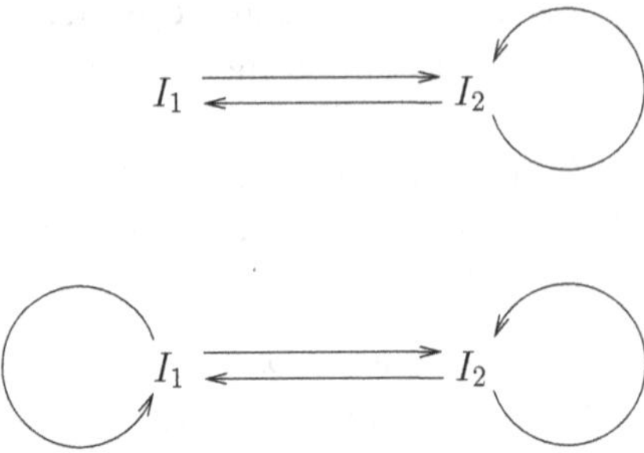

Figure 7.1. The two possible Markov graphs for period three.

The last assertion follows from the preceding paragraph together with part (1). □

A *partition* of an interval I is a (finite or infinite) collection of closed subintervals $\{I_k\}$, with pairwise disjoint interiors, whose union is I. The *Markov graph* of f associated to the partition $\{I_k\}$ is the directed graph with vertices I_k, and a directed edge from I_i to I_j if and only if I_i f-covers I_j. By Lemma 7.3.2, any loop of length n in the Markov graph of f forces the existence of a periodic point of (not necessarily minimal) period n.

As a warmup to the proof of the full Sharkovsky Theorem, we prove that the existence of a periodic point of minimal period three implies the existence of periodic points of all periods. This result was rediscovered in 1975 by Li and Yorke, and popularized in their paper *Period three implies chaos* (Li and Yorke, 1975).

Let x be a point of period three. Replacing x with $f(x)$ or $f^2(x)$ if necessary, we may assume that $x < f(x)$ and $x < f^2(x)$. Then there are two cases: (1) $x < f(x) < f^2(x)$; or (2) $x < f^2(x) < f(x)$. In the first case, we let $I_1 = [x, f(x)]$ and $I_2 = [f(x), f^2(x)]$. The associated Markov graph is one of the two graphs shown in Figure 7.1.

For $k \geqslant 2$, the path $I_1 \to I_2 \to I_2 \to \cdots \to I_2 \to I_1$ of length k implies the existence of a periodic point y of period k with the itinerary $I_1, I_2, I_2, \ldots, I_2, I_1$. If the minimal period of y is less than k, then $y \in I_1 \cap I_2 = \{f(x)\}$. But $f(x)$ does not have the specified itinerary for $k \neq 3$, so the minimal period of y is k. A similar argument applies to case (2), and this proves the Sharkovsky Theorem for $n = 3$.

To prove the full Sharkovsky Theorem it is convenient to use a subgraph of the Markov graph defined as follows. Let $P = \{x_1, x_2, \ldots, x_n\}$ be a periodic orbit of (minimal) period $n > 1$, where $x_1 < x_2 < \cdots < x_n$. Let $I_j = [x_j, x_{j+1}]$. The *P-graph* of f is the directed graph with vertices I_j, and a directed edge from I_j to I_k if and only if $I_k \subset [f(x_j), f(x_{j+1})]$. Since $f(I_j) \supset [f(x_j), f(x_{j+1})]$, it follows that the P-graph is a subgraph of the Markov graph associated to the same partition. In particular, any loop in the P-graph is also a loop in the Markov graph. The P-graph has the virtue that it is completely determined by the ordering of the periodic orbit, and is independent of the behavior of the map on the intervals I_j. For example, in Figure 7.1, the top graph is the unique P-graph for a periodic orbit of period three with ordering $x < f(x) < f^2(x)$.

LEMMA 7.3.3. *The P-graph of f contains a trivial loop, i.e. there is a vertex I_j with a directed edge from I_j to itself.*

Proof. Let $j = \max\{i\colon f(x_i) > x_i\}$. Then $f(x_j) > x_j$ and $f(x_{j+1}) \leqslant x_{j+1}$, so $f(x_j) \geqslant x_{j+1}$ and $f(x_{j+1}) \leqslant x_j$. Thus $[f(x_j), f(x_{j+1})]) \supset [x_j, x_{j+1}]$. □

We will renumber the vertices of the P-graph (but not the points of P) so that $I_1 = [x_j, x_{j+1}]$, where $j = \max\{i\colon f(x_i) > x_i\}$. By the proof of the preceding lemma, I_1 is a vertex with a directed edge from itself to itself.

For any two points $x_i < x_k$ in P, define

$$\hat{f}([x_i, x_k]) = \bigcup_{l=i}^{k-1} [f(x_l), f(x_{l+1})].$$

In particular, $\hat{f}(I_k) = [f(x_k), f(x_k + 1)]$. If $\hat{f}(I_k) \supset I_l$, we say that I_k $\hat{f}$-covers I_l. Since we will only be using P-graphs throughout the remainder of this section, we also redefine the notation $I_k \to I_l$ to mean that I_k $\hat{f}$-covers I_l.

PROPOSITION 7.3.4. *Any vertex of the P-graph can be reached from I_1.*

Proof. The nested sequence $I_1 \subset \hat{f}(I_1) \subset \hat{f}^2(I_1) \subset \cdots$ must eventually stabilize, since $\hat{f}^k(I_1)$ is an interval whose endpoints are in the orbit of x. Then for k sufficiently large, $\mathcal{O}(x) \cap \hat{f}^k(I_1)$ is an invariant subset of $\mathcal{O}(x)$, and is therefore equal to $\mathcal{O}(x)$. It follows that $\hat{f}^k(I_1) = [x_1, x_n]$, so any vertex of the P-graph can be reached from I_1. □

LEMMA 7.3.5. *Suppose the P-graph has no directed edge from any interval I_k, $k \neq 1$, to I_1. Then n is even, and f has a periodic point of period 2.*

Proof. Let $J_0 = [x_1, x_j]$ and $J_1 = [x_{j+1}, x_{n-1}]$, where $j = \max\{i\colon f(x_i) > x_i\}$ (the case $j = 1$ is not excluded a priori). Then $\hat{f}(J_0) \not\subset J_0$ (since $f(x_j) > x_j$) and $\hat{f}(J_0) \not\supset I_1$, so $\hat{f}(J_0) \subset J_1$ since $\hat{f}(J_0)$ is connected. Likewise, $\hat{f}(J_1) \subset J_0$. Now $\hat{f}(J_0) \cup \hat{f}(J_1) \supset \mathcal{O}(x)$, so $\hat{f}(J_0) = J_1$ and $\hat{f}(J_1) = J_0$. Thus J_0 f-covers

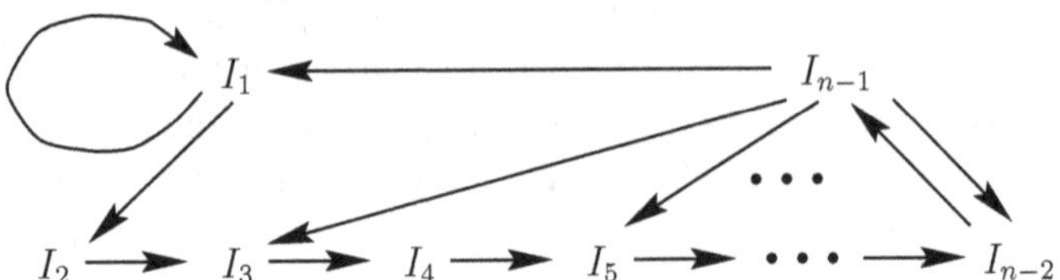

Figure 7.2. The P-graph for Lemmas 7.3.6 and 7.3.9.

J_1 and J_1 f-covers J_0, so f has a periodic point of minimal period 2, and $n = |\mathcal{O}(x)| = 2|\mathcal{O}(x) \cap J_0|$ is even. □

LEMMA 7.3.6. *Suppose $n > 1$ is odd and f has no non-fixed periodic points of smaller odd period. Then there is a numbering of the vertices of the P-graph so that the graph contains the following edges, and no others (see Figure 7.2):*

1. $I_1 \to I_1$ *and* $I_{n-1} \to I_1$;
2. $I_i \to I_{i+1}$, *for* $i = 1, \dots, n-2$;
3. $I_{n-1} \to I_{2i+1}$, *for* $0 \leqslant i < (n-1)/2$.

Proof. By Lemmas 7.3.4 and 7.3.5, there is a non-trivial loop in the P-graph starting from I_1. By choosing a shortest such loop and renumbering the vertices of the graph, we may assume that we have a loop

$$I_1 \to I_2 \to \cdots \to I_k \to I_1 \tag{7.1}$$

in the P-graph, $k \leqslant n-1$. The existence of this loop implies that f has a periodic point of minimal period k. The path

$$I_1 \to I_1 \to I_2 \to \cdots \to I_k \to I_1$$

implies the existence of a periodic point of minimal period $k+1$. By the minimality of n, we conclude that $k = n-1$, which proves statement (1).

Let $I_1 = [x_j, x_{j+1}]$. Note that $\hat{f}(I_1)$ contains I_1 and I_2, but no other I_i, since otherwise we would have a shorter path than (7.1). Similarly, if $1 \leqslant i < (n-2)$, then $\hat{f}(I_i)$ cannot contain I_k for $k > i+1$. Thus $\hat{f}(I_1) = [x_j, x_{j+2}]$, or $\hat{f}(I_1) = [x_{j-1}, x_{j+1}]$. Suppose the latter holds (the other case is similar). Then $I_2 = [x_{j-1}, x_j]$, $f(x_{j+1}) = x_{j-1}$, and $f(x_j) = x_{j+1}$. If $2 < (n-1)$, then $\hat{f}(I_2)$ can contain at most I_2 and I_3, so $f(x_{j-1}) = x_{j+2}$. Continuing in this way (see Figure 7.3), we find that the intervals of the partition are ordered on the interval I as follows:

$$I_{n-1},\ I_{n-3},\ \dots\ I_2,\ I_1,\ I_3, \dots\ I_{n-2}.$$

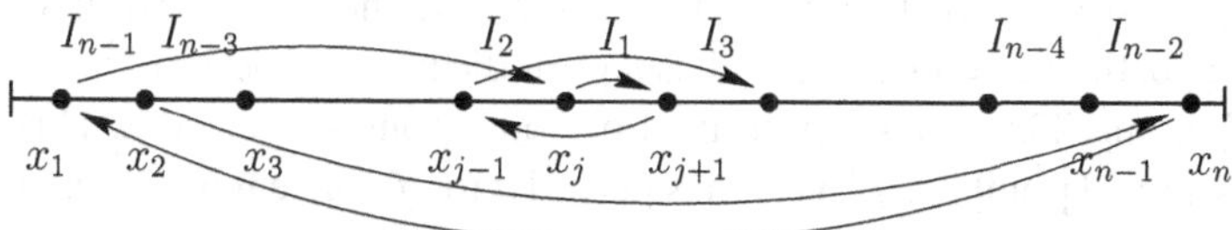

Figure 7.3. The action of f from Lemma 7.3.6 on x_k is shown by arrows.

Moreover, $f(x_2) = x_n$, $f(x_n) = x_1$, and $f(x_1) = x_j$, so $\hat{f}(I_{n-1}) = [x_j, x_{n-1}]$, and $\hat{f}(I_{n-1})$ contains all the odd-numbered intervals, which completes the proof of the lemma. □

COROLLARY 7.3.7. *If n is odd, then f has a periodic point of minimal period q for any $q > n$ and for any even integer $q < n$.*

Proof. Let $m > 1$ be the minimal odd period of a non-fixed periodic point. By the preceding lemma, there are paths of the form

$$I_1 \to I_1 \to \cdots \to I_1 \to I_2 \to \cdots \to I_{m-1} \to I_1$$

of any length $q \geqslant m$. For $q = 2i < m$, the path

$$I_{m-1} \to I_{m-2i} \to I_{m-2i+1} \to \cdots \to I_{m-1}$$

gives a periodic point of period q. The verification that these periodic points have minimal period q is left as an exercise (Exercise 7.3.3). □

LEMMA 7.3.8. *If n is even, then f has a periodic point of minimal period* 2.

Proof. Let m be the smallest even period of a non-fixed periodic point, and let I_1 be an interval of the associated partition that $\hat{f}$-covers itself. If no other interval $\hat{f}$-covers I_1, then Lemma 7.3.5 implies that $m = 2$.

Suppose then that some other interval $\hat{f}$-covers I_1. In the proof of Lemma 7.3.6, we used the hypothesis that n is odd only to conclude the existence of such an interval. Thus the same argument as in the proof of that lemma implies that the P-graph contains the paths

$$I_1 \to I_2 \to \cdots \to I_{n-1} \to I_1 \text{ and } I_{n-1} \to I_{2i} \text{ for } 0 \leqslant i < n/2.$$

Then $I_{n-1} \to I_{n-2} \to I_{n-1}$ implies the existence of a periodic point of minimal period 2. □

Conclusion of the proof of the first assertion of the Sharkovsky Theorem. There are two cases to consider.

1. $n = 2^k, k > 0$. If $q \prec n$, then $q = 2^l$ with $0 \leqslant l < k$. The case $l = 0$ is trivial. If $l > 0$, then $g = f^{q/2} = f^{2^{l-1}}$ has a periodic point of period 2^{k-l+1}, so by Lemma 7.3.8, g has a non-fixed periodic point of period 2. This point is a fixed point for f^q, i.e. it has period q for f. Since it is not fixed by g, its minimal period is q.
2. $n = p2^k$, p odd. The map f^{2^k} has a periodic point of minimal period p, so by Corollary 7.3.7, f^{2^k} has periodic points of minimal period m for all $m \geqslant p$ and all even $m < p$. Thus f has periodic points of minimal period $m2^k$ for all $m \geqslant p$ and all even $m < p$. In particular, f has a periodic point of minimal period 2^{k+1}, so by case 1, f has periodic points of minimal period 2^i for $i = 0, \dots, k$. □

The next lemma finishes the proof of the Sharkovsky Theorem.

LEMMA 7.3.9. *For any* $\alpha \in \mathbb{N}_{\mathrm{Sh}}$, *there is a continuous map* $f\colon [0,1] \to [0,1]$ *such that* $\mathrm{MinPer}(f) = S(\alpha)$.

Proof. We distinguish three cases:
1. $\alpha \in \mathbb{N}, \alpha$ odd;
2. $\alpha \in \mathbb{N}, \alpha$ even;
3. $\alpha = 2^\infty$.

Case 1. Suppose $n \in \mathbb{N}$ is odd, and $\alpha = n$. Choose points $x_0, \dots, x_{n-1} \in [0,1]$ such that

$$0 = x_{n-1} < \cdots < x_4 < x_2 < x_0 < x_1 < x_3 < \cdots < x_{n-2} = 1,$$

and let $I_1 = [x_0, x_1]$, $I_2 = [x_2, x_0]$, $I_3 = [x_1, x_3]$, etc. Let $f\colon [0,1] \to [0,1]$ be the unique map defined by:
1. $f(x_i) = x_{i+1}, i = 0, \dots, n-2$, and $f(x_{n-1}) = x_0$;
2. f is linear (or affine, to be precise) on each interval $I_j, j = 1, \dots, n-1$.

Then x_0 is periodic of period n, and the associated P-graph is shown in Figure 7.2. Any path that avoids I_1 has even length. Loops of length less than n must be of the following form:
1. $I_i \to I_{i+1} \to \cdots \to I_{n-1} \to I_{2j+1} \to I_{2j+2} \to \cdots \to I_i$ for $i > 1$; or
2. $I_{n-1} \to I_{2i+1} \to \cdots \to I_{n-1}$; or
3. $I_1 \to I_1 \to \cdots \to I_1 \to I_1$.

Paths of type 1 or 2 have even length, so no point in $int(I_j), j = 2, \dots, n-1$, can have odd period $k < n$. Since $f(I_1) = I_1 \cup I_2, |f'| > 1$ on I_1, so every non-fixed point in $\mathrm{int}(I_1)$ must move away from the (unique) fixed point in I_1, and therefore eventually enters I_2. Once a point enters I_2 it must enter every I_j

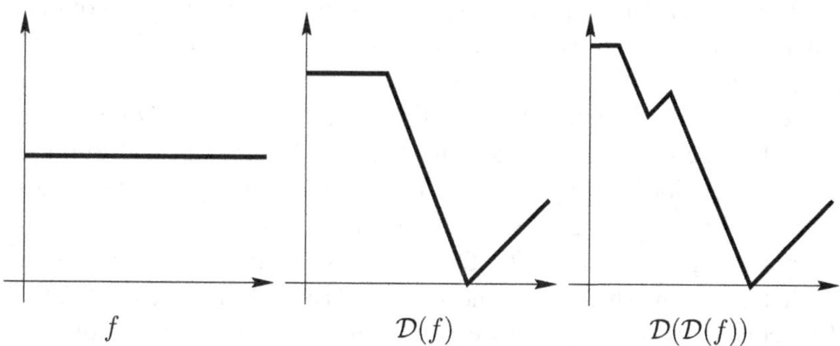

Figure 7.4. Graphs of $\mathcal{D}^k(f)$ for $f \equiv 1/2$.

before it returns to $\text{int}(I_1)$. Thus there is no non-fixed periodic point in I_1 of period less than n. It follows that no point has odd period less than n. This finishes the proof of the theorem for n odd.

Case 2. Suppose $n \in \mathbb{N}$ is even, and $\alpha = n$. For $f\colon [0,1] \to [0,1]$, define a new function $\mathcal{D}\colon [0,1] \to [0,1]$ by

$$\mathcal{D}(f)(x) = \begin{cases} \frac{2}{3} + \frac{1}{3} f(3x) & x \in [0, \frac{1}{3}] \\ (2 + f(1))(\frac{2}{3} - x) & x \in [\frac{1}{3}, \frac{2}{3}] \\ x - \frac{2}{3} & x \in [\frac{2}{3}, 1] \end{cases} .$$

The operator $\mathcal{D}(f)$ is sometimes called the *doubling operator*, because $\text{MinPer}(\mathcal{D}(f)) = 2\,\text{MinPer}(f) \cup \{1\}$, i.e. $\mathcal{D}$ doubles the periods of a map. To see this, let $g = \mathcal{D}(f)$, and let $I_1 = [0, 1/3]$, $I_2 = [1/3, 2/3]$, and $I_3 = [2/3, 1]$. For $x \in I_1, g^2(x) = f(3x)/3$, so $g^{2k}(x) = f^k(3x)/3$. Thus $g^{2k}(x) = x$ if and only if $f^k(3x) = 3x$, so $\text{MinPer}(g) \supset 2\,\text{MinPer}(f)$ (see Figure 7.4).

On the interval I_2, $|g'| \geqslant 2$, so there is a unique repelling fixed point in $(1/3, 2/3)$, and every other point eventually leaves this interval and never returns since $g(I_1 \cup I_3) \cap I_2 = \emptyset$. Thus no non-fixed point in I_2 is periodic.

Finally, any periodic point in I_3 enters I_1, so its period is in $2\,\text{MinPer}(f)$, which verifies our claim that $\text{MinPer}(\mathcal{D}(f)) = 2\,\text{MinPer}(f) \cup \{1\}$.

Since n is even, we can write $n = p2^k$, where p is odd and $k > 0$. Let f be a map whose minimum odd period is p (see Case 1). Then $\text{MinPer}(\mathcal{D}^k(f)) = 2^k\,\text{MinPer}(f) \cup \{2^{k-1}, 2^{k-2}, \ldots, 1\}$, which settles Case 2 of the lemma.

Case 3. Suppose $\alpha = 2^\infty$. Let $g_k = \mathcal{D}^k(\text{Id})$, where Id is the identity map. Then, by the induction and the remarks in the proof of Case 2, $\text{MinPer}(g_k) =$

$\{2^{k-1}, 2^{k-2}, \ldots, 1\}$. The sequence $\{g_k\}_{k\in\mathbb{N}}$ converges uniformly to a continuous map $g_\infty\colon [0,1] \to [0,1]$, and $g_\infty = g_k$ on $[2/3^k, 1]$ (Exercise 7.3.4). It follows that $\mathrm{MinPer}(g_\infty) \supset S(2^\infty)$.

Let x be a periodic point of g_∞. If $0 \notin \mathcal{O}(x)$, then $\mathcal{O}(x) \subset [2/3^k, 1]$ for k sufficiently large, so x is a periodic point of g_k and has even period. Suppose then that 0 is periodic with period p. If $p \succ 2^\infty$, then there is $q \in \mathbb{N}$ such $p \succ q \succ 2^\infty$. By the first part of the Sharkovsky Theorem, g_∞ has a periodic point y with minimal period q. Since $0 \in \mathcal{O}(y)$, we conclude by the preceding argument that q is even, which contradicts $q \succ 2^\infty$. Thus $\mathrm{MinPer}(g_\infty) = S(2^\infty)$.

This concludes the proof of Lemma 7.3.9, and thus the proof of Theorem 7.3.1. □

Exercise 7.3.1. Let σ be a permutation of $\{1, \ldots, n-1\}$. Show that there is a continuous map $f\colon [0,1] \to [0,1]$ with a periodic point x of period n such that $x < f^{\sigma(1)} < \cdots < f^{\sigma(n-1)}$.

Exercise 7.3.2. Show that there are maps $f, g\colon [0,1] \to [0,1]$, each with a periodic point of period n (for some n) such that the associated P-graphs are not isomorphic. (Note that for $n = 3$, all P-graphs are isomorphic.)

Exercise 7.3.3. Verify that the periodic points in the proof of Corollary 7.3.7 have *minimal* period q.

Exercise 7.3.4. Show that the sequence $\{g_k\}_{k\in\mathbb{N}}$ defined near the end of the proof of Lemma 7.3.9 converges uniformly, and the limit g_∞ satisfies $g_\infty = g_k$ on $[2/3^k, 1]$.

7.4 Combinatorial theory of piecewise-monotone mappings[1]

Let $I = [a, b]$ be a compact interval. A continuous map $f\colon I \to I$ is *piecewise-monotone* if there are points $a = c_0 < c_1 < \cdots < c_l < c_{l+1} = b$ such that f is strictly monotone on each interval $I_i = [c_{i-1}, c_i]$, $i = 1, \ldots, l+1$. We always assume that each interval $[c_{i-1}, c_i]$ is a maximal interval on which f is monotone, so the orientation of f reverses at the *turning points* $c_1, \ldots, c_l$. The intervals I_i are called *laps* of f.

Note that any piecewise-monotone map $f\colon I \to I$ can be extended to a piecewise-monotone map of a larger interval J in such a way that $f(\partial J) \subset \partial J$. Thus we assume (without losing much generality) that $f(\partial I) \subset \partial I$. If f has l

[1] Our arguments in this section follow in part those of Collet and Eckmann (1980) and Milnor and Thurston (1988).

turning points and $f(\partial I) \subset \partial I$, then f is *l-modal.* If f has exactly one turning point, then f is *unimodal.*

The *address* of a point $x \in I$ is the symbol c_j if $x = c_j$ for some $j \in \{1, \ldots, l\}$, or the symbol I_j if $x \in I_j$ and $x \notin \{c_1, \ldots, c_l\}$. Note that c_0 and c_{l+1} are *not* included as addresses. The *itinerary* of x is the sequence $i(x) = (i_k(x))_{k \in \mathbb{N}_0}$, where $i_k(x)$ is the address of $f^k(x)$. Let

$$\Sigma = \{I_1, \ldots, I_{l+1}, c_1, \ldots, c_l\}^{\mathbb{N}_0}.$$

Then $i\colon I \to \Sigma$, and $i \circ f = \sigma \circ i$, where σ is the one-sided shift on Σ.

Example. Any quadratic map $q_\mu(x) = \mu x(1-x), 0 < \mu \leqslant 4$, is a unimodal map of $I = [0, 1]$, with turning point $c_1 = 1/2$, $I_1 = [0, 1/2]$, $I_2 = [1/2, 1]$. If $0 < \mu < 2$, then $f(I) \subset [0, 1/2)$, so the only possible itineraries are $(I_1, I_1, \ldots)$, $(c_1, I_1, I_1, \ldots)$, and $(I_2, I_1, I_1, \ldots)$. Note that the map $i\colon [0, 1] \to \Sigma$ is not continuous at c_1.

If $\mu = 2$, then the possible itineraries are $(I_1, I_1, \ldots)$, $(c_1, c_1, \ldots)$, and $(I_2, I_1, I_1, \ldots)$. If $2 < \mu < 3$, there is an attracting fixed point $(\mu - 1)/\mu \in (1/2, 2/3)$. Thus the possible itineraries are:

$$(I_1, I_1, \ldots)$$
$$(c_1, I_2, I_2, \ldots)$$
$$(I_2, I_2, \ldots)$$
$$(I_1, \ldots, I_1, I_2, I_2, \ldots)$$
$$(I_1, \ldots, I_1, C_1, I_2, I_2, \ldots)$$

any of the above preceded by I_2.

LEMMA 7.4.1. *The itinerary $i(x)$ is eventually periodic if and only if the iterates of x converge to a periodic orbit of f.*

Proof. If $i(x)$ is eventually periodic, then by replacing x by one of its forward iterates, we may assume that $i(x)$ is periodic, of period p. If $i_j(x) = c_j$ for some j, then c_j is periodic and we are done. Thus we may assume that $f^k(x)$ is contained in the interior of a lap of f for each k. For $j = 0, \ldots, p-1$, let J_j be the smallest closed interval containing $\{f^k(x)\colon k = j \bmod p\}$. Since the itinerary is periodic of period p, each J_i is contained in a single lap, so $f\colon J_j \to J_{j+1}$ is strictly monotone. It follows that $f^p\colon J_0 \to J_0$ is strictly monotone.

Suppose $f^p\colon J_0 \to J_0$ is increasing. If $f^p(x) \geqslant x$, then by induction, $f^{kp}(x) \geqslant f^{(k-1)p}(x)$ for all $k > 0$, so $\{f^{kp}(x)\}$ converges to a point $y \in J_0$, which is fixed for f^p. A similar argument holds if $f^p(x) < x$.

If $f^p: J_0 \to J_0$ is decreasing, then $f^{2p}: J_0 \to J_0$ is increasing, and by the argument in the preceding paragraph, the sequence $\{f^{2kp}(x)\}$ converges to a fixed point of f^{2p}.

Conversely, suppose that $f^{kq}(x) \to y$ as $k \to \infty$, where $f^q(y) = y$. If the orbit of y does not contain any turning points, then eventually x has the same itinerary as y. The case where $\mathcal{O}(y)$ does contain a turning point is left as an exercise (Exercise 7.4.1). □

Let ϵ be a function defined on $\{I_1, \dots, I_l, c_1, \dots, c_l\}$ such that $\epsilon(I_1) = \pm 1$, $\epsilon(I_k) = (-1)^{k+1}\epsilon(I_1)$, and $\epsilon(c_k) = 1$ for $k = 0, \dots, l$. Associated to ϵ is a *signed lexicographic ordering* $\prec$ on Σ, defined as follows. For $s \in \Sigma$, define

$$\tau_n(s) = \prod_{0 \leqslant k < n} \epsilon(s_k).$$

We order the symbols $\{\pm I_j, \pm c_k\}$ by

$$-I_{l+1} < -c_l < -I_l < \cdots < -I_1 < I_1 < c_1 < I_2 < \cdots < c_l < I_{l+1}.$$

Given $s = (s_i), t = (t_i) \in \Sigma$, we say $s \prec t$ if and only if $s_0 < t_0$, or there is $n > 0$ such that $s_i = t_i$ for $i = 0, \dots, n-1$, and $\tau_n(s)s_n < \tau_n(t)t_n$. The proof that $\prec$ is an ordering is left as an exercise.

Associated to an l-modal map f is a natural signed lexicographic ordering with $\epsilon(I_k) = 1$ if f is increasing on I_k and $\epsilon(I_k) = -1$ otherwise, and $\epsilon(c_k) = 1$, for $k = 1, \dots, l$. For $x \in I$, we define $\tau_n(x) = \tau_n(i(x))$. Note that if $\{x, f(x), \dots, f^{n-1}(x)\}$ contains no turning points, then $\tau_n(x)$ is the orientation of f^n at x: positive (i.e. increasing) if and only if $\tau_n(x) = 1$.

LEMMA 7.4.2. *For $x, y \in I$, if $x < y$, then $i(x) \preceq i(y)$. Conversely, if $i(x) \prec i(y)$, then $x < y$.*

Proof. Suppose $i(x) \neq i(y)$, $i_k(x) = i_k(y)$ for $k = 0, \dots, n-1$, and $i_n(x) \neq i_n(y)$. Then there is no turning point in the intervals $[x, y]$, $f([x, y]), \dots, f^{n-1}([x, y])$, so f^n is monotone on $[x, y]$, and is increasing if and only if $\tau_n(i(x)) = 1$. Thus $x < y$ if and only if $\tau_n(x) f^n(x) < \tau_n(y) f^n(y)$, and the latter holds if and only if $\tau_n(x) i_n(x) < \tau_n(y) i_n(y)$ since $i_n(x) \neq i_n(y)$. □

LEMMA 7.4.3. *Let $I(x) = \{y\colon i(y) = i(x)\}$. Then*

1. *$I(x)$ is an interval (which may consist of a single point).*
2. *If $I(x) \neq \{x\}$, then $f^n(I(x))$ does not contain any turning points for $n \geqslant 0$. In particular, every power of f is strictly monotone on $I(x)$.*
3. *Either the intervals $I(x), f(I(x)), f^2(I(x)), \dots$ are pairwise disjoint, or the iterates of every point in $I(x)$ converge to a periodic orbit of f.*

Proof. Lemma 7.4.2 implies immediately that $I(x)$ is an interval. To prove part (2), suppose there is $y \in I(x)$ such that $f^n(y)$ is a turning point. If $I(x)$ is not a single point, then there is some point $z \in I(x)$ such that $f^n(y) \neq f^n(z)$ since f^n is not constant on any interval. Thus $i_n(z) \neq i_n(y) = f(y)$, which contradicts the fact that $y, z \in I(x)$. Thus $I(x)$ must be a single point.

To prove part (3), suppose the intervals $I(x), f(I(x)), f^2(I(x)), \ldots$ are not pairwise disjoint. Then there are integers $n \geqslant 0, p > 0$, such that $f^n(I(x)) \cap f^{n+p}(I(x)) \neq \emptyset$. Then $f^{n+kp}(I(x)) \cap f^{n+(k+1)p}(I(x)) \neq \emptyset$ for all $k \geqslant 1$. It follows that $L = \bigcup_{k\geqslant 1} f^{kp}(I(x))$ is a non-empty interval that contains no turning points and is invariant by f^p. Since f^p is strictly monotone on L, for any $y \in L$, the sequence $\{f^{2kp}(y)\}$ is monotone, and converges to a fixed point of f^{2p}. □

An interval $J \subset I$ is *wandering* if the intervals $J, f(J), f^2(J), \ldots$ are pairwise disjoint and $f^n(J)$ does not converge to a periodic orbit of f. Recall that if x is an attracting periodic point, then the *basin of attraction* $BA(x)$ of x is the set of all points whose ω-limit set is $\mathcal{O}(x)$.

COROLLARY 7.4.4. *Suppose f does not have wandering intervals, attracting periodic points, or intervals of periodic points. Then $i\colon I \to \Sigma$ is an injection, and therefore a bijective order-preserving map onto its image.*

Proof. To prove that i is injective we need only show that $I(x) = \{x\}$ for every $x \in I$. If not, then by the proof of Lemma 7.4.3, either $I(x)$ is wandering or there is an interval L with non-empty interior and $p > 0$ such that f^p is monotone on L, $f^p(L) \subset L$, and the iterates of any point in L converge to a periodic orbit of f of period $2p$. The former case is excluded by hypothesis. In the latter case, by Exercise 7.4.2, either L contains an interval of periodic points, or some open interval in L converges to a single periodic point, contrary to the hypothesis. So $I(x) = \{x\}$. □

Our next goal is to characterize the subset $i(I) \subset \Sigma$. As we indicated above, the map $i\colon I \to \Sigma$ is not continuous. Nevertheless, for any $x \in I$ and $k \in \mathbb{N}_0$, there is $\delta > 0$ such that $i_k(y)$ is constant on $(x, x+\delta)$ and on $(x-\delta, x)$ (but not necessarily the same on both intervals). Thus the limits $i(x^+) = \lim_{y\to x^+} i(y)$ and $i(x^-) = \lim_{y\to x^-} i(y)$ exist. Moreover, $i(x^+)$ and $i(x^-)$ are both contained in $\{I_1, \ldots, I_l\}^{\mathbb{N}_0} \subset \Sigma$. For $j = 1, \ldots, l$, we define the jth *kneading invariant* of f to be $\nu_j = i(c_j^+)$. For convenience we also define sequences $\nu_0 = i(c_0) = i(c_0^+)$ and $\nu_{l+1} = i(c_{l+1}) = i(c_{l+1}^-)$. Note that ν_0 and ν_{l+1} are eventually periodic of period 1 or 2 since by hypothesis the set $\{c_0, c_{l+1}\}$ is invariant. In fact, there are only four possibilities for the pair ν_0, ν_{l+1}, corresponding to the four possibilities for $f|\partial I$.

LEMMA 7.4.5. *For any $x \in I$, $i(x)$ satisfies the following:*

1. $\sigma^n i(x) = i(c_k)$ *if* $f^n(x) = c_k$.
2. $\sigma \nu_k \preceq \sigma^{n+1} i(x) \preceq \sigma \nu_{k+1}$ *if* $f^n(x) \in I_{k+1}$ *and* f *is increasing on* I_{k+1}.
3. $\sigma \nu_k \succeq \sigma^{n+1} i(x) \succeq \sigma \nu_{k+1}$ *if* $f^n(x) \in I_{k+1}$ *and* f *is decreasing on* I_{k+1}.

Moreover, if f has no wandering intervals, attracting periodic points, or intervals of periodic points, then the inequalities in conditions (2) and (3) are strict.

Proof. The first assertion is obvious. To prove the second, suppose that $f^n(x) \in I_{k+1}$ and f is increasing on I_{k+1}. Then for $y \in (c_k, f^n(x))$, we have $f(c_k) < f(y) < f^{n+1}(x)$, so

$$i(f(c_k)) \preceq i(f(y)) \preceq i(f^{n+1}(x)) = \sigma^{n+1} i(x).$$

Since $\nu_k = \lim_{y \to c_k^+} i(y)$, we conclude that $\sigma \nu_k \preceq \sigma^{n+1} i(x)$. The other inequalities are proved in a similar way.

If f has no wandering intervals, attracting periodic points, or intervals of periodic points, then Corollary 7.4.4 implies that i is injective, so $\preceq$ can be replaced by $\prec$ everywhere in the preceding paragraph. □

The following immediate corollary of Lemma 7.4.5 gives an admissibility criterion for kneading invariants.

COROLLARY 7.4.6. *If* $\sigma^n(\nu_j) = (I_{k+1}, \dots)$, *then*

1. $\sigma \nu_k \preceq \sigma^{n+1} \nu_j \preceq \sigma \nu_{k+1}$ *if* f *is increasing on* I_{k+1};
2. $\sigma \nu_k \succeq \sigma^{n+1} \nu_j \succeq \sigma \nu_{k+1}$ *if* f *is decreasing on* I_{k+1}.

Let $f \colon I \to I$ be an l-modal map with kneading invariants $\nu_1, \dots, \nu_l$, and let ν_0, ν_{l+1} be the itineraries of the endpoints of I. Define Σ_f to be the set of all sequences $t = (t_n) \in \Sigma$ satisfying the following:

1. $\sigma^n t = i(c_k)$ if $t_n = c_k, k \in \{0, \dots, l\}$.
2. $\sigma \nu_k \prec \sigma^{n+1} t \prec \sigma \nu_{k+1}$ if $t_n = I_{k+1}$ and $\epsilon(I_{k+1}) = +1$.
3. $\sigma \nu_k \succ \sigma^{n+1} t \succ \sigma \nu_{k+1}$ if $t_n = I_{k+1}$ and $\epsilon(I_{k+1}) = -1$.

Similarly, we define $\hat{\Sigma}_f$ to be the set of sequences in Σ satisfying the conditions (1)–(3) with $\prec$ replaced by $\preceq$.

THEOREM 7.4.7. *Let $f \colon I \to I$ be an l-modal map with kneading invariants $\nu_1, \dots, \nu_l$, and let ν_0, ν_{l+1} be the itineraries of the endpoints. Then $i(I) \subset \hat{\Sigma}_f$. Moreover, if f has no wandering intervals, attracting periodic points or intervals of periodic points, then $i(I) = \Sigma_f$, and $i \colon I \to \Sigma_f$ is an order-preserving bijection.*

Proof. Lemma 7.4.5 implies that $i(I) \subset \hat{\Sigma}_f$, and $i(I) \subset \Sigma_f$ if there are no wandering intervals, attracting periodic points or intervals of periodic points.

Suppose f has no wandering intervals, attracting periodic points or intervals of periodic points. Let $t = (t_n) \in \Sigma_f$, and suppose $t \notin i(f)$. Then

$$L_t = \{x \in I : i(x) \prec t\}, \qquad R_t = \{x \in I : i(x) \succ t\}$$

are disjoint intervals, and $I = L_t \cup R_t$.

We claim that L_t and R_t are non-empty. The proof of this claim breaks into four cases according to the four possibilities for $f|\partial I$. We prove it in the case $f(c_0) = f(c_{l+1}) = c_0$. Then $\nu_0 = i(c_0^+) = (I_1, I_1, \ldots)$, $\nu_{l+1} = i(c_{l+1}^-) = (I_{l+1}, I_1, I_1, \ldots)$, $\epsilon(I_1) = 1$ and $\epsilon(I_{l+1}) = -1$. Note that $t \neq i(c_0) = \nu_0$ and $t \neq i(c_{l+1}) = \nu_{l+1}$ since $t \notin i(f)$. Thus $\nu_0 \prec t$, so $c_0 \in L_t$. If $t_0 < I_{l+1}$, then $t \prec \nu_{l+1}$ so $c_{l+1} \in R_t$ and we are done. So suppose $t_0 = I_{l+1}$. If $t_1 > I_1$, then $t \prec \nu_{l+1}$, and again we are done. If $t_1 = I_1$, then condition (2) implies that $\sigma \nu_0 \prec \sigma^2 t$, which implies in turn that $t \prec \nu_{l+1}$. Thus $\nu_{l+1} \in R_t$.

Let $a = \sup L_t$. We will show that $a \notin L_t$. Suppose for a contradiction that $a \in L_t$. Since $x \notin L_t$ for all $x > a$, we conclude that $i(a) \prec t \preceq i(a^+)$. This implies that the orbit of a contains a turning point. Let $n \geqslant 0$ be the smallest integer such that $i_n(a) = c_k$ for some $k \in \{1, \ldots, l\}$. Then $i_j(a) = t_j = i_j(a^+)$ for $j = 1, \ldots, n-1$, and $i_n(a^+) = I_k$ or $i_n(a^+) = I_{k+1}$. Suppose the latter holds. Then f^n is increasing on a neighborhood of a. Since $i(a) \prec t \preceq i(a^+)$ and $i_j(a) = t_j = i_j(a^+)$ for $j = 0, \ldots, n-1$, it follows that

$$i(c_k) = \sigma^n(i(a)) \prec \sigma^n(t) \prec \sigma^n(i(a^+)) = \nu_k,$$

and $c_k \leqslant t_n \leqslant I_{k+1}$.

If $t_n = c_k$, then by (1), $\sigma^n(t) = i(c_k)$, so $t = i(a)$, contradicting the fact that $t \notin i(f)$. Thus we may assume that $t_n = I_{k+1}$. If f is increasing on I_{k+1}, then condition (2) implies that $\sigma^{n+1}(t) \succ \sigma \nu_k$. But $\sigma^n(t) \preceq \sigma^n(i(a^+))$, $\tau_n(t) = +1$ and $t_n = i_n(a^+)$ imply that

$$\sigma^{n+1}(t) \preceq \sigma^{n+1}(i(a^+) = \sigma(\nu_k).$$

Similarly, if f is decreasing on I_{k+1}, then condition (3) implies that $\sigma^{n+1} \prec \sigma \nu_k$, which contradicts $\sigma^n(t) \preceq \sigma^n(i(a^+))$, $\tau_{n+1}(t) = -1$ and $t_n = i_n(a^+)$.

We have shown that the case $i_n(a^+) = I_{k+1}$ leads to a contradiction. Similarly, the case $i_n(a^+) = I_k$ leads to a contradiction. Thus $a \notin L_t$. By similar arguments, $\inf R_t \notin R_t$, which contradicts the fact that I is the disjoint union of L_t and R_t. Thus $t \in i(I)$, so $i(I) = \Sigma_f$.

Lemma 7.4.2 now implies that $i: I \to \Sigma_f$ is an order-preserving bijection. □

COROLLARY 7.4.8. *Let f and g be l-modal maps of I with no wandering intervals, no attracting periodic points and no intervals of periodic points. If f*

and g have the same kneading invariants and endpoint itineraries, then f and g are topologically conjugate.

Proof. Let i_f and i_g be the itinerary maps of f and g, respectively. Then $i_f^{-1} \circ i_g \colon I \to \Sigma(\nu_0, \nu_1, \ldots, \nu_{l+1}) \to I$ is an order-preserving bijection, and therefore a homeomorphism, which conjugates f and g. □

REMARK 7.4.9. *One can show that the following extension of Corollary 7.4.8 is also true. Let f and g be l-modal maps of I, and suppose f has no wandering intervals, no attracting periodic points and no intervals of periodic points. If f and g have the same kneading invariants and endpoint itineraries, then f and g are topologically semiconjugate.*

Example. Consider the unimodal quadratic map $f \colon [-1,1] \to [-1,1]$, $f(x) = -2x^2 + 1$. This map is conjugate to the quadratic map $q_4 \colon [0,1] \to [0,1]$, $q_4(x) = 4x(1-x)$, via the homeomorphism $h \colon [-1,1] \to [0,1]$, $h(x) = (1/2)(x+1)$. The orbit of the turning point $c = 0$ of f is $0, 1, -1, -1, \ldots$, so the kneading invariant is $\nu = (I_2, I_2, I_1, I_1, \ldots)$.

Now let $I = [-1,1]$ and consider the *tent map* $T \colon I \to I$ defined by

$$T(x) = \begin{cases} 2x+1 & x \leqslant 0 \\ -2x+1 & x > 0 \end{cases}.$$

The homeomorphism $\phi \colon I \to I, \phi(x) = (2/\pi)\sin^{-1}(x)$, conjugates f to T.

For any $n > 0$, the map f^{n+1} maps each of the intervals $[k/2^n, (k+1)/2^n]$, $k = -2^n, \ldots, 2^n$, homeomorphically onto I. Thus the forward iterates of any open set cover I, or equivalently, the backward orbit of any point in I is dense in I. It follows from the next lemma that T has no wandering intervals, attracting periodic points, or intervals of periodic points, so any unimodal map with the same kneading invariants as T is semiconjugate to T. In particular, any unimodal map $g \colon [a,b] \to [a,b]$ with $g(a) = g(b) = a$ and $g(c) = b$ is semiconjugate to T.

LEMMA 7.4.10. *Let $I = [a,b]$ be an interval, and $f \colon I \to I$ a continuous map with $f(\partial I) \subset \partial I$. Suppose that every backward orbit is dense in I and that f has a fixed point x_0 not in ∂I. Then f has no wandering intervals, no intervals of periodic points and no attracting periodic points.*

Proof. Let $U \subset I$ be an open interval. Fix $x \in U$. By density of $\cup f^{-n}(x)$, there is $n > 0$ such that $f^{-n}(x) \cap U \neq \emptyset$. Then $f^n(U) \cap U \neq \emptyset$ so U is not a wandering interval.

Suppose $z \in I$ is an attracting periodic point. Then the basin of attraction $BA(z)$ is a forward-invariant set with non-empty interior. Since backward orbits are dense, $BA(z)$ is a dense open subset of I and therefore intersects the backward orbit of x_0. Thus $z = x_0$. On the other hand, the backward orbits of a and b are dense, and therefore intersect $BA(z)$, which is a contradiction. Thus there can be no attracting periodic point.

Any point in $\mathrm{Per}(f)$ has finitely many preimages in $\mathrm{Per}(f)$, so if $\mathrm{Per}(f)$ had non-empty interior, the backward orbit of a point in $\mathrm{Per}(f)$ would not be dense in $\mathrm{Per}(f)$. Thus f has no intervals of periodic points. □

The final result of this section is a realization theorem, which asserts that any "admissible" set of sequences in Σ is the set of kneading invariants of an l-modal map.

Note that for an l-modal map f, the endpoint itineraries are determined completely by the orientation of f on the first and last laps of f. Thus, given l and a function ϵ as in the definition of signed lexicographic orderings, we can define natural "endpoint itineraries" ν_0 and ν_{l+1} as sequences in the symbol space $\{I_1, I_{l+1}\}$.

THEOREM 7.4.11. *Let $\nu_1, \dots, \nu_l \in \{I_1, \dots, I_{l+1}\}^{\mathbb{N}_0}$, and $\epsilon(I_j) = \epsilon_0(-1)^j$, where $\epsilon_0 = \pm 1$. Let $\prec$ be the signed lexicographic ordering on $\Sigma = \{I_1, \dots, I_{l+1}, c_1, \dots, c_l\}^{\mathbb{N}_0}$ associated to ϵ. Let ν_0, ν_{l+1} be the "endpoint itineraries" determined uniquely by ϵ and l. If $\{\nu_0, \dots, \nu_{l+1}\}$ satisfies the admissibility criterion of Corollary 7.4.6, then there is a continuous l-modal map $f\colon [0,1] \to [0,1]$ with kneading invariants $\nu_1, \dots, \nu_{l+1}$.*

Proof. Define an equivalence relation $\sim$ on Σ by the rule $t \sim s$ if and only if $t = s$, or $\sigma(t) = \sigma(s)$ and $t_0 = I_k, s_0 = I_{k\pm1}$. To paraphrase: t and s are equivalent if and only if they differ at most in the first position, and then only if the first positions are adjacent intervals. (Thus, for example, $i(c_k^-) \sim i(c_k^+)$ for a turning point of an l-modal map.)

We will define a sequence of l-modal maps $f_N, N \in \mathbb{N}_0$, whose kneading invariants agree up to order N with $\nu_1, \dots, \nu_l$. The desired map f will be the limit in the C^0 topology of these maps.

Let $p_j^0 = c_j$, $j = 0, \dots, l+1$. Choose points $p_j^1 \in [0,1]$, $j = 0, \dots, l+1$, such that:

1. if $\sigma^m(\nu_i) \sim \sigma^n(\nu_j)$ then $p_i^m = p_j^n$;
2. $p_i^m < p_j^n$ if and only if $\sigma^m(\nu_i) \prec \sigma^n(\nu_j)$ and $\sigma^m(\nu_i) \not\sim \sigma^n(\nu_j)$; and
3. the new points are equidistributed in each of the intervals $[p_j^0, p_{j+1}^0]$, $j = 0, \dots, l+1$.

Define $f_1\colon [0,1] \to [0,1]$ to be the piecewise-linear map specified by $f(p_j^0) = p_j^1$. Note that $p_j^1 < p_{j+1}^1$ if and only if $\sigma\nu_j < \sigma\nu_{j+1}$, which happens if and only if $\epsilon(I_{j+1}) = +1$. Thus f_1 is l-modal.

For $N > 0$ we define inductively points $p_j^N \in [0,1]$, $j = 0, \ldots, l+1$ satisfying conditions (1) and (2) for all $n, m \leqslant N$ and $j = 0, \ldots, l+1$, and so that in any subinterval defined by the points $\{p_j^n : 0 < n < N,\ 0 \leqslant j \leqslant l+1\}$, the new points $\{p_j^N\}$ in that interval are equidistributed. Then we define the map $f_N\colon I \to I$ to be the piecewise-linear map connecting the points (p_j^n, p_j^{n+1}), $j = 0, \ldots, l+1, n = 0, \ldots, N-1$. It follows (Exercise 7.4.5) that:

1. f_N is l-modal for each $N > 0$;
2. $\{f_N\}$ converges in the C^0 topology to an l-modal map f with turning points $c_1, \ldots, c_l$; and
3. the kneading invariants of f are $\nu_1, \ldots, \nu_l$. □

Exercise 7.4.1. Finish the proof of Lemma 7.4.1.

Exercise 7.4.2. Let L be an interval and $f\colon L \to L$ a strictly monotone map. Show that either L contains an interval of periodic points, or some open interval in L converges to a single periodic point.

Exercise 7.4.3. Work out the ordering on the set of itineraries of the quadratic map q_μ for $2 < \mu < 3$.

Exercise 7.4.4. Show that the tent map has exactly 2^n periodic points of period n, and the set of periodic points is dense in $[-1, 1]$.

Exercise 7.4.5. Verify the last three assertions in the proof of Theorem 7.4.11.

7.5 The Schwarzian derivative

Let f be a C^3 function defined on an interval $I \subset \mathbb{R}$. If $f'(x) \neq 0$, we define the *Schwarzian derivative* of f at x to be

$$Sf(x) = \frac{f'''(x)}{f'(x)} - \frac{3}{2}\left(\frac{f''(x)}{f'(x)}\right)^2.$$

If x is an isolated critical point of f, we define $Sf(x) = \lim_{y\to x} Sf(y)$ if the limit exists.

For the quadratic map $q_\mu(x) = \mu x(1-x)$, we have $Sq_\mu(x) = -6/(1-2x)^2$ for $x \neq 1/2$, and $Sf(1/2) = -\infty$. We also have $S\exp(x) = -1/2$ and $S\log(x) = 1/2x^2$.

LEMMA 7.5.1. *The Schwarzian derivative has the following properties:*

1. $S(f \circ g) = (Sf \circ g)(g')^2 + Sg$;
2. $S(f^n) = \sum_{i=0}^{n-1} Sf(f^i(x)) \cdot ((f^i)'(x))^2$;
3. *If* $Sf < 0$, *then* $S(f^n) < 0$ *for all* $n > 0$.

The proof is left as an exercise (Exercise 7.5.3).

A function with negative Schwarzian derivative satisfies the following *Minimum Principle.*

LEMMA 7.5.2 (Minimum Principle). *Let* I *be an interval and* $f\colon I \to I$ *a* C^3 *map with* $f'(x) \neq 0$ *for all* $x \in I$. *If* $Sf < 0$, *then* $|f'(x)|$ *does not attain a local minimum in the interior of* I.

Proof. Let z be a critical point of f'. Then $f''(z) = 0$, which implies that $f'''(z)/f'(z) < 0$ since $Sf < 0$. Thus $f'''(z)$ and $f'(z)$ have opposite signs. If $f'(z) < 0$, then $f'''(z) > 0$ and z is a local minimum of f', so z is a local maximum of $|f'|$. Similarly, if $f'(z) > 0$, z is also a local maximum of $|f'|$. Since f' is never zero on I, this implies that $|f'|$ does not have a local minimum on I. □

THEOREM 7.5.3 (Singer, 1978). *Let* I *be a closed interval (possibly unbounded) and* $f\colon I \to I$ *a* C^3 *map with negative Schwarzian derivative. If* f *has* n *critical points, then* f *has at most* $n+2$ *attracting periodic orbits.*

Proof. Let z be an attracting periodic point of period m. Let $W(z)$ be the maximal interval about z such that $f^{mn}(y) \to z$ as $n \to \infty$ for all $y \in U$. Then $W(z)$ is open (in I), and $f^m(W(z)) \subset W(z)$.

Suppose that $W(z)$ is bounded and does not contain a point in ∂I, so $W(z) = (a, b)$, for some $a < b \in \mathbb{R}$. We claim that f^m has a critical point in $W(z)$. By maximality of $W(z)$, f^m must preserve the set of endpoints of $W(z)$. If $f^m(a) = f^m(b)$, then f^m must have a maximum or minimum in $W(z)$, and therefore a critical point in $W(z)$. If $f^m(a) \neq f^m(b)$, then f^m must permute a and b. Suppose $f^m(a) = a$ and $f^m(b) = b$. Then $(f^m)' \geqslant 1$ on ∂U, since otherwise a or b would be an attracting fixed point for f^m whose basin of attraction overlaps U. By the Minimum Principle, if f^m has no critical points in U, then $(f^m)' > 1$ on U, which contradicts $f^m(W(z)) = W(z)$, so f^m has a critical point in $W(z)$. If $f^m(a) = b$ and $f^m(b) = a$, then applying the preceding argument to f^{2m}, we conclude that f^{2m} has a critical point in $W(z)$. Since $f^m(W(z)) = W(z)$ it follows that f^m also has a critical point in $W(z)$.

By the chain rule, if $p \in W(z)$ is a critical point of f^m, then one of the points $p, f(p), \dots, f^{m-1}(p)$ is a critical point of f. Thus we have shown that

$W(z)$ is either unbounded, or meets ∂I, or there is a critical point of f whose orbit meets $W(z)$. Since there are only n critical points, and only two boundary points (or unbounded ends) of I, the theorem is proved. □

COROLLARY 7.5.4. *For any $\mu > 4$, the quadratic map $q_\mu\colon \mathbb{R} \to \mathbb{R}$ has at most one (finite) attracting periodic orbit.*

Proof. The proof of Theorem 7.5.3 shows that if z is an attracting periodic point, then $W(z)$ is either unbounded or contains the critical point of q_μ. Since ∞ is an attracting periodic point, the basin of attraction of z must be bounded, and therefore must contain the critical point. □

We now discuss a relation between the Schwarzian derivative and length distortion that is used in producing absolutely continuous invariant measures for maps of the interval with negative Schwarzian derivative[2].

Let f be a piecewise-monotone real-valued function defined on a bounded interval I. Suppose $J \subset I$ is a subinterval such that $I \setminus J$ consists of disjoint non-empty intervals L and R. Denote by $|F|$ the length of an interval F. Define the *cross-ratios*:

$$C(I,J) = \frac{|I|\cdot|J|}{|J\cup L|\cdot|J\cup R|}, \qquad D(I,J) = \frac{|I|\cdot|J|}{|L|\cdot|R|}.$$

If f is monotone on I, set

$$A(I,J) = \frac{C\big(f(I), f(J)\big)}{C(I,J)}, \qquad B(I,J) = \frac{D\big(f(I), f(J)\big)}{D(I,J)}.$$

The group $\mathcal{M}$ of real *Möbius transformations* consists of maps of the extended real line $\mathbb{R}\cup\{\infty\}$ of the form $\phi(x) = (ax+b)/(cx+d)$, where $a, b, c, d \in \mathbb{R}$ and $ad - bc \neq 0$. Möbius transformations have Schwarzian derivative equal to 0 and preserve the cross-ratios C and D (Exercise 7.5.4). The group of Möbius transformations is *simply transitive* on triples of points in the extended real line, i.e. given any three distinct points $a, b, c \in \mathbb{R}\cup\{\infty\}$, there is a unique Möbius transformation $\phi \in \mathcal{M}$ such that $\phi(0) = a, \phi(1) = b$ and $\phi(\infty) = c$ (Exercise 7.5.5). Möbius transformations are also called *linear fractional transformations*.

PROPOSITION 7.5.5. *Let f be a C^3 real-valued function defined on a compact interval I such that f has negative Schwarzian derivative and $f'(x) \neq 0$, $x \in I$. Let $J \subset I$ be a closed subinterval that does not contain the endpoints of I. Then $A(I,J) > 1$ and $B(I,J) > 1$.*

[2] Our exposition here follows to a large extent van Strien (1988) and de Melo and van Strien (1993).

Proof. Since every Möbius transformation has Schwarzian derivative 0 and preserves C and D, we may assume, by composing f on the left and on the right with appropriate Möbius transformations and using Lemma 7.5.1, that $I = [0,1]$, $J = [a,b]$ with $0 < a < b < 1$, $f(0) = 0$, $f(a) = a$, and $f(1) = 1$. By Lemma 7.5.2, $|f'|$ does not have a local minimum in $[0,1]$, and hence f cannot have fixed points except 0, a and 1. Therefore $f(x) < x$ if $0 < x < a$ and $f(x) > x$ if $a < x < 1$; in particular, $f(b) > b$. We have:

$$B(I,J) = \frac{|f(1) - f(0)| \cdot |f(b) - f(a)|}{|f(a) - f(0)| \cdot |f(1) - f(b)|} \cdot \left(\frac{|1-0| \cdot |b-a|}{|a-0| \cdot |1-b|} \right)^{-1}$$

$$= \frac{1 \cdot (f(b) - a) \cdot a \cdot (1-b)}{a \cdot (1 - f(b)) \cdot 1 \cdot (b-a)} > 1.$$

This proves the second inequality. The first one is left as an exercise (Exercise 7.5.6). □

The following proposition, which we do not prove, describes bounded distortion properties of maps with negative Schwarzian derivative on intervals without critical points.

PROPOSITION 7.5.6 (van Strien, 1988; de Melo and van Strien, 1993). *Let $f\colon [a,b] \to \mathbb{R}$ be a C^3 map. Assume that $Sf < 0$ and $f'(x) \neq 0$, for all $x \in [a,b]$. Then*

1. $|f'(a)| \cdot |f'(b)| \geqslant \big(|f(b) - f(a)|/(b-a)\big)^2$;
2. $\dfrac{|f'(x)| \cdot |f(b) - f(a)|}{b-a} \geqslant \dfrac{|f(x) - f(a)|}{x-a} \cdot \dfrac{|f(b) - f(x)|}{b-x}$ *for every* $x \in (a,b)$.

Exercise 7.5.1. Prove that if $f\colon I \to \mathbb{R}$ is a C^3 diffeomorphism onto its image and $g(x) = \frac{d}{dx} \log |f'(x)|$, then

$$Sf(x) = g'(x) - \frac{1}{2}(g(x))^2 = -2\sqrt{|f'(x)|} \cdot \frac{d^2}{dx^2} \frac{1}{\sqrt{|f'(x)|}}.$$

Exercise 7.5.2. Show that any polynomial with distinct real roots has negative Schwarzian derivative.

Exercise 7.5.3. Prove Lemma 7.5.1.

Exercise 7.5.4. Prove that each Möbius transformation has Schwarzian derivative 0 and preserves the cross-ratios C and D.

Exercise 7.5.5. Prove that the action of the group of Möbius transformations on the extended real line is simply transitive on triples of points.

Exercise 7.5.6. Prove the remaining inequality of Proposition 7.5.5.

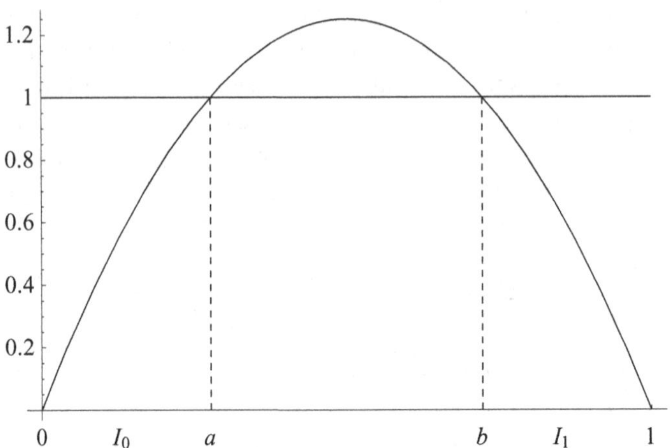

Figure 7.5. Quadratic map.

7.6 Real quadratic maps

In §1.5, we introduced the one-parameter family of real quadratic maps $q_\mu(x) = \mu x(1-x)$, $\mu \in \mathbb{R}$. We showed that for $\mu > 1$, the orbit of any point outside $I = [0,1]$ converges monotonically to $-\infty$. Thus the interesting dynamics is concentrated on the set

$$\Lambda_\mu = \{x \in I \mid q_\mu^n(x) \in I, \ \forall n \geqslant 0\}.$$

THEOREM 7.6.1. *Let $\mu > 4$. Then Λ_μ is a Cantor set, i.e. a perfect, nowhere-dense subset of* $[0,1]$. *The restriction $q_\mu|_{\Lambda_\mu}$ is topologically conjugate to the one-sided shift $\sigma : \Sigma_2^+ \to \Sigma_2^+$.*

Proof. Let $a = 1/2 - \sqrt{1/4 - 1/\mu}$ and $b = 1/2 + \sqrt{1/4 - 1/\mu}$ be the two solutions of $q_\mu(x) = 1$, and let $I_0 = [0,a]$, $I_1 = [b,1]$. Then $q_\mu(I_0) = q_\mu(I_1) = I$, and $q_\mu((a,b)) \cap I = \emptyset$ (see Figure 7.5). Observe that the images $q_\mu^n(1/2)$ of the critical point $1/2$ lie outside I and tend to $-\infty$. Therefore the two inverse branches $f_0 : I \to I_0$ and $f_1 : I \to I_1$ and their compositions are well-defined. For $k \in \mathbb{N}$, denote by W_k the set of all words of length k in the alphabet $\{0,1\}$. For $w = \omega_1\omega_2\cdots\omega_k \in W_k$ and $j \in \{0,1\}$, set $I_{wj} = f_j(I_w)$, and $g_w = f_{\omega_k} \circ \cdots \circ f_{\omega_2} \circ f_{\omega_1}$, so that $I_w = g_w(I)$.

LEMMA 7.6.2. $\lim_{k\to\infty} \max_{w\in W_k} \max_{x\in I} |g_w'(x)| = 0.$

Proof. If $\mu > 2 + \sqrt{5}$, then $1 > |f_j'(1)| = \mu\sqrt{1 - 4/\mu} \geqslant |f_j'(x)|$ for every $x \in I$, $j = 0, 1$, and the lemma follows.

For $4 < \mu < 2 + \sqrt{5}$, the lemma follows from Theorem 8.5.10 (see also Theorem 8.5.11). □

Lemma 7.6.2 implies that the length of the interval I_w tends to 0 as the length of w tends to infinity. Therefore, for each $\omega = \omega_1\omega_2\cdots \in \Sigma_2^+$ the intersection $\bigcap_{n\in\mathbb{N}} I_{\omega_1\cdots\omega_n}$ consists of exactly one point $h(\omega)$. The map $h\colon \Sigma_2^+ \to \Lambda_\mu$ is a homeomorphism conjugating the shift σ and $q_\mu|_{\Lambda_\mu}$ (Exercise 7.6.2). □

Exercise 7.6.1. Prove that if $\mu > 4$ and $1/2 - \sqrt{1/4 - 1/\mu} < x < 1/2 + \sqrt{1/4 - 1/\mu}$, then $q_\mu^n(x) \to -\infty$ as $n \to \infty$.

Exercise 7.6.2. Prove that the map $h\colon \Sigma_2^+ \to \Lambda_\mu$ in the proof of Theorem 7.6.1 is a homeomorphism and that $q_\mu \circ h = h \circ \sigma$.

7.7 Bifurcations of periodic points[3]

The family of real quadratic maps $q_\mu(x) = \mu x(1 - x)$ (§§1.5, 7.6) is an example of a (one-dimensional) parametrized family of dynamical systems. Although the specific quantitative behavior of a dynamical system depends on the parameter, it is often the case that the qualitative behavior remains unchanged for certain ranges of the parameter. A parameter value where the qualitative behavior changes is called a *bifurcation* value of the parameter. For example, in the family of quadratic maps, the parameter value $\mu = 3$ is a bifurcation value because the stability of the fixed point $1 - 1/\mu$ changes from repelling to attracting. The parameter value $\mu = 1$ is a bifurcation value because for $\mu < 1$, 0 is the only fixed point, and for $\mu > 1$, q_μ has two fixed points.

A bifurcation is called *generic* if the same bifurcation occurs for all nearby families of dynamical systems, where "nearby" is defined with respect to an appropriate topology (usually the C^2 or C^3 topology). For example, the bifurcation value $\mu = 3$ is generic for the family of quadratic maps. To see this, note that for μ close to 3, the graph of q_μ crosses the diagonal transversely at the fixed point $x_\mu = 1 - 1/\mu$, and the magnitude of $q'_\mu(x_\mu)$ is less than 1 for $\mu < 3$ and greater than 1 for $\mu > 3$. If f_μ is another family of maps C^1-close to q_μ, then the graph of $f_\mu(x)$ must also cross the diagonal at a point y_μ near x_μ, and the magnitude of $f'_\mu(y_\mu)$ must cross 1 at some parameter value close to 3. Thus f_μ has the same kind of bifurcation as q_μ. Similar reasoning shows that the bifurcation value $\mu = 1$ is also generic.

[3] The exposition in this section follows to a certain extent that of Robinson (1995).

Generic bifurcations are the primary ones of interest. The notion of genericity depends on the dimension of the parameter space (i.e. a bifurcation may be generic for a one-parameter family, but not for a two-parameter family). Bifurcations that are generic for one-parameter families of dynamical systems are called *codimension-one* bifurcations. In this section we describe codimension-one bifurcations of fixed and periodic points for one-dimensional maps.

We begin with a non-bifurcation result. If the graph of a differentiable map f intersects the diagonal transversely at a point x_0, then the fixed point x_0 persists under a small C^1 perturbation of f.

PROPOSITION 7.7.1. *Let $U \subset \mathbb{R}^m$ and $V \subset \mathbb{R}^n$ be open subsets and let $f_\mu : U \to \mathbb{R}^m$, $\mu \in V$, be a family of C^1 maps such that:*

1. *the map $(x, \mu) \mapsto f_\mu(x)$ is a C^1 map;*
2. *$f_{\mu_0}(x_0) = x_0$ for some $x_0 \in U$ and $\mu_0 \in V$;*
3. *1 is not an eigenvalue of $df_{\mu_0}(x_0)$.*

Then there are open sets $U' \subset U, V' \subset V$ with $x_0 \in U', \mu_0 \in V'$ and a C^1 function $\xi : V' \to U'$ such that for each $\mu \in V'$, $\xi(\mu)$ is the only fixed point of f_μ in U'.

Proof. The proposition is an immediate consequence of the Implicit Function Theorem applied to the map $(x, \mu) \mapsto f_\mu(x) - x$ (Exercise 7.7.1). □

Proposition 7.7.1 shows that if 1 is not an eigenvalue of the derivative, then the fixed point does not bifurcate into multiple fixed points and does not disappear. The next proposition shows that periodic points cannot appear in a neighborhood of a hyperbolic fixed point.

PROPOSITION 7.7.2. *Under the assumption (and notation) of Proposition 7.7.1, suppose in addition that x_0 is a hyperbolic fixed point of f_{μ_0}, i.e. no eigenvalue of $df_{\mu_0}(x_0)$ has absolute value 1. Then for each $k \in \mathbb{N}$ there are neighborhoods $U_k \subset U'$ of x_0 and $V_k \subset V'$ of μ_0 such that $\xi(\mu)$ is the only fixed point of f_μ^k in U_k.*

If, in addition, x_0 is an attracting fixed point of f_{μ_0}, i.e. all eigenvalues of $df_{\mu_0}(x_0)$ are strictly less than 1 in absolute value, then the neighborhoods U_k and V_k can be chosen independent of k.

Proof. Since no eigenvalue of $df_{\mu_0}(x_0)$ has absolute value 1, it follows that 1 is not an eigenvalue of $df_{\mu_0}^k(x_0)$, so the first statement follows from Proposition 7.7.1.

The second statement is left as an exercise (Exercise 7.7.2). □

Propositions 7.7.1 and 7.7.2 show that, for differentiable one-dimensional maps, bifurcations of fixed or periodic points can occur only if the absolute value of the derivative is 1. For one-dimensional maps there are only two types of generic bifurcations: the *saddle-node bifurcation* (or *fold bifurcation*) may occur if the derivative at a periodic point is 1, and the *period-doubling bifurcation* (or *flip bifurcation*) may occur if the derivative at a periodic point is -1. We describe these bifurcations in the next two propositions. See Chow and Hale (1982) or Hale and Koçak (1991) for a more extensive discussion of bifurcation theory, or Golubitsky and Guillemin (1973) for a thorough exposition of the closely related topic of singularities of differentiable maps.

PROPOSITION 7.7.3 (Saddle-node bifurcation). *Let $I, J \subset \mathbb{R}$ be open intervals and $f: I \times J \to \mathbb{R}$ be a C^2 map such that*

1. $f(x_0, \mu_0) = x_0$ *and* $\frac{\partial f}{\partial x}(x_0, \mu_0) = 1$ *for some* $x_0 \in I$ *and* $\mu_0 \in J$;
2. $\frac{\partial^2 f}{\partial x^2}(x_0, \mu_0) < 0$ *and* $\frac{\partial f}{\partial \mu}(x_0, \mu_0) > 0$.

Then there are $\epsilon, \delta > 0$ and a C^2 function $\alpha: (x_0 - \epsilon, x_0 + \epsilon) \to (\mu_0 - \delta, \mu_0 + \delta)$ such that:

1. $\alpha(x_0) = \mu_0, \alpha'(x_0) = 0, \alpha''(x_0) = -\frac{\partial^2 f}{\partial x^2}(x_0, \mu_0) \Big/ \frac{\partial f}{\partial \mu}(x_0, \mu_0) > 0.$
2. *Each $x \in (x_0 - \epsilon, x_0 + \epsilon)$ is a fixed point of $f(\cdot, \alpha(x))$, i.e. $f(x, \alpha(x)) = x$, and $\alpha^{-1}(\mu)$ is exactly the fixed point set of $f(\cdot, \mu)$ in $(x_0 - \epsilon, x_0 + \epsilon)$ for $\mu \in (\mu_0 - \delta, \mu_0 + \delta)$.*
3. *For each $\mu \in (\mu_0, \mu_0 + \delta)$, there are exactly two fixed points $x_1(\mu) < x_2(\mu)$ of $f(\cdot, \mu)$ in $(x_0 - \epsilon, x_0 + \epsilon)$ with $\frac{\partial f}{\partial x}(x_1(\mu), \mu) > 1$ and $0 < \frac{\partial f}{\partial x}(x_2(\mu), \mu) < 1$; $\alpha(x_i(\mu)) = \mu$ for $i = 1, 2$.*
4. *$f(\cdot, \mu)$ does not have fixed points in $(x_0 - \epsilon, x_0 + \epsilon)$ for $\mu \in (\mu_0 - \delta, \mu_0)$.*

REMARK 7.7.4. *The inequalities in the second hypothesis of Proposition 7.7.3 correspond to one of the four possible generic cases when the two derivatives do not vanish. The other three cases are similar (Exercise 7.7.3).*

Proof. Consider the function $g(x, \mu) = f(x, \mu) - x$ (see Figure 7.6). Observe that

$$\frac{\partial g}{\partial \mu}(x_0, \mu_0) = \frac{\partial f}{\partial \mu}(x_0, \mu_0) > 0.$$

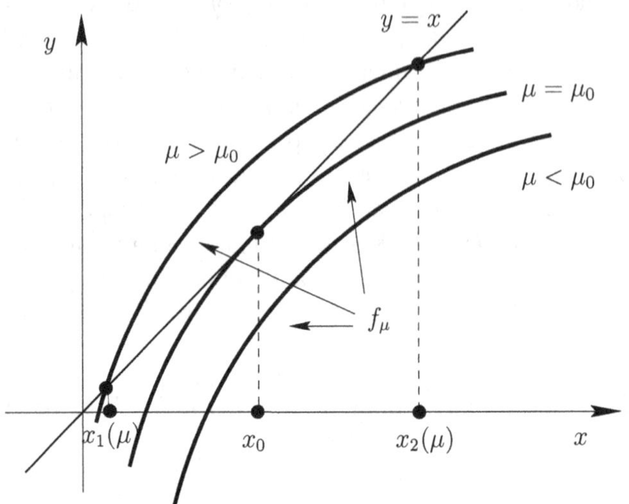

Figure 7.6. Saddle-node bifurcation.

Therefore, by the Implicit Function Theorem, there are $\epsilon, \delta > 0$ and a C^2 function $\alpha\colon (x_0 - \epsilon, x_0 + \epsilon) \to J$ such that $g(x, \alpha(x)) = 0$ for each $x \in (x_0 - \epsilon, x_0 + \epsilon)$ and there are no other zeros of g in $(x_0 - \epsilon, x_0 + \epsilon) \times (\mu_0 - \epsilon, \mu_0 + \epsilon)$. A direct calculation shows that α satisfies statement (1). Since $\alpha''(x_0) > 0$, statements (3) and (4) are satisfied for ϵ and δ sufficiently small (Exercise 7.7.4). □

PROPOSITION 7.7.5 (Period-doubling bifurcation). *Let I, $J \subset \mathbb{R}$ be open intervals and $f\colon I \times J \to \mathbb{R}$ be a C^3 map such that:*

1. *$f(x_0, \mu_0) = x_0$ and $\frac{\partial f}{\partial x}(x_0, \mu_0) = -1$ for some $x_0 \in I$ and $\mu_0 \in J$, so that by Proposition 7.7.1, there is a curve $\mu \mapsto \xi(\mu)$ of fixed points of $f(\cdot, \mu)$ for μ close to μ_0.*
2. $\eta = \left.\dfrac{d}{d\mu}\right|_{\mu=\mu_0} \dfrac{\partial f}{\partial x}(\xi(\mu), \mu) < 0.$
3. $\zeta = \dfrac{\partial^3 f(f(x_0, \mu_0), \mu_0)}{\partial x^3} = -2\dfrac{\partial^3 f}{\partial x^3}(x_0, \mu_0) - 3\left(\dfrac{\partial^2 f}{\partial x^2}(x_0, \mu_0)\right)^2 < 0.$

Then there are ϵ, $\delta > 0$ and C^3 functions $\xi\colon (\mu_0 - \delta, \mu_0 + \delta) \to \mathbb{R}$ with $\xi(\mu_0) = x_0$ and $\alpha\colon (x_0 - \epsilon, x_0 + \epsilon) \to \mathbb{R}$ with $\alpha(x_0) = \mu_0$, $\alpha'(x_0) = 0$ and $\alpha''(x_0) = -2\eta/\zeta > 0$ such that:

1. *$f(\xi(\mu), \mu) = \xi(\mu)$ and $\xi(\mu)$ is the only fixed point of $f(\cdot, \mu)$ in $(x_0 - \epsilon, x_0 + \epsilon)$ for $\mu \in (\mu_0 - \delta, \mu_0 + \delta)$;*

2. *$\xi(\mu)$ is an attracting fixed point of $f(\cdot,\mu)$ for $\mu_0-\delta<\mu<\mu_0$ and is a repelling fixed point for $\mu_0<\mu<\mu_0+\delta$;*
3. *for each $\mu\in(\mu_0,\mu_0+\delta)$, the map $f(\cdot,\mu)$ has, in addition to the fixed point $\xi(\mu)$, exactly two attracting period-2 points $x_1(\mu), x_2(\mu)$ in $(x_0-\epsilon,x_0+\epsilon)$; moreover, $\alpha(x_i(\mu))=\mu$ and $x_i(\mu)\to x_0$ as $\mu\searrow\mu_0$, for $i=1,2$;*
4. *for each $\mu\in(\mu_0-\delta,\mu_0]$, the map $f\big(f(\cdot,\mu),\mu\big)$ has exactly one fixed point $\xi(\mu)$ in $(x_0-\epsilon,x_0+\epsilon)$.*

REMARK 7.7.6. *The stability of the fixed point $\xi(\mu)$ and of the periodic points $x_1(\mu)$ and $x_2(\mu)$ depend on the signs of the derivatives in the third and fourth hypotheses of Proposition 7.7.5. This proposition deals with only one of the four possible generic cases when the derivatives do not vanish. The other three cases are similar and we do not consider them here (Exercise 7.7.5).*

Proof. Since $\frac{\partial f}{\partial x}(x_0,\mu_0)=-1\neq 1$, we can apply the Implicit Function Theorem to $f(x,\mu)-x=0$ to obtain a differentiable function ξ such that $f(\xi(\mu),\mu)=\xi(\mu)$ for μ close to μ_0 and $\xi(\mu_0)=x_0$. This proves statement (1).

Differentiating $f(\xi(\mu),\mu)=\xi(\mu)$ with respect to μ gives

$$\frac{d}{d\mu}f(\xi(\mu),\mu)=\frac{\partial f}{\partial\mu}(\xi(\mu),\mu)+\frac{\partial f}{\partial x}(\xi(\mu),\mu)\cdot\xi'(\mu)=\xi'(\mu),$$

and hence

$$\xi'(\mu)=\frac{\frac{\partial f}{\partial\mu}(\xi(\mu),\mu)}{1-\frac{\partial f}{\partial x}(\xi(\mu),\mu)},\qquad \xi'(\mu_0)=\frac{1}{2}\frac{\partial f}{\partial\mu}(x_0,\mu_0).$$

Therefore

$$\frac{d}{d\mu}\bigg|_{\mu=\mu_0}\frac{\partial f}{\partial x}(\xi(\mu),\mu)=\frac{\partial^2 f}{\partial\mu\partial x}(x_0,\mu_0)+\frac{1}{2}\frac{\partial f}{\partial\mu}(x_0,\mu_0)\cdot\frac{\partial^2 f}{\partial x^2}(x_0,\mu_0)=\eta$$

and assumption (2) yields statement (2).

To prove statements (3) and (4) consider the change of variables $y=x-\xi(\mu), 0=x_0-\xi(\mu_0)$ and the function $g(y,\mu)=f(f(y+\xi(\mu),\mu),\mu)-\xi(\mu)$. Observe that fixed points of $f(f(\cdot,\mu),\mu)$ correspond to solutions of $g(y,\mu)=y$. Moreover,

$$g(0,\mu)\equiv 0,\qquad \frac{\partial g}{\partial y}(0,\mu_0)=1,\qquad \frac{\partial^2 g}{\partial y^2}(0,\mu_0)=0,$$

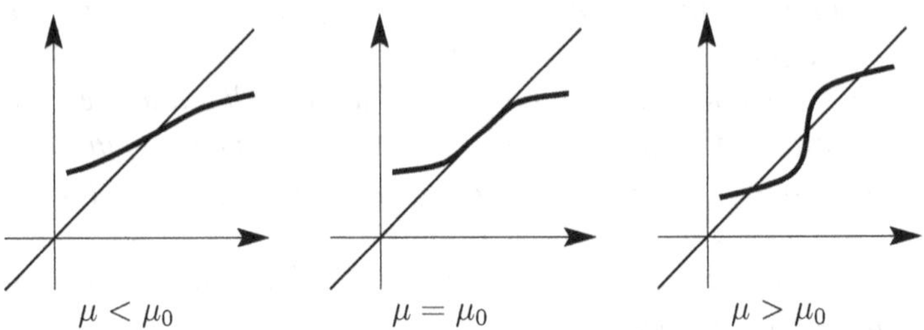

Figure 7.7. Period doubling bifurcation, the graph of the second iterate.

i.e. the graph of the second iterate of $f(\cdot, \mu_0)$ is tangent to the diagonal at (x_0, μ_0) with second derivative 0 (see Figure 7.7). A direct calculation shows that, by assumption (3), the third derivative does not vanish:

$$\frac{\partial^3 g}{\partial y^3}(0, \mu_0) = -2\,\frac{\partial^3 f}{\partial x^3}(x_0, \mu_0) - 3\left(\frac{\partial f^2}{\partial x^2}(x_0, \mu_0)\right)^2 = \zeta < 0.$$

Therefore

$$g(y, \mu_0) = y + \frac{1}{3!}\zeta y^3 + o(y^3). \tag{7.2}$$

Since $\xi(\mu)$ is a fixed point of $f(\cdot, \mu)$, we have that $g(0, \mu) \equiv 0$ in an interval about μ_0. Therefore there is a differentiable function h such that $g(y, \mu) = y \cdot h(y, \mu)$ and to find the period 2 points of $f(\cdot, \mu)$ different from $\xi(\mu)$ we must solve the equation $h(y, \mu) = 1$. From (7.2) we obtain

$$h(y, \mu_0) = 1 + \frac{1}{3!}\zeta y^2 + o(y^2),$$

i.e. $h(0, \mu_0) = 1$, $\dfrac{\partial h}{\partial y}(0, \mu_0) = 0$, and $\dfrac{\partial^2 h}{\partial y^2}(0, \mu_0) = \zeta/3$. On the other hand,

$$\begin{aligned}\frac{\partial h}{\partial \mu}(0, \mu_0) &= \lim_{y\to 0}\frac{1}{y}\frac{\partial g}{\partial \mu}(y, \mu_0) = \frac{\partial^2 g}{\partial \mu \partial y}(0, \mu_0)\\ &= \frac{d}{d\mu}\left(\frac{\partial f}{\partial x}(\xi(\mu), \mu)\right)^2\Big|_{\mu=\mu_0} = -2\eta > 0.\end{aligned}$$

By the Implicit Function Theorem, there is $\epsilon > 0$ and a differentiable function $\beta\colon (-\epsilon, \epsilon) \to \mathbb{R}$ such that $h(y, \beta(y)) = 1$ for $|y| < \epsilon$ and $\beta(0) = \mu_0$. Differentiating $h(y, \beta(y)) = 1$ with respect to y we obtain that $\beta'(0) = 0$. The second differentiation yields $\beta''(0) = \zeta/6\eta > 0$. Therefore $\beta(y) > 0$ for $y \neq 0$ and the new period 2 orbit appears only for $\mu > \mu_0$.

Note that since $g(\cdot, \mu)$ has three fixed points near x_0 for μ close to μ_0, and the middle one, $\xi(\mu)$, is unstable, the other two must be stable. In fact, a direct calculation shows that

$$\frac{\partial g}{\partial y}(y, \beta(y)) = \frac{\partial g}{\partial y}(0, \mu_0) + \frac{1}{2!}\frac{\partial^2 g}{\partial y^2}(0, \mu_0)y + \frac{1}{3!}\frac{\partial^3 g}{\partial y^3}(0, \mu_0)y^2 + o(y^2)$$
$$= 1 + \frac{\zeta}{6}y^2 + o(y^2).$$

Since $\zeta < 0$, the period 2 orbit is stable. □

Exercise 7.7.1. Prove Proposition 7.7.1.

Exercise 7.7.2. Prove the second statement of Proposition 7.7.2.

Exercise 7.7.3. State the analog of Proposition 7.7.3 for the remaining three generic cases when the derivatives from assumption (3) do not vanish.

Exercise 7.7.4. Prove statements (3) and (4) of Proposition 7.7.3.

Exercise 7.7.5. State the analog of Proposition 7.7.5 for the remaining three generic cases when the derivatives from assumptions (3) and (4) do not vanish.

Exercise 7.7.6. Prove that a period-doubling bifurcation occurs for the family $f_\mu(x) = 1 - \mu x^2$ at $\mu_0 = 3/4, x_0 = 2/3$.

7.8 The Feigenbaum phenomenon

M. Feigenbaum studied the family

$$f_\mu(x) = 1 - \mu x^2, \quad 0 < \mu \leqslant 2,$$

of unimodal maps of the interval $[-1, 1]$ (Feigenbaum, 1979). For $\mu < 3/4$, the unique attracting fixed point of f_μ is

$$x_\mu = \frac{\sqrt{1 + 4\mu} - 1}{2\mu}.$$

The derivative $f'_\mu(x_\mu) = 1 - \sqrt{1 + 4\mu}$ is greater than -1 for $\mu < 3/4$, equals -1 for $\mu = 3/4$, and is less than -1 for $\mu > 3/4$. A period-doubling bifurcation occurs at $\mu = 3/4$ (Exercise 7.7.6). For $\mu > 3/4$, the map f_μ has an attracting period-2 orbit. Numerical studies show that there is an increasing sequence of bifurcation values μ_n at which an attracting periodic orbit of period 2^n for f_μ loses stability and an attracting periodic orbit of period 2^{n+1} is born. The sequence μ_n converges, as $n \to \infty$, to a limit μ_∞ and

$$\lim_{n\to\infty} \frac{\mu_\infty - \mu_{n-1}}{\mu_\infty - \mu_n} = \delta = 4.669201609\cdots. \tag{7.3}$$

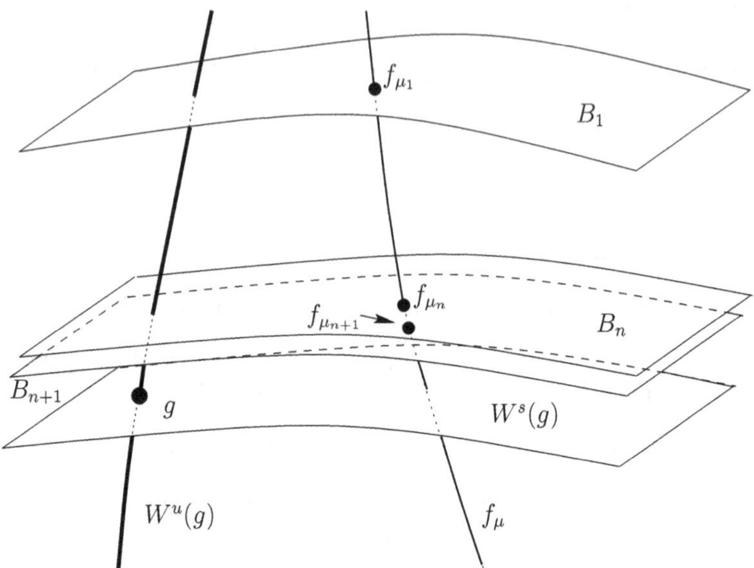

Figure 7.8. Fixed point and stable and unstable manifolds for the Feigenbaum map Φ.

The constant δ is called the *Feigenbaum constant*. Numerical experiments show that the Feigenbaum constant appears for many other one-parameter families.

The Feigenbaum phenomenon can be explained as follows. Consider the infinite-dimensional space $\mathcal{A}$ of real analytic maps $\psi\colon [-1,1] \to [-1,1]$ with $\psi(0) = 1$ and the map $\Phi\colon \mathcal{A} \to \mathcal{A}$ given by the formula:

$$\Phi(\psi)(x) = \frac{1}{\lambda}\psi \circ \psi(\lambda x), \ \lambda = \psi(1). \tag{7.4}$$

A fixed point g of Φ (which Feigenbaum estimated numerically) is an even function satisfying the Cvitanović–Feigenbaum equation

$$g \circ g(\lambda x) - \lambda g(x) = 0. \tag{7.5}$$

The function g is a hyperbolic fixed point of Φ. The stable manifold $W^s(g)$ has codimension one, and the unstable manifold $W^u(g)$ has dimension one and corresponds to a simple eigenvalue $\delta = 4.669201609\cdots$ of the derivative $d\Phi_g$. The codimension-one bifurcation set B_1 of maps ψ for which an attracting fixed point loses stability and an attracting period two orbit is born, intersects $W^u(g)$ transversely. The preimage $B_n = \Phi^{1-n}(B_1)$ is the bifurcation set of maps for which an attracting orbit of period 2^{n-1} is replaced by an attracting orbit of period 2^n (Exercise 7.8.1). Figure 7.8 is a graphical depiction of the process underlying the Feigenbaum phenomenon.

By the infinite-dimensional version of the Inclination Lemma 5.7.2, the codimension-one bifurcation sets B_n accumulate to $W^s(g)$. Let f_μ be a one-parameter family of maps that intersects $W^s(g)$ transversely and let μ_n be the sequence of period-doubling bifurcation parameters, $f_{\mu_n} \in B_n$. Using the Inclination Lemma, one can show that the sequence μ_n satisfies (7.3). O.E. Lanford established the correctness of this model through a computer-assisted proof (Lanford, 1984).

Exercise 7.8.1. Prove that if ψ has an attracting periodic orbit of period $2k$, then $\Phi(\psi)$ has an attracting periodic orbit of period k.

CHAPTER EIGHT

Complex dynamics

In this chapter, we consider *rational* maps $R(z) = P(z)/Q(z)$ of the *Riemann sphere* $\overline{\mathbb{C}} = \mathbb{C} \cup \{\infty\}$, where P and Q are complex polynomials. These maps exhibit many interesting dynamical properties, and lend themselves to the computer-aided drawing of fractals and other fascinating pictures in the complex plane. For more thorough exposition of the dynamics of rational maps see Beardon (1991) and Carleson and Gamelin (1993).

8.1 Complex analysis on the Riemann sphere

We assume that the reader is familiar with the basic ideas of complex analysis (see, for example, Berenstein and Gay (1991) or Conway (1995)).

Recall that a function f from a domain $D \subset \mathbb{C}$ to $\overline{\mathbb{C}}$ is said to be *meromorphic* if it is analytic except at a discrete set of singularities, all of which are poles. In particular, rational functions are meromorphic.

The Riemann sphere is the one-point compactification of the complex plane, $\overline{\mathbb{C}} = \mathbb{C} \cup \{\infty\}$. The space $\overline{\mathbb{C}}$ has the structure of a complex manifold, given by the standard coordinate system on $\mathbb{C}$ and the coordinate $z \mapsto \frac{1}{z}$ on $\overline{\mathbb{C}} \setminus \{0\}$. If M and N are complex manifolds, then a map $f \colon M \to N$ is *analytic* if for every point $\zeta \in M$, there are complex coordinate neighborhoods U of ζ and V of $f(\zeta)$ such that $f \colon U \to V$ is analytic in the coordinates on U and V. An analytic map into $\overline{\mathbb{C}}$ is said to be *meromorphic*. This terminology is somewhat confusing, because in the modern sense (as maps of manifolds) meromorphic functions are analytic, while in the classical sense (as functions on $\mathbb{C}$), meromorphic functions are generally not analytic. Nevertheless, the terminology is so entrenched that it cannot be avoided.

Many of the proofs in this chapter follow the corresponding arguments from Carleson and Gamelin (1993).

It is easy to see that a map $f\colon \overline{\mathbb{C}} \to \overline{\mathbb{C}}$ is analytic (and meromorphic) if and only if both $f(z)$ and $f(1/z)$ are meromorphic (in the classical sense) on $\mathbb{C}$. It is known that every analytic map from the Riemann sphere to itself is a rational map. Note that the constant map $f(z) = \infty$ is considered to be analytic.

The group of Möbius transformations

$$\left\{z \to \frac{az+b}{cz+d} : a, b, c, d \in \mathbb{C};\ ad - bc = 1\right\}$$

acts on the Riemann sphere and is *simply transitive* on triples of points, i.e. for any three distinct points $x, y, z \in \overline{\mathbb{C}}$, there is a unique Möbius transformation that carries x, y, z to $0, 1, \infty$, respectively (see §7.5).

Suppose $f : \overline{\mathbb{C}} \to \overline{\mathbb{C}}$ is a meromorphic map and ζ is a periodic point of minimal period k. If $\zeta \neq \infty$, the *multiplier* of ζ is the derivative $\lambda(\zeta) = (f^k)'(\zeta)$. If $\zeta = \infty$, the multiplier of ζ is $g'(0)$, where $g(z) = 1/f^k(1/z)$. The periodic point ζ is *attracting* if $0 < |\lambda(\zeta)| < 1$, *superattracting* if $\lambda(\zeta) = 0$, *repelling* if $|\lambda(\zeta)| > 1$, *rationally neutral* if $\lambda(\zeta)^m = 1$ for some $m \in \mathbb{N}$, and *irrationally neutral* if $|\lambda(\zeta)| = 1$ but $\lambda(\zeta)^m \neq 1$ for every $m \in \mathbb{N}$. One can prove that a periodic point is attracting or superattracting if and only if it is a (topologically) attracting periodic point in the sense of Chapter 1; similarly for repelling periodic points. The orbit of an attracting or superattracting periodic point is said to be an attracting or superattracting periodic orbit, respectively.

For an attracting or superattracting fixed point ζ of a meromorphic map f, we define the *basin of attraction* $\mathrm{BA}(\zeta)$ as the set of points $z \in \overline{\mathbb{C}}$ for which $f^n(z) \to \zeta$ as $n \to \infty$. Since the multiplier of ζ is less than 1, there is a neighborhood U of ζ which is contained in $\mathrm{BA}(\zeta)$ and $\mathrm{BA}(\zeta) = \bigcup_{n\in\mathbb{N}} f^{-n}(U)$. The set $\mathrm{BA}(\zeta)$ is open. The connected component of $\mathrm{BA}(\zeta)$ containing ζ is called the *immediate basin of attraction*, and is denoted $\mathrm{BA}^\circ(\zeta)$.

If ζ is an attracting or superattracting periodic point of period k, then the basin of attraction of the periodic orbit is the set of all points z for which $f^{nk}(z) \to f^j(\zeta)$ as $n \to \infty$ for some $j \in \{0, 1, \dots, k\}$, and is denoted $\mathrm{BA}(\zeta)$. The union of the connected components of $\mathrm{BA}(\zeta)$ containing a point in the orbit of ζ is called the *immediate basin of attraction*, and is denoted $\mathrm{BA}^\circ(\zeta)$.

A point ζ is a critical point (or branch point) of a meromorphic function f if f is not 1-1 on a neighborhood of ζ. A critical point ζ has *order* m if f is $(m+1)$-to-1 on $U \setminus \{\zeta\}$ for a sufficiently small neighborhood U of ζ. Equivalently, ζ is a critical point of order m if ζ is a zero of f' (in local coordinates) of multiplicity m. If ζ is a critical point, then $f(\zeta)$ is called a critical value.

For a rational map $R = P/Q$, with P and Q relatively prime polynomials of degree p and q, respectively, the *degree* of R is $\deg(R) = \max(p, q)$. If R has degree d, then the map $R: \overline{\mathbb{C}} \to \overline{\mathbb{C}}$ is a *branched covering of degree* d, i.e. any $\xi \in \overline{\mathbb{C}}$ that is not a critical value has exactly d preimages; in fact, every point has exactly d preimages if critical points are counted with multiplicity. Since the number of preimages of a generic point is a topological invariant of R, the degree is invariant under conjugation by a Möbius transformation.

The rational maps of degree 1 are the *Möbius* transformations. A rational map is a polynomial if and only if the only preimage of ∞ is ∞.

The *postcritical set* of a rational map R is the union of the forward orbits of all critical points of R, and is denoted $CL(R)$.

PROPOSITION 8.1.1. *Let R be a rational map of degree d. Then the number of critical points, counted with multiplicity, is $2d - 2$. If $CL(R)$ has exactly two points then R is conjugate by a Möbius transformation to z^d or z^{-d}.*

Proof. By composing with a Möbius transformation we may assume that $R(\infty) = 0$ and that ∞ is neither a critical point nor a critical value. Then $R(\infty) = 0$ and the fact that ∞ is not a critical point imply that

$$R(z) = \frac{\alpha z^{d-1} + \cdots}{\beta z^d + \cdots}$$

where $\alpha \neq 0$ and $\beta \neq 0$. Then

$$R'(z) = -\frac{\alpha\beta z^{2d-2} + \cdots}{(\beta z^d + \cdots)^2},$$

and the critical points of R are the zeros of the numerator (since ∞ is not a critical value).

The proof of the second assertion is left as an exercise (Exercise 8.1.5). □

A family F of meromorphic functions in a domain $D \subset \overline{\mathbb{C}}$ is *normal* if every sequence from F contains a subsequence that converges uniformly on compact subsets of D in the standard spherical metric on $\overline{\mathbb{C}} \approx S^2$. A family F is *normal at a point* $z \in \overline{\mathbb{C}}$ if it is normal in a neighborhood of z.

The *Fatou set* $F = F(R) \subset \overline{\mathbb{C}}$ of a rational map $R: \overline{\mathbb{C}} \to \overline{\mathbb{C}}$ is the set of points $z \in \overline{\mathbb{C}}$ such that the family of forward iterates $\{R^n\}_{n\in\mathbb{N}}$ is normal at z. The *Julia set* $J = J(R)$ is the complement of the Fatou set. Both $F(R)$ and $J(R)$ are completely invariant under R (see Proposition 8.5.1). Points belonging to the same component of $F(R)$ have the same asymptotic behavior. As we will see later, the Fatou set contains all basins of attraction and the Julia set is the closure of the set of all repelling periodic points. The "interesting" dynamics is concentrated on the Julia set. The case when $J(R)$ is a hyperbolic set is reasonably well understood (Theorem 8.5.10).

Exercise 8.1.1. Prove that any Möbius transformation is conjugate by another Möbius transformation to either $z \mapsto az$ or $z \mapsto z + a$.

Exercise 8.1.2. Prove that a non-constant rational map R is conjugate to a polynomial by a Möbius transformation if and only if $R^{-1}(z) = \{z\}$ for some $z \in \overline{\mathbb{C}}$.

Exercise 8.1.3. Find all Möbius transformations that commute with $q_0(z) = z^2$.

Exercise 8.1.4. Let R be a rational map such that $R(\infty) = \infty$ and let f be a Möbius transformation such that $f(\infty)$ is finite. Define the multiplier $\lambda_R(\infty)$ of R at ∞ to be the multiplier of $f \circ Rf^{-1}$ at $f(\infty)$. Prove that $\lambda_R(\infty)$ does not depend on the choice of f.

Exercise 8.1.5. Prove the second assertion of Proposition 8.1.1.

Exercise 8.1.6. Let R be a non-constant rational map. Prove that

$$\deg(R) - 1 \leqslant \deg(R') \leqslant 2\deg(R)$$

with equality on the left if and only if R is a polynomial, and with equality on the right if and only if all poles of R are simple and finite.

8.2 Examples

The global dynamics of a rational map R depends heavily on the behavior of the critical points of R under its iterates. In most of the examples below the Fatou set consists of finitely many components that are basins of attraction. Some of the assertions in the following examples will be proved in later sections of this chapter. Proofs of most of the assertions that are not proved here can be found in Carleson and Gamelin (1993).

Let $q_a\colon \overline{\mathbb{C}} \to \overline{\mathbb{C}}$ be the quadratic map $q_a(z) = z^2 - a$ and denote by S^1 the unit circle $\{z \in \mathbb{C} : |z| = 1\}$. The critical points of q_a are 0 and ∞, and the critical values are $-a$ and ∞; if $a \neq 0$, the only superattracting periodic (fixed) point is ∞. In the examples below we observe drastically different global dynamics depending on whether the critical point lies in the basin of a finite attracting periodic point, or in the basin of ∞, or in the Julia set.

1. $q_0(z) = z^2$. There is a superattracting fixed point at 0, whose basin of attraction is the open disk $\Delta_1 = \{z \in \mathbb{C} \mid |z| < 1\}$, and a superattracting fixed point at ∞, whose basin of attraction is the exterior of S^1.

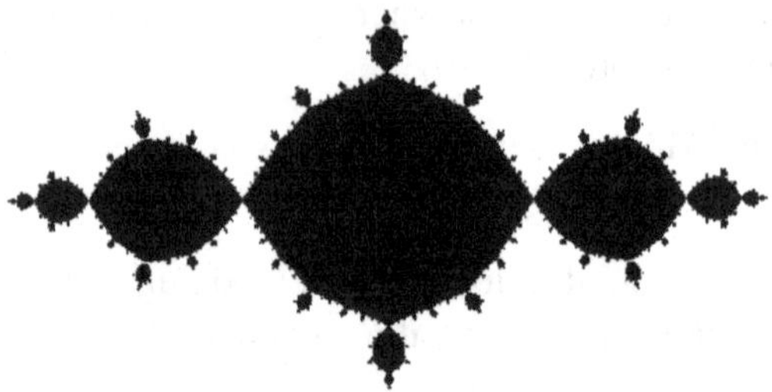

Figure 8.1. The Julia set for $a = -1$.

There is also a repelling fixed point at 1 and, for each $n \in \mathbb{N}$, there are 2^n repelling periodic points of period n on S^1. The Julia set is S^1, the Fatou set is the complement of S^1. The map q_0 acts on S^1 by $\phi \mapsto 2\phi \bmod 2\pi$ (where ϕ is the angular coordinate of a point $z \in S^1$). If U is a neighborhood of $\zeta \in S^1 = J(q_0)$, then $\bigcup_{n\in\mathbb{N}_0} q_0^n(U) = \mathbb{C} \setminus \{0\}$.

2. $q_\epsilon(z) = z^2 + \epsilon$, $0 < \epsilon \ll 1$. There is an attracting fixed point near 0, a superattracting fixed point at ∞ and, for each $n \in \mathbb{N}$, 2^n repelling periodic points near S^1. The Julia set $J(q_\epsilon)$ is a closed continuous q_ϵ-invariant curve which is C^0 close to S^1, is not differentiable at a dense set of points and has Hausdorff dimension greater than 1. The basins of attraction of the fixed points near 0 and at ∞ are, respectively the interior and exterior of $J(q_\epsilon)$. The critical point and critical value lie in the immediate basin of attraction of the attracting fixed point near 0.

   The same properties hold true for maps of the form $f(z) = z^2 + \epsilon P(z)$, where P is a polynomial and ϵ is small enough.
3. $q_1(z) = z^2 - 1$. Note that $q_1(0) = -1$, $q_1(-1) = 0$. Therefore, 0 and -1 are superattracting periodic points of period 2. On the real line the repelling fixed point $(1 - \sqrt{5})/2$ separates the basins of attraction of 0 and -1. The Julia set $J(q_1)$ contains two simple closed curves σ_0 and σ_{-1} which surround 0 and -1 and bound their immediate basins of attraction. The only preimage of -1 is 0, hence the only preimage of σ_{-1} is σ_0. However, 0 has two preimages, $+1$ and -1. Therefore σ_0 has two preimages, σ_1 and a closed curve σ_1 surrounding 1. Continuing in this manner and using the complete invariance of the Julia set (Proposition 8.5.1) we conclude that $J(q_1)$ contains infinitely many closed curves. Their interiors are components of the Fatou set. Figure 8.1 shows the Julia set for q_1[1].

[1] The illustrations in this chapter were produced with *mandelspawn, see http://www.araneus.fi/gson/mandelspawn/*.

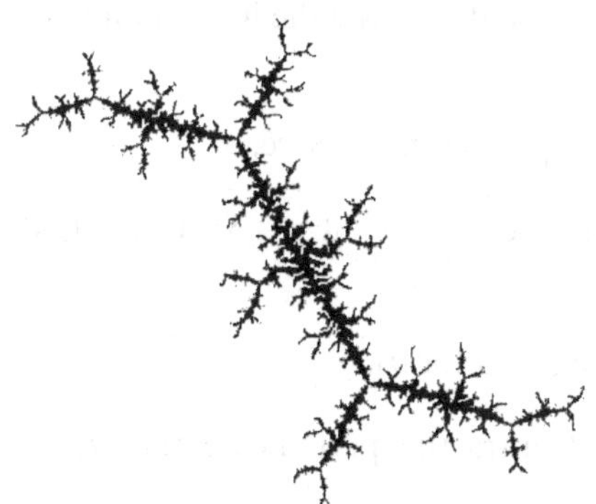

Figure 8.2. The Julia set for $a = -i$.

4. $q_{-i} = z^2 + i$. The critical point 0 is eventually periodic: $q^2_{-i}(0) = i - 1$, and $i - 1$ is a repelling periodic point of period 2. The only attracting periodic fixed point is ∞. The Fatou set consists of one component and coincides with $\mathrm{BA}(\infty)$. The Julia set is a *dendrite*, i.e. a compact, path-connected, locally connected, nowhere dense subset of $\mathbb{C}$ that does not separate $\mathbb{C}$. Figure 8.2 shows the Julia set for q_{-i}.
5. $q_2(z) = z^2 - 2$. The change of variables $z = \zeta + \zeta^{-1}$ conjugates q_2 on $\overline{\mathbb{C}} \setminus [-2, 2]$ with $\zeta \mapsto \zeta^2$ on the exterior of S^1. Hence $J(q_2) = [-2, 2]$ and $F(q_2) = \overline{\mathbb{C}} \setminus [-2, 2]$ is the basin of attraction of ∞. The image of the critical point 0 is $-2 \in J(q_2)$. The change of variables $y = (2 - x)/4$ conjugates the action of q_2 on the real axis to $y \mapsto 4y(1 - y)$. The only attracting periodic point is ∞.
6. $q_4(z) = z^2 - 4$. The only attracting periodic point is ∞; the critical value -4 lies in the (immediate) basin of attraction of ∞, $J(q_4)$ is a Cantor set on the real axis; $\mathrm{BA}(\infty)$ is the complement of $J(q_4)$.
7. This example illustrates the connection between the dynamics of rational maps and issues of convergence for the *Newton method*. Let $Q(z) = (z - a)(z - b)$ with $a \neq b$. To find the roots a and b using the Newton method one iterates the map

$$f(z) = z - \frac{Q(z)}{Q'(z)} = z - \frac{1}{\frac{1}{z-a} + \frac{1}{z-b}}.$$

The change of variables $\zeta = (z - a)/(z - b)$ sends a to 0, b to ∞, ∞ to 1, the line $l = \{(a + b)/2 + ti(a - b) : t \in \mathbb{R}\}$ to the unit circle, and conjugates Q with $\zeta \mapsto \zeta^2$. Therefore the Newton method for Q converges to a and b if the initial point lies in the half-planes of l containing a

and b, respectively; the Newton method diverges if the initial point lies on l.

Exercise 8.2.1. Prove the properties of q_0 described above.

Exercise 8.2.2. Let U be a neighborhood of a point $z \in S^1$. Prove that $\bigcup_{n\in\mathbb{N}} q_0(U) = \mathbb{C} \setminus \{0\}$.

Exercise 8.2.3. Check the above conjugacies for q_2.

Exercise 8.2.4. Prove that ∞ is the only attracting periodic point of q_4.

Exercise 8.2.5. Let $|a| > 2$ and $|z| \geqslant |c|$. Prove that $q_a^n(z) \to \infty$ as $n \to \infty$.

Exercise 8.2.6. Prove the statements in example (7).

8.3 Normal families

The theory of normal families of meromorphic functions is a keystone in the study of complex dynamics. The principal result, Theorem 8.3.2, is due to Montel.

PROPOSITION 8.3.1. *Suppose F is a family of analytic functions in a domain D and suppose that for every compact subset $K \subset D$ there is $C(K) > 0$ such that $|f(z)| < C(K)$ for all $z \in K$ and $f \in F$. Then F is a normal family.*

Proof. Let $K \subset D$ be compact, and let $\delta = \frac{1}{2} \min_{z\in K} \operatorname{dist}(z, \partial D)$. Let K_δ be the closure of the δ-neighborhood of K. Then $K_\delta \subset D$ and, for $z \in K$, the circle $\gamma = \gamma(z)$ of radius δ centered at z is contained in K_δ. By the Cauchy formula,

$$f'(z) = \frac{1}{2\pi} \int_\gamma \frac{f(\xi)}{(\xi - z)^2}\, d\xi$$

so $|f'(z)| < C(K_\delta)/\delta$ for every $f \in F$ and $z \in K$. Thus the family F is equicontinuous on K, and therefore normal by the Arzela–Ascoli Theorem. □

We say that a family F of functions on a domain D *omits* a point a if $f(z) \neq a$ for every $f \in F$ and $z \in D$.

THEOREM 8.3.2 (Montel). *Suppose that a family F of meromorphic functions in a domain $D \subset \overline{\mathbb{C}}$ omits three distinct points $a, b, c \in \overline{\mathbb{C}}$. Then F is normal in D.*

Proof. Since D is covered by disks, we may assume without loss of generality that D is a disk. By applying a Möbius transformation, we may also assume that $a = 0$, $b = 1$ and $c = \infty$. Let Δ_1 be the unit disk. By the Uniformization Theorem (Ahlfors, 1973), there is an analytic covering map $\phi : \Delta_1 \to (\mathbb{C} \setminus \{0, 1\})$ (ϕ is called the *modular function*). For every function $f\colon D \to (\mathbb{C} \setminus \{0, 1\})$ there is a lift $\tilde{f}\colon D \to \Delta_1$ such that $\phi \circ \tilde{f} = f$. The family $\tilde{F} = \{\tilde{f} : f \in F\}$ is bounded, and therefore, by Proposition 8.3.1, normal. The normality of F follows immediately. □

Exercise 8.3.1. Let f be a meromorphic map defined on a domain $D \subset \overline{\mathbb{C}}$ and let $k > 1$. Show that the family $\{f^n\}_{n\in\mathbb{N}}$ is normal on D if and only if the family $\{f^{kn}\}_{n\in\mathbb{N}}$ is normal.

8.4 Periodic points

THEOREM 8.4.1. *Let ζ be an attracting fixed point of a meromorphic map $f\colon \overline{\mathbb{C}} \to \overline{\mathbb{C}}$. Then there is a neighborhood $U \ni \zeta$ and an analytic map $\phi\colon U \to \mathbb{C}$ that conjugates f and $z \mapsto \lambda(\zeta)z$, i.e. $\phi(f(z)) = \lambda(\zeta)\phi(z)$ for all $z \in U$.*

Proof. We abbreviate $\lambda(\zeta) = \lambda$. Conjugating by a translation (or by $z \mapsto 1/z$ if $\zeta = \infty$), we replace ζ by 0. Then on any sufficiently small neighborhood of 0, say $\Delta_{1/2} = \{z \mid |z| < 1/2\}$, there is a $C > 0$ such that $|f(z) - \lambda z| \leqslant C|z|^2$. Hence for every $\epsilon > 0$ there is a neighborhood U of 0 such that $|f(z)| < (|\lambda| + \epsilon)|z|$, for all $z \in U$, and, assuming that $|\lambda| + \epsilon < 1$,

$$|f^n(z)| < (|\lambda| + \epsilon)^n|z|\,.$$

Set $\phi_n(z) = \lambda^{-n} f^n(z)$. Then, for $z \in U$,

$$|\phi_{n+1}(z) - \phi_n(z)| = \left|\frac{f(f^n(z)) - \lambda f^n(z)}{\lambda^{n+1}}\right| \leqslant \frac{C(|\lambda| + \epsilon)^{2n}|z|^2}{|\lambda|^{n+1}},$$

and hence the sequence ϕ_n converges uniformly if $(|\lambda| + \epsilon)^2 < |\lambda|$ and $|z| < 1$.

By construction, $\phi_n(f(z)) = \lambda\phi_{n+1}(z)$. Therefore the limit $\phi = \lim_{n\to\infty} \phi_n$ is the required conjugation. □

COROLLARY 8.4.2. *Let ζ be a repelling fixed point of a meromorphic map f. Then there is a neighborhood U of ζ and an analytic map $\phi\colon U \to \overline{\mathbb{C}}$ that conjugates f and $z \mapsto \lambda(\zeta)z$, i.e. $\phi(f(z)) = \lambda(\zeta)\phi(z)$ for $z \in U$.*

Proof. Apply Theorem 8.4.1 to the branch g of f^{-1} with $g(\zeta) = \zeta$. □

PROPOSITION 8.4.3. *Let ζ be a fixed point of a meromorphic map f. Assume that $\lambda = f'(\zeta)$ is not 0 and is not a root of 1 and suppose that an* *analytic map ϕ conjugates f and $z \mapsto \lambda z$. Then ϕ is unique up to multiplication by a constant.*

Proof. Again, we assume $\zeta = 0$. If there are two conjugating maps ϕ and ψ, then $\eta = \phi^{-1} \circ \psi$ conjugates $z \mapsto \lambda z$ with itself, i.e. $\eta(\lambda z) = \lambda \eta(z)$. If $\eta = a_1 z + a_2 z^2 + \cdots$, then $a_n \lambda^n = \lambda a_n$ and $a_n = 0$ for $n > 1$. □

LEMMA 8.4.4. *Any rational map R of degree >1 has infinitely many periodic points.*

Proof. Observe that the number of solutions of $R^n(z) - z = 0$ (counted with multiplicity) tends to ∞ as $n \to \infty$. Therefore, if R has only finitely many periodic points, their multiplicities cannot be bounded in n.

On the other hand, if ζ is a multiple root of $R^n(z) - z = 0$, then $R^{n\prime}(\zeta) = 1$ and $R^n(z) = \zeta + (z - \zeta) + a(z - \zeta)^m + \cdots$ for some $a \neq 0$ and $m \geqslant 2$. By induction, $R^{nk}(z) = \zeta + (z - \zeta) + ka(z - \zeta)^m + \cdots$ for $k \in \mathbb{N}$. Therefore, ζ has the same multiplicity as a fixed point of R^n and of R^{nk}. □

PROPOSITION 8.4.5. *Let f be a meromorphic map of $\overline{\mathbb{C}}$. If ζ is an attracting or superattracting periodic point of f, then the family $\{f^n\}_{n \geqslant 0}$ is normal in* $\mathrm{BA}(\zeta)$.

If ζ is a repelling periodic point of f, then the family $\{f^n\}$ is not normal at ζ.

Proof. Exercise 8.4.1. □

THEOREM 8.4.6. *Let ζ be an attracting periodic point of a rational map R. Then the immediate basin of attraction* $\mathrm{BA}^\circ(\zeta)$ *contains a critical point of f.*

Proof. Consider first the case when ζ is a fixed point. Suppose that $\mathrm{BA}^\circ(\zeta)$ does not contain a critical point. For a small enough $\epsilon > 0$, there is a branch g of R^{-1} which is defined in the open ϵ-disk D_ϵ about ζ and satisfies $g(\zeta) = \zeta$. The map $g : D_\epsilon \to \mathrm{BA}^\circ(\zeta)$ is a diffeomorphism onto its image, and therefore $g(D_\epsilon)$ is simply connected and does not contain a critical point. Thus g extends uniquely to a map on $g(D_\epsilon)$. By induction, g extends uniquely to $g^n(D_\epsilon)$, which is a simply connected subset of $\mathrm{BA}^\circ(\zeta)$. The sequence $\{g^n\}$ is normal on D_ϵ since it omits infinitely many periodic points of R different from ζ (Lemma 8.4.4). (Note that if R is a polynomial, then $\{g^n\}$ omits a neighborhood of ∞ and Lemma 8.4.4 is not needed.) On the other hand, $|g'(\zeta)| > 1$ and hence $g^{n\prime}(\zeta) \to \infty$ as $n \to \infty$, therefore the family $\{g^n\}$ is not normal; a contradiction.

If ζ is a periodic point of order n, then the preceding argument shows that the immediate basin of attraction of ζ for the map R^n contains a critical point of R^n. Since the components of $BA^\circ(\zeta)$ are permuted by R, it follows from the chain rule that one of the components contains a critical point of R. □

COROLLARY 8.4.7. *A rational map has at most* $2d - 2$ *attracting and super-attracting periodic orbits.*

Proof. The corollary follows immediately from Theorem 8.4.6 and Proposition 8.1.1. □

More delicate analysis that is beyond the scope of this book leads to the following theorem.

THEOREM 8.4.8 (Shishikura, 1987). *The total number of attracting, super-attracting and neutral periodic orbits of a rational map of degree d is at most* $2d - 2$.

The upper bound $6d - 6$ was obtained by Fatou.

Exercise 8.4.1. Prove Proposition 8.4.5.

Exercise 8.4.2. Let $D \subset \overline{\mathbb{C}}$ be a domain whose complement contains at least three points and let $f\colon D \to D$ be a meromorphic map with an attracting fixed point $z_0 \in D$. Prove that the sequence of iterates f^n converges in D to z_0 uniformly on compact sets.

Exercise 8.4.3. Prove that every rational map $R \neq \mathrm{Id}$ of degree $d \geqslant 1$ has $d + 1$ fixed points in $\overline{\mathbb{C}}$ counted with multiplicity.

8.5 The Julia set

Recall that the Fatou set $F(R)$ of a rational map R is the set of points $z \in \overline{\mathbb{C}}$ such that the family of forward iterates R^n, $n \in \mathbb{N}$, is normal at z. The Julia set $J(R)$ is the complement of $F(R)$. The Julia set of a rational map is closed by definition, and non-empty by Lemma 8.4.4, Proposition 8.4.5 and Theorem 8.4.8. If U is a connected component of $F(R)$, then $R(U)$ is also a connected component of $F(R)$ (Exercise 8.5.1).

Suppose $V \neq BA(\infty)$ is a component of $BA(\infty)$. Then $R^n(V) \subset BA^\circ(\infty)$ for some $n > 0$. Moreover, $R^n(V)$ is both open and closed in $BA^\circ(\infty)$ since $R^n(V) = R^n(V \cup J(R)) \setminus J(R)$. It follows that $R^n(V) = BA^\circ(\infty)$.

PROPOSITION 8.5.1. *Let* $R\colon \overline{\mathbb{C}} \to \overline{\mathbb{C}}$ *be a rational map. Then* $F(R)$ *and* $J(R)$ *are completely invariant, i.e.* $R^{-1}(F(R)) = F(R)$ *and* $R(F(R)) = F(R)$ *and similarly for* $J(R)$.

Proof. Let $\zeta = R(\xi)$. Then R^{n_k} converges in a neighborhood of ζ if and only if R^{n_k+1} converges in a neighborhood of ξ. □

PROPOSITION 8.5.2. *Let $R\colon \overline{\mathbb{C}} \to \overline{\mathbb{C}}$ be a rational map. Then either $J(R) = \overline{\mathbb{C}}$ or $J(R)$ has no interior.*

Proof. Suppose $U \subset J(R)$ is non-empty and open in $\overline{\mathbb{C}}$. Then the family $\{R^n\}_{n\in\mathbb{N}}$ is not normal on U and, in particular, by Theorem 8.3.2, $\bigcup_n R^n(U)$ omits at most two points in $\overline{\mathbb{C}}$. Since $J(R)$ is invariant and closed, $J(R) = \overline{\mathbb{C}}$. □

Let $R\colon \overline{\mathbb{C}} \to \overline{\mathbb{C}}$ be a rational map and U an open set such that $U \cap J(R) \neq \emptyset$. The family of iterates $\{R^n\}_{n\in\mathbb{N}_0}$ is not normal in U, so it omits at most two points in $\overline{\mathbb{C}}$. The set E_U of omitted points is called the exceptional set of R on U. The exceptional set of R is the set $E = \cup E_U$, where the union is over all open sets U with $U \cap J(R) \neq \emptyset$. A point in E is called an *exceptional point* of R.

PROPOSITION 8.5.3. *Let R be a rational map of degree greater than 1. Then the exceptional set of R contains at most two points. If the exceptional set consists of a single point, then R is conjugate by a Möbius transformation to a polynomial. If it consists of two points, then R is rationally conjugate to z^m or $1/z^m$, for some $m > 1$. The exceptional set is disjoint from $J(R)$.*

Proof. If E_U is empty for every U with $U \cap J(R) \neq \emptyset$, there is nothing to show.

Suppose $\{R^n\}_{n\in\mathbb{N}_0}$ omits two points z_0, z_1 on U for some open set U with $U \cap J(R) \neq \emptyset$. Then after conjugating by the rational map $\phi(z) = (z - z_1)/(z - z_0)$, R becomes a rational map whose family of iterates omits only 0 and ∞ on the set $\phi(U)$. Thus there are no solutions of $R(z) = \infty$ except possibly 0 and ∞. If $R(0) \neq \infty$, then R has no poles, so it is a polynomial, and is therefore equal to $z^m, m > 0$, since $R(z) = 0$ has no non-zero solutions. If $R(0) = \infty$, then R has a unique pole at 0; since there are no finite solutions of $R(z) = 0$, it follows that $R(z) = 1/z^m$. We have shown that R is conjugate to z^m, $|m| > 1$, if the exceptional set of some open set U has two points. In this case the exceptional set is $\{0, \infty\}$.

Suppose that $\{R^n\}_{n\in\mathbb{N}_0}$ omits at most a single point on U for every open set U with $U \cap J(R) \neq \emptyset$. Fix such a set U with $E_U \neq \emptyset$ and let z_0 be the omitted point. Replacing R with its conjugate by the rational map $\phi(z) = 1/(z - z_0)$, we may take $z_0 = \infty$. Since $\{\infty\}$ is omitted, R has no poles, and is therefore a polynomial. Thus R omits ∞ on every open subset $U \subset \mathbb{C}$, and (by hypothesis) omits only a single point on U if $U \cap J(R) \neq \emptyset$, so ∞ is the only exceptional point of R.

In either case, $J(R)$ does not contain any exceptional points. □

PROPOSITION 8.5.4. *Let $R\colon \overline{\mathbb{C}} \to \overline{\mathbb{C}}$ be a rational map of degree >1. For any point $\zeta \notin E$, $J(R)$ is contained in the closure of the set of backward iterates of ζ. In particular, $J(R)$ is the closure of the set of backward iterates of any point in $J(R)$.*

Proof. Let $z \in J(R)$, and let U be a neighborhood of z. Then $\overline{\mathbb{C}} \setminus E \subset \bigcup_{n\in\mathbb{N}} R^n(U)$. Hence for any $\zeta \notin E$, some preimage of ζ is in U. □

COROLLARY 8.5.5. *The Julia set of a rational map of degree >1 is perfect, i.e. it does not have isolated points.*

Proof. Exercise 8.5.3. □

PROPOSITION 8.5.6. *Let $R\colon \overline{\mathbb{C}} \to \overline{\mathbb{C}}$ be a rational map of degree >1. Then $J(R)$ is the closure of the set of repelling periodic points.*

Proof. We will show that $J(R)$ is contained in the closure of the set $\mathrm{Per}(R)$ of the periodic points of R. The result will follow since $J(R)$ is perfect and there are only finitely many non-repelling periodic points.

Suppose $\zeta \in J(R)$ has a neighborhood U that contains no periodic points, no poles and no critical values of R. Since the degree of R is >1, there are distinct branches f and g of R^{-1} in U, and $f(z) \neq g(z)$, $f(z) \neq R^n(z)$ and $g(z) \neq R^n(z)$ for all $n \geqslant 0$ and all $z \in U$. Hence the family

$$h_n(z) = \frac{R^n(z) - f(z)}{R^n(z) - g(z)} \cdot \frac{z - g(z)}{z - f(z)}, \quad n \in \mathbb{N},$$

omits $0, 1$ and ∞ in U and therefore is normal by Theorem 8.3.2. Since R^n can be expressed in terms of h_n, the family $\{R^n\}$ is also normal in U, a contradiction. Therefore $J(R) \subset \overline{\mathrm{Per}(R)}$. □

The following proposition shows that the Julia set possesses self-similarity.

PROPOSITION 8.5.7. *Let $R\colon \overline{\mathbb{C}} \to \overline{\mathbb{C}}$ be a rational map of degree >1, and let U be a neighborhood of a point $\zeta \in J(R)$. Then $\bigcup_{n\in\mathbb{N}} R^n(U) = \overline{\mathbb{C}} \setminus E$, and $J(R) \subset R^n(U)$ for some $n \in \mathbb{N}$.*

Proof. If E contains two points, then by Proposition 8.5.3, R is conjugate to z^m, $|m| > 1$, and the proof is left as an exercise (Exercise 8.5.4).

Suppose E is empty or consists of a single point. If the latter, we may and do assume that the omitted point is ∞ and R is a polynomial. Since repelling periodic points are dense in $J(R)$, we may choose a neighborhood $V \subset U$ and $n > 0$ such that $R^n(V) \supset V$. The family $\{R^{nk}\}_{k\in\mathbb{N}}$ on V does not omit any points in $\mathbb{C}$, and ∞ is omitted if and only if R is a polynomial, in which case $\infty \notin J(R)$. Hence $J(R) \subset \bigcup_n R^n(V)$. Since $J(R)$ is compact and $R^{nk}(V) \subset R^{n(k-1)}(V)$, the proposition follows. □

Let $P\colon \overline{\mathbb{C}} \to \overline{\mathbb{C}}$ be a polynomial. Then $P(\infty) = \infty$ and locally near ∞ there are $\deg P$ branches of P^{-1}. The complete preimage of any connected domain containing ∞ is connected since $\infty = P^{-1}(\infty)$ must belong to every connected component of the preimage. Therefore $\mathrm{BA}(\infty)$ is connected, i.e. $\mathrm{BA}(\infty) = \mathrm{BA}^\circ(\infty)$.

LEMMA 8.5.8. *Let $f\colon \overline{\mathbb{C}} \to \overline{\mathbb{C}}$ be a meromorphic function and suppose ζ is an attracting periodic point. Then every component of $BA^\circ(\zeta)$ is simply connected.*

Proof. Since f cyclically permutes the components of $BA^\circ(\zeta)$, we may replace f by f^n where n is the minimal period of ζ and assume that ζ is fixed. After conjugating by a Möbius transformation, we may assume that ζ is finite.

Let γ be a smooth simple closed curve in $BA^\circ(\zeta)$, and let D be the simply connected region (in $\mathbb{C}$) that it bounds. Suppose $D \not\subset BA^\circ(\zeta)$. Let δ be the distance from ζ to the boundary of $BA^\circ(\zeta)$, and let U be the disk of radius $\delta/2$ around ζ. Because ζ is attracting, and γ is a compact subset of $BA^\circ(\zeta)$, there is $n > 0$ such that $f^n(\gamma) \subset U$. Let $g(z) = f^n(z) - \zeta$. Then $|g(z)| < \delta/2$ on γ, but $|g(z)| > \delta$ for some $z \in D$ since $f^n(D) \not\subset BA^\circ(\zeta)$. This contradicts the Maximum Principle for analytic functions. Thus $D \subset BA^\circ(\zeta)$, and $BA^\circ(\zeta)$ is simply connected. □

PROPOSITION 8.5.9. *Let $R\colon \overline{\mathbb{C}} \to \overline{\mathbb{C}}$ be a rational map of degree >1. If U is any completely invariant component of $F(R)$, then $J(R) = \overline{U} \setminus U$, and $J(R) = \partial U$ if $F(R)$ is not connected. Every other component of $F(R)$ is simply connected. There are at most two completely invariant components. If R is a polynomial, then $BA(\infty)$ is completely invariant.*

Proof. Suppose U is a completely invariant component of $F(R)$. Then, by Proposition 8.5.4, $J(R)$ is contained in the closure of U and also of $F(R) \setminus U$ if the latter is non-empty. This proves the first assertion. Since $J(R) \cup U = \overline{U}$ is connected, every component of the complement in $\overline{\mathbb{C}}$ is simply connected (by a basic result of homotopy theory).

Suppose there is more than one completely invariant component of $F(R)$. Then, by the preceding paragraph, each must be simply connected. Let U be such a component. Then $R\colon U \to U$ is a branched covering of degree d, so there must be $d-1$ critical points, counted with multiplicity. Since the total number of critical points is $2d-2$ (Proposition 8.1.1), this implies that there are at most 2 completely invariant components.

If R is a polynomial, then $\mathrm{BA}(\infty)$ is completely invariant (Exercise 8.5.1). □

THEOREM 8.5.10 (Fatou). *Let R be a rational map of degree >1. Suppose that all critical points of R tend to attracting periodic points of R under the forward iterates of R. Then $J(R)$ is a hyperbolic set for R, i.e. there are $a>1$ and $n \in \mathbb{N}$ such that $|R^{n\prime}(z)| \geqslant a$ for every $z \in J(R)$.*

Proof. If $CL(R)$ has exactly two points then R is conjugate to z^d or z^{-d} (Proposition 8.1.1), and the theorem follows by a direct computation.

We assume then that $CL(R)$ has at least three points. Let $U = \overline{\mathbb{C}} \setminus \overline{CL(R)}$; then $R^{-1}(U) \subset U$. By the Uniformization Theorem (Ahlfors, 1973), there is an analytic covering map $\phi : \Delta_1 \to U$. Let $g : \Delta_1 \to \Delta_1$ be the lift of a locally defined branch of R^{-1}, so $R \circ \phi \circ g = \phi$.

The family $\{\phi \circ g^n\}$ is normal since it omits $CL(R)$. Let f be the uniform limit of a sequence $\phi \circ g^{n_k}$. Let $z_0 \in \phi^{-1}(J(R))$, and let $O \subset \Delta_1$ be a neighborhood of z_0 such that $\phi(O)$ does not contain any attracting periodic points of R. Since $J(R)$ is invariant (Proposition 8.5.1) and closed, $f(z_0) \in J(R)$. If $f'(z_0) \neq 0$, then $f(O)$ contains a neighborhood of $f(z_0)$, and hence (by Proposition 8.5.9) contains a point $z_1 \in \mathrm{BA}(\xi)$, where ξ is an attracting periodic point. Since $\phi \circ g^{n_k} \to f$, the value z_1 is taken on by every $\phi \circ g^{n_k}$ with k large enough. This implies that $R^{n_k}(z_1) \in \phi(O)$ for k sufficiently large, which contradicts the fact that $z_1 \in \mathrm{BA}(\xi)$ and $\xi \notin \phi(O)$. Therefore, $f'(z_0) = 0$, so f is constant on $\phi^{-1}(J(R))$. It follows that $R^{n_k\prime} = 1/(g^{n_k})'$ goes to infinity uniformly on $J(P)$, which proves the theorem. □

THEOREM 8.5.11 (Fatou). *Let $P\colon \overline{\mathbb{C}} \to \overline{\mathbb{C}}$ be a polynomial such that $P^n(c) \to \infty$ as $n \to \infty$ for every critical point c. Then the Julia set $J(P)$ is totally disconnected, i.e. $J(P)$ is a Cantor set.*

Proof. Let D be a disk centered at 0 that contains $J(P)$, and choose N large enough that P^N carries all critical points outside of $\overline{D}$. Then for $n \geqslant N$, branches of P^{-n} are globally defined on D. Fix $z_0 \in J(P)$ and let g_n be the branch of P^{-n} with $g_n(P^n(z_0)) = z_0$, for $n \geqslant N$. The family $F = \{g_n\}_{n \geqslant N}$ is uniformly bounded on $\overline{D}$, and is therefore normal on $\overline{D}$. Let f be the uniform limit of a sequence in F. Since P is hyperbolic on $J(P)$ (Theorem 8.5.10), f must be constant on $J(P)$, and therefore constant on $\overline{D}$ since f is analytic and $J(P)$ has no isolated points. If $y \neq z_0$ is any other point of $J(P)$, then

$y \notin g_n(D)$ for n sufficiently large since the diameter of $g_n(D)$ converges to zero. The set $g_n(D) \cap J(P)$ is both open and closed in $J(P)$ because ∂D does not interset $J(P)$. Thus z_0 and y are in different components of $J(P)$, so $J(P)$ is totally disconnected. □

PROPOSITION 8.5.12. *Let $P\colon \overline{\mathbb{C}} \to \overline{\mathbb{C}}$ be a polynomial such that no critical point lies in* BA(∞). *Then $J(P)$ is connected.*

Proof. $BA(\infty)$ is simply connected (Lemma 8.5.8) and completely invariant. If $F(P)$ has only one component, then $J(P)$ is the complement in $\overline{\mathbb{C}}$ of $BA(\infty)$, and is therefore connected by a fundamental result of algebraic topology.

We assume then that $F(P)$ has at least two components. We conjugate by a Möbius transformation that carries ∞ to 0, and one of the other components of $F(P)$ to a neighborhood of ∞. We obtain a rational map R such that 0 is a superattracting fixed point and $BA(0)$ is a bounded, simply connected, completely invariant component of $F(R)$ that contains no critical points. Let g_n be the branch of R^n on $BA(0)$ with $g_n(0) = 0$. Let γ be the unit circle. Then $g_n(\gamma)$ converges to $J(R)$, so $J(R)$ is connected. □

There are many other results about the Fatou and Julia sets that are beyond the scope of this book. For example, results of Wolff (1926) and Denjoy (1926), and Douady and Hubbard (1982), show that if a component of the Fatou set is eventually mapped back to itself, then its closure contains either an attracting periodic point or a neutral periodic point. A result of Sullivan (1985) shows that the Fatou set has no "wandering" components, i.e. no orbit in the set of components is infinite.

Exercise 8.5.1. Show that if U is a connected component of $F(R)$, then $R(U)$ is also a connected component of $F(R)$. Show that if P is a polynomial, then $BA(\infty)$ is connected, and is therefore completely invariant.

Exercise 8.5.2. Show that, for $m > 1$, the Julia set of $z \mapsto z^m$ is the unit circle S^1, BA(∞) is the exterior of S^1 and the α-limit set of every $z \neq 0$ is S^1.

Exercise 8.5.3. Prove Corollary 8.5.5.

Exercise 8.5.4. Prove Proposition 8.5.7 for $R(z) = z^m$, $|m| > 1$.

Exercise 8.5.5. Let P be a polynomial of degree at least 2. Prove that $P^n \to \infty$ on the component of $F(P)$ which contains ∞.

Exercise 8.5.6. Show that if R is a rational map of degree >1, and $F(R)$ has only finitely many components, then it has either 0, 1, or 2 components.

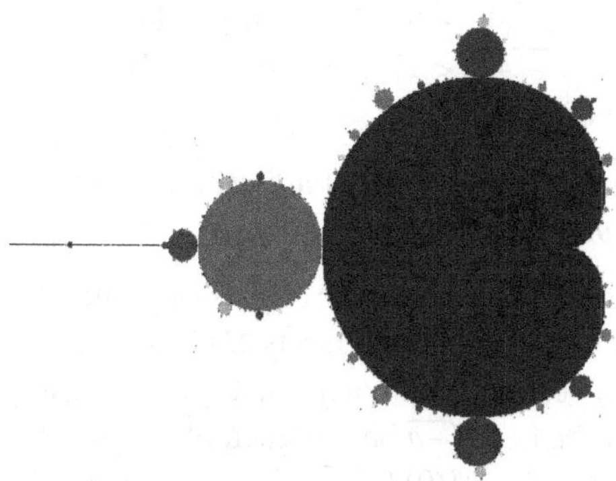

Figure 8.3. The Mandelbrot set.

8.6 The Mandelbrot set

For a general quadratic function $q(z) = \alpha z^2 + \beta z + \gamma$ with $\alpha \neq 0$, the change of variables $\zeta = (z - \beta/2)/\alpha$ maps the critical point to 0 and conjugates q with $q_a(z) = z^2 + a$. Since the conjugation is unique, the maps q_a, $a \in \mathbb{C}$, are in one-to-one correspondence with conjugacy classes of quadratic maps. If $q_a^n(0) \to \infty$, then $J(q_a)$ is totally disconnected (see Theorem 8.5.11). Otherwise, the orbit $\{q_a^n(0)\}_{n\in\mathbb{N}}$ is bounded and $J(q_a)$ is connected (Proposition 8.5.12).

The *Mandelbrot set M* is the set of parameter values a for which the orbit of 0 is bounded, or equivalently, $M = \{a \in \mathbb{C} : 0 \notin \mathrm{BA}(\infty) \text{ for } q_a\}$. The Mandelbrot set is shown in Figure 8.3.

THEOREM 8.6.1 (Douady and Hubbard, 1982). *$M = \{a \in \mathbb{C} : |q_a^n(0)| \leqslant 2$ for all $n \in \mathbb{N}\}$. M is closed and simply connected.*

Proof. Let $|a| > 2$. We have $|q_a(0)| = |a| > 2$, $|q_a^2(0)| = |q_a(a)| \geqslant |a^2| - |a| = |a|(|a| - 1)$ and $|q_a^n(0)| \geqslant |a|(|a| - 1)^{n-1}$ for $n \in \mathbb{N}$ (Exercise 8.6.1). Therefore $a \notin M$. If $|a| \leqslant 2$ and $|q_a^n(0)| = 2 + \alpha$ for some $n \in \mathbb{N}$ and $\alpha > 0$, then $|q_a^{n+1}(0)| \geqslant (2 + \alpha)^2 - 2 > 2 + 4\alpha$ and $|q_a^{n+k}(0)| \geqslant 2 + 4^k\alpha \to \infty$ as $k \to \infty$. Therefore $a \notin M$. The first and second statements follow.

If D is a bounded component of $\mathbb{C} \setminus M$, then $\max_{a\in\overline{D}} |q_a^n(0)| > 2$ for some $n \in \mathbb{N}$ and, by the maximum principle, $|q_a^n(0)| > 2$ for some $a \in \partial D \subset M$. This contradicts the first assertion of the theorem. Thus $\mathbb{C} \setminus M$ has no bounded components, has only one unbounded component containing ∞, and is therefore connected. Hence M is simply connected. □

The fixed points of q_a are $z_a^\pm = (1 \pm \sqrt{1-4a})/2$ with multipliers $\lambda^\pm = 1 \pm \sqrt{1-4a}$. The set $\{a \in \mathbb{C} : |1 \pm \sqrt{1-4a}| < 1\}$ is a subset of M (Exercise 8.6.3) and is called the *main cardioid* of M.

PROPOSITION 8.6.2. *Every point in ∂M is an accumulation point of the set of values of a for which q_a has a superattracting cycle.*

Proof. Since 0 is the only critical point of q_a, a periodic orbit is superattracting if and only if it contains 0. Let D be a disk that intersects ∂M and does not contain 0, and suppose that 0 is not a periodic point of q_a for any $a \in D$. Then $(q_a^n(0))^2 \neq -a$ for all $a \in D$ and $n \in \mathbb{N}$. Let $\sqrt{-a}$ be a branch of the inverse of $z \mapsto z^2$ defined on D, and define $f_n(a) = q_a^n(0)/\sqrt{-a}$ for $n \in \mathbb{N}$ and $a \in D$. Then the family $\{f_n\}_{n\in\mathbb{N}}$ omits $0, 1$ and ∞ on D, and is therefore normal in D. On the other hand, since D intersects ∂M, it contains both points a for which $f_n(a)$ is bounded and points a for which $f_n(a) \to \infty$, and hence the family $\{f_n\}$ is not normal on D. Thus 0 must be periodic for q_a for some $a \in D$. □

Exercise 8.6.1. Prove by induction that if $|a| > 2$, then $|q_a^n(0)| \geqslant |a|(|a|-1)^{n-1}$ for $n \in \mathbb{N}$.

Exercise 8.6.2. Prove that the intersection of M with the real axis is $[-2, 1/4]$.

Exercise 8.6.3. Prove that the main cardioid is contained in M.

Exercise 8.6.4. Prove that the set of values a in $\mathbb{C}$, for which q_a has an attracting periodic point of period 2, is the disk of radius $1/4$ centered at -1 (it is tangent to the main cardioid). Prove that this set is contained in M.

CHAPTER NINE

Measure theoretic entropy

In this chapter we give a short introduction to *measure-theoretic entropy*, also called *metric entropy*, for measure-preserving transformations. This invariant was introduced by Kolmogorov (1958, 1959), to classify Bernoulli automorphisms and developed further by Sinai (1959) for general measure-preserving dynamical systems. The measure-theoretic entropy has deep roots in thermodynamics, statistical mechanics and information theory. We explain the interpretation of entropy from the perspective of information theory at the end of the first section.

9.1 Entropy of a partition

Throughout this chapter $(X, \mathfrak{A}, \mu)$ is a Lebesgue space with $\mu(X) = 1$. We use the notation of Chapter 4. A (finite) *partition* of X is a finite collection ζ of essentially disjoint measurable sets C_i (called *elements* or *atoms* of ζ) whose union covers X mod 0. We say that a partition ζ' is a *refinement* of ζ and write $\zeta \leqslant \zeta'$ (or $\zeta' \geqslant \zeta$) if every element of ζ' is contained mod 0 in an element of ζ. Partitions ζ and ζ' are *equivalent* if each is a refinement of the other. We will deal with equivalence classes of partitions. The *common refinement* $\zeta \vee \zeta'$ of partitions ζ and ζ' is the partition into intersections $C_\alpha \cap C'_\beta$, where $C_\alpha \in \zeta$ and $C'_\beta \in \zeta'$; it is the smallest partition which is $\geqslant \zeta$ and ζ'. The *intersection* $\zeta \wedge \zeta'$ is the largest measurable partition which is $\leqslant \zeta$ and ζ'. The trivial partition consisting of a single element X is denoted by $\boldsymbol{\nu}$.

Although many definitions and statements in this chapter hold for infinite partitions, we discuss only finite partitions.

For $A, B \subset X$, let $A \vartriangle B = (A \setminus B) \cup (B \setminus A)$. Let $\xi = \{C_i \colon 1 \leqslant i \leqslant m\}$ and $\eta = \{D_j \colon 1 \leqslant j \leqslant n\}$ be finite partitions. By adding null sets if necessary, we may assume that $m = n$. Define

$$d(\xi, \eta) = \min_{\sigma \in S_m} \sum_{i=1}^{m} \mu(C_i \vartriangle D_{\sigma(i)}),$$

where the minimum is taken over all permutations of m elements. The axioms of distance are satisfied by d (Exercise 9.1.1).

Partitions ζ and ζ' are *independent*, and we write $\zeta \perp \zeta'$, if $\mu(C \cap C') = \mu(C) \cdot \mu(C')$ for all $C \in \zeta$ and $C' \in \zeta'$.

For a transformation T and partition $\xi = \{C_1, \ldots, C_m\}$, let $T^{-1}(\xi) = \{T^{-1}(C_1), \ldots, T^{-1}(C_m)\}$.

To motivate the definition of entropy below consider a Bernoulli automorphism of Σ_m with probabilities $q_i > 0, q_1 + \cdots + q_m = 1$ (see §4.4). Let ξ be the partition of Σ_m into m sets $C_i = \{\omega \in \Sigma_m \colon \omega_0 = i\}$, $\mu(C_i) = q_i$. Set $\eta_n = \vee_{k=0}^{n-1} \sigma^{-k}(\xi)$ and let $\eta_n(\omega)$ denote the element of η_n containing ω. For $\omega \in \Sigma_m$, let $f_i^n(\omega)$ be the relative frequency of symbol i in the word $\omega_1 \cdots \omega_n$. Since σ is ergodic with respect to μ, by the Birkhoff Ergodic Theorem 4.5.5, for every $\epsilon > 0$ there are $N \in \mathbb{N}$ and a subset $A_\epsilon \subset \Sigma_m$ with $\mu(A_\epsilon) > 1 - \epsilon$ such that $|f_i^n(\omega) - q_i| < \epsilon$ for each $\omega \in A_\epsilon$ and $n \geqslant N$. Therefore, if $\omega \in A_\epsilon$, then

$$\mu(\eta_n(\omega)) = \prod_{i=1}^{m} q_i^{(q_i+\epsilon_i)n} = 2^{n \sum_{i=1}^{m} (q_i+\epsilon_i) \log q_i},$$

where $|\epsilon_i| < \epsilon$ and from now on log denotes logarithm base 2 with $0 \log 0 = 0$. It follows that for μ-a.e. $\omega \in \Sigma_m$,

$$\lim_{n \to \infty} \frac{1}{n} \log \mu(\eta_n(\omega)) = \sum_{i=1}^{m} q_i \log q_i,$$

and hence the number of elements of η_n with approximately correct frequency of symbols $1, \ldots, m$ grows exponentially as 2^{nh}, where $h = -\sum_{i=1}^{m} q_i \log q_i$.

For a partition $\zeta = \{C_1, \ldots, C_n\}$ define the *entropy of* ζ by

$$H(\zeta) = -\sum_{i=1}^{n} \mu(C_i) \log \mu(C_i)$$

(recall that $0 \log 0 = 0$). Note that $-x \log x$ is a strictly concave continuous function on $[0, 1]$, i.e. if $x_i \geqslant 0, \lambda_i \geqslant 0, i = 1, \ldots, n$, and $\sum_i \lambda_i = 1$, then

$$-\left(\sum_{i=1}^{n} \lambda_i x_i\right) \cdot \log \sum_{i=1}^{n} \lambda_i x_i \geqslant -\sum_{i=1}^{n} \lambda_i x_i \log x_i \tag{9.1}$$

with equality if and only if all x_i's are equal. For $x \in X$, let $m(x, \zeta)$ denote the measure of the element of ζ containing x. Then

$$H(\zeta) = -\int_X \log m(x, \zeta)\, d\mu.$$

PROPOSITION 9.1.1. *Let ξ and η be finite partitions. Then:*

1. *$H(\xi) \geqslant 0$, and $H(\xi) = 0$ if and only if $\xi = \nu$;*
2. *if $\xi \leqslant \eta$, then $H(\xi) \leqslant H(\eta)$, and equality holds if and only if $\xi = \eta$;*
3. *if ξ has n elements, then $H(\xi) \leqslant \log n$, and equality holds if and only if each element of ξ has measure $1/n$;*
4. *$H(\xi \vee \eta) \leqslant H(\xi) + H(\eta)$ with equality if and only if $\xi \perp \eta$.*

Proof. We leave the first three statements as exercises (Exercise 9.1.2). To prove the last statement, let μ_i, ν_j and κ_{ij} be the measures of the elements of ξ, η and $\xi \vee \eta$, respectively, so that $\sum_j \kappa_{ij} = \mu_i$ and $\sum_i \kappa_{ij} = \nu_j$. It follows from (9.1) that

$$-\nu_j \log \nu_j \geqslant -\sum_i \mu_i \frac{\kappa_{ij}}{\mu_i} \cdot \log \frac{\kappa_{ij}}{\mu_i} = -\sum_i \kappa_{ij} \log \kappa_{ij} + \sum_i \kappa_{ij} \log \mu_i$$

and summation over j finishes the proof of the inequality. The equality is achieved if and only if $x_i = \kappa_{ij}/\mu_i$ does not depend on i for each j, which is equivalent to the independence of ξ and η. □

The entropy of a partition has a natural interpretation as the "average information of the elements of the partition." For example, suppose X represents the set of all possible outcomes of an experiment, and μ is the probability distribution of the outcomes. To extract information from the experiment, we devise a measuring scheme that effectively partitions X into finitely many observable subsets, or *events*, $C_1, C_2, \dots, C_n$. We define the information of an event C to be $I(C) = -\log \mu(C)$. This is a natural choice given that the information should have the following properties:

1. The information is a non-negative and decreasing function of the probability of an event; the lower the probability of an event, the greater the informational content of observing that event.
2. The information of the trivial event X is 0.
3. For independent events C and D, the information is additive, i.e. $I(C \cap D) = I(C) + I(D)$.

Up to a constant, $-\log \mu(C)$ is the only such function.

With this definition of information, the entropy of a partition is simply the average information of the elements of the partition.

Exercise 9.1.1. Prove: (i) $d(\xi, \eta) \geqslant 0$ with equality if and only if $\xi = \eta \mod 0$, and (ii) $d(\xi, \zeta) \leqslant d(\xi, \eta) + d(\eta, \zeta)$.

Exercise 9.1.2. Prove the first three statements of Proposition 9.1.1.

Exercise 9.1.3. For $n \in \mathbb{N}$, let $\mathcal{P}_n$ be the space of equivalence classes of finite partitions with n elements with metric d. Prove that the entropy is a continuous function on $\mathcal{P}_n$.

9.2 Conditional entropy

For measurable subsets $C, D \subset X$ with $\mu(D) > 0$, set $\mu(C|D) = \mu(C \cap D)/\mu(D)$. Let $\xi = \{C_i \colon i \in I\}$ and $\eta = \{D_j \colon j \in J\}$ be partitions. The *conditional entropy of* ξ *with respect to* η is defined by the formula:

$$H(\xi|\eta) = -\sum_{j \in J} \mu(D_j) \sum_{i \in I} \mu(C_i|D_j) \log \mu(C_i|D_j).$$

The quantity $H(\xi|\eta)$ is the average entropy of the partition induced by ξ on an element of η. If $C(x) \in \xi$ and $D(x) \in \eta$ are the elements containing x, then

$$H(\xi|\eta) = -\int_X \log \mu(C(x)|D(x))\, d\mu.$$

The following proposition gives several simple properties of conditional entropy.

PROPOSITION 9.2.1. *Let ξ, η and ζ be finite partitions. Then*

1. *$H(\xi|\eta) \geqslant 0$ with equality if and only if $\xi \leqslant \eta$;*
2. *$H(\xi|\nu) = H(\xi)$;*
3. *if $\eta \leqslant \zeta$, then $H(\xi|\eta) \geqslant H(\xi|\zeta)$;*
4. *if $\eta \leqslant \zeta$, then $H(\xi \vee \eta|\zeta) = H(\xi|\zeta)$;*
5. *if $\xi \leqslant \eta$, then $H(\xi|\zeta) \leqslant H(\eta|\zeta)$ with equality if and only if $\xi \vee \zeta = \eta \vee \zeta$;*
6. *$H(\xi \vee \eta|\zeta) = H(\xi|\zeta) + H(\eta|\xi \vee \zeta)$ and $H(\xi \vee \eta) = H(\xi) + H(\eta|\xi)$;*
7. *$H(\xi|\eta \vee \zeta) \leqslant H(\xi|\zeta)$;*
8. *$H(\xi|\eta) \leqslant H(\xi)$ with equality if and only if $\xi \perp \eta$.*

Proof. To prove part (6), let $\xi = \{A_i\}, \eta = \{B_j\}, \zeta = \{C_k\}$. Then

$$\begin{aligned}
H(\xi \vee \eta|\zeta) &= -\sum_{i,j,k} \mu(A_i \cap B_j \cap C_k) \cdot \log \frac{\mu(A_i \cap B_j \cap C_k)}{\mu(C_k)} \\
&= -\sum_{i,j,k} \mu(A_i \cap B_j \cap C_k) \cdot \log \frac{\mu(A_i \cap C_k)}{\mu(C_k)} \\
&\quad -\sum_{i,j,k} \mu(A_i \cap B_j \cap C_k) \cdot \log \frac{\mu(A_i \cap B_j \cap C_k)}{\mu(A_i \cap C_k)} \\
&= H(\xi|\zeta) + H(\eta|\xi \vee \zeta)
\end{aligned}$$

and the first equality follows. The second equality follows from the first one with $\zeta = \nu$.

The remaining statements of Proposition 9.2.1 are left as exercises (Exercise 9.2.1). □

For finite partitions ξ and η, define

$$\rho(\xi, \eta) = H(\xi|\eta) + H(\eta|\xi).$$

The function ρ, which is called the *Rokhlin metric*, defines a metric on the space of equivalence classes of partitions (Exercise 9.2.2).

PROPOSITION 9.2.2. *For every $\epsilon > 0$ and $m \in \mathbb{N}$ there is $\delta > 0$ such that if ξ and η are finite partitions with at most m elements and $d(\xi, \eta) < \delta$, then $\rho(\xi, \eta) < \epsilon$.*

Proof (Katok and Hasselblatt, 1995, Proposition 4.3.5.) Let $\xi = \{C_i: 1 \leqslant i \leqslant m\}$, $\eta = \{D_i: 1 \leqslant i \leqslant m\}$ be two partitions, and let $d(\xi, \eta) = \sum_{i=1}^{m} \mu(C_i \vartriangle D_i) = \delta$. We will estimate $H(\eta|\xi)$ in terms of δ and m.

If $\mu(C_i) > 0$, set $\alpha_i = \mu(C_i \setminus D_i)/\mu(C_i)$. Then

$$-\mu(C_i \cap D_i) \log \frac{\mu(C_i \cap D_i)}{\mu(C_i)} \leqslant -\mu(C_i)(1 - \alpha_i) \log(1 - \alpha_i)$$

and, by Proposition 9.1.1(3) applied to the partition of $C_i \setminus D_i$ induced by η,

$$-\sum_{j \neq i} \mu(C_i \cap D_j) \log \frac{\mu(C_i \cap D_j)}{\mu(C_i)} \leqslant -\mu(C_i)\alpha_i(\log \alpha_i - \log(m-1)).$$

Therefore, since $\log x$ is concave,

$$-\sum_j \mu(C_i \cap D_j) \log \frac{\mu(C_i \cap D_j)}{\mu(C_i)}$$
$$\leqslant \mu(C_i)\left((1-\alpha_i)\log\frac{1}{1-\alpha_i} + \alpha_i \log \frac{m-1}{\alpha_i}\right) \leqslant \mu(C_i)\log m.$$

It follows that

$$H(\eta|\xi) \leqslant \sum_{\mu(C_i)<\sqrt{\delta}} \mu(C_i)\log m$$
$$+ \sum_{\mu(C_i)\geqslant\sqrt{\delta}} \mu(C_i)(-(1-\alpha_i)\log(1-\alpha_i) - \alpha_i\log\alpha_i + \alpha_i\log(m-1)).$$

The first term does not exceed $\sqrt{\delta}\, m\log m$. To estimate the second term, observe that $\alpha_i\mu(C_i) \leqslant \delta$. Hence, if $\mu(C_i) \geqslant \sqrt{\delta}$, then $\alpha_i \leqslant \sqrt{\delta}$. Since the function $f(x) = -x\log x - (1-x)\log(1-x)$ is increasing on $(0, 1/2)$, for a small δ, the second term does not exceed $f(\sqrt{\delta}) + \sqrt{\delta}\log(m-1)$, and

$$H(\eta|\xi) \leqslant f(\sqrt{\delta}) + \sqrt{\delta}(m\log m + \log(m-1)).$$

Since $f(x) \to 0$ as $x \to 0$, the proposition follows. □

Exercise 9.2.1. Prove the remaining statements of Proposition 9.2.1.

Exercise 9.2.2. Prove that (i) $\rho(\xi,\eta) \geqslant 0$ with equality if and only if $\xi = \eta$ mod 0, and (ii) $\rho(\xi,\zeta) \leqslant \rho(\xi,\eta) + \rho(\eta,\zeta)$.

9.3 Entropy of a measure-preserving transformation

Let T be a measure-preserving transformation of a measure space $(X, \mathfrak{A}, \mu)$ and $\zeta = \{C_\alpha : \alpha \in I\}$ be a partition of X with finite entropy. For $k, n \in \mathbb{N}$, set $T^{-k}(\zeta) = \{T^{-k}(C_\alpha) : \alpha \in I\}$ and

$$\zeta^n = \zeta \vee T^{-1}(\zeta) \vee \cdots \vee T^{-n+1}(\zeta).$$

Since $H(T^{-k}(\zeta)) = H(\zeta)$ and $H(\xi \vee \eta) \leqslant H(\xi) + H(\eta)$, we have that $H(\zeta^{m+n}) \leqslant H(\zeta^m) + H(\zeta^n)$. By subadditivity (Exercise 2.5.3), the limit

$$h(T,\zeta) = \lim_{n\to\infty} \frac{1}{n} H(\zeta^n)$$

exists, and is called the *metric* (or *measure-theoretic*) *entropy* of T *relative to* ζ. Note that $h(T,\zeta) \leqslant H(\zeta)$.

PROPOSITION 9.3.1. $h(T,\zeta) = \lim_{n\to\infty} H(\zeta | T^{-1}(\zeta^n))$.

Proof. Since $H(\xi|\eta) \geqslant H(\xi|\zeta)$ for $\eta \leqslant \zeta$, the sequence $H(\zeta|T^{-1}(\zeta^n))$ is non-increasing in n. Since $H(T^{-1}\xi) = H(\xi)$ and $H(\xi \vee \eta) = H(\xi) + H(\eta|\xi)$, we get

$$\begin{aligned} H(\zeta^n) &= H(T^{-1}(\zeta^{n-1}) \vee \zeta) \\ &= H(\zeta^{n-1}) + H(\zeta|T^{-1}(\zeta^{n-1})) \\ &= H(\zeta^{n-2}) + H(\zeta|T^{-1}(\zeta^{n-2})) + H(\zeta|T^{-1}(\zeta^{n-1})) \\ &= \cdots = H(\zeta) + \sum_{k=1}^{n-1} H(\zeta|T^{-1}(\zeta^k)). \end{aligned}$$

Dividing by n and passing to the limit as $n \to \infty$ finishes the proof. □

Proposition 9.3.1 means that $h(T, \zeta)$ is the average information added by the present state on condition that all past states are known.

PROPOSITION 9.3.2. *Let ξ and η be finite partitions. Then*

1. $h(T, T^{-1}(\xi)) = h(T, \xi)$; *if T is invertible, then* $h(T, T(\xi)) = h(T, \xi)$;
2. $h(T, \xi) = h\left(T, \bigvee_{i=0}^{n} T^{-i}(\xi)\right)$ *for $n \in \mathbb{N}$; if T is invertible, then* $h(T, \xi) = h\left(T, \bigvee_{i=-n}^{n} T^{-i}(\xi)\right)$ *for $n \in \mathbb{N}$;*
3. $h(T, \xi) \leqslant h(T, \eta) + H(\xi|\eta)$; *if $\xi \leqslant \eta$, then* $h(T, \xi) \leqslant h(T, \eta)$;
4. $|h(T, \xi) - h(T, \eta)| \leqslant \rho(\xi, \eta) = H(\xi|\eta) + H(\eta|\xi)$ *(the Rokhlin inequality);*
5. $h(T, \xi \vee \eta) \leqslant h(T, \xi) + h(T, \eta)$.

Proof. To prove statement (3) observe that, by the second statement of Proposition 9.2.1(6), $H(\xi^n) \leqslant H(\xi^n \vee \eta^n) = H(\eta^n) + H(\xi^n|\eta^n)$. We apply Proposition 9.2.1(6) n times to get

$$\begin{aligned} H(\xi^n|\eta^n) &= H(\xi \vee T^{-1}(\xi^{n-1})|\eta^n) \\ &= H(\xi|\eta^n) + H(T^{-1}(\xi^{n-1})|\xi \vee \eta^n) \\ &\leqslant H(\xi|\eta) + H(T^{-1}(\xi^{n-1})|\eta^n) \\ &\leqslant H(\xi|\eta) + H(T^{-1}(\xi)|T^{-1}(\eta)) + H(T^{-2}(\xi^{n-2})|\eta^n) \\ &\cdots \\ &\leqslant nH(\xi|\eta). \end{aligned}$$

Therefore

$$\frac{1}{n}H(\xi^n) \leqslant \frac{1}{n}H(\eta^n) + H(\xi|\eta)$$

and statement (3) follows.

The remaining statements of Proposition 9.3.2 are left as exercises (Exercise 9.3.2). □

The *metric* (or *measure-theoretic*) *entropy* is the supremum of the entropies $h(T, \zeta)$ over all finite measurable partitions ζ of X.

If two measure-preserving transformations are isomorphic (i.e. if there exists a measure-preserving conjugacy), then their measure-theoretic entropies are equal. If the entropies are different, the transformations are not isomorphic.

We will need the following lemma.

LEMMA 9.3.3. *Let η be a finite partition and let ζ_n be a sequence of finite partitions such that $d(\zeta_n, \eta) \to 0$. Then there are finite partitions $\xi_n \leqslant \zeta_n$ such that $H(\eta|\xi_n) \to 0$.*

Proof. Let $\eta = \{D_j\colon 1 \leqslant j \leqslant m\}$. For each j choose a sequence $C_j^n \in \zeta_n$ such that $\mu(D_j \triangle C_j^n) \to 0$. Let ξ_n consist of C_j^n, $1 \leqslant j \leqslant m$, and $C_{m+1}^n = X \setminus \bigcup_{j+1}^m C_j^n$. Then $\mu(C_j^n) \to \mu(D_j)$and $\mu(C_{m+1}^n) \to 0$. We have

$$\begin{aligned} H(\eta|\xi_n) = &-\sum_{i=1}^{m} \mu(C_i^n \cap D_i) \cdot \log \frac{\mu(C_i^n \cap D_i)}{\mu(C_i^n)} \\ &-\sum_{j=1}^{m} \mu(C_{m+1}^n \cap D_j) \cdot \log \frac{\mu(C_{m+1}^n \cap D_j)}{\mu(C_{m+1}^n)} \\ &-\sum_{i=1}^{m} \sum_{j \neq i} \mu(C_i^n \cap D_j) \cdot \log \frac{\mu(C_i^n \cap D_j)}{\mu(C_i^n)}. \end{aligned}$$

The first sum tends to 0 since $\mu(C_i^n \cap D_i) \to \mu(C_i^n)$. The second and third sums tend to 0 since $\mu(C_i^n \cap D_j) \to 0$ for $j \neq i$. □

A sequence (ζ_n) of finite partitions is called *refining* if $\zeta_{n+1} \geqslant \zeta_n$ for $n \in \mathbb{N}$.

A sequence (ζ_n) of finite partitions is called *generating* if for every finite partition ξ and every $\delta > 0$ there is $n_0 \in \mathbb{N}$ such that for every $n \geqslant n_0$ there is a partition ξ_n with $\xi_n \leqslant \bigvee_{i=1}^n \zeta_i$ and $d(\xi_n, \xi) < \delta$, or equivalently if every measurable set can be approximated by a union of elements of $\bigvee_{i=1}^n \zeta_i$ for a large enough n.

Every Lebesgue space has a generating sequence of finite partitions (Exercise 9.3.3). If X is a compact metric space with a non-atomic Borel measure μ, then a sequence of finite partitions ζ_n is generating if the maximal diameter of elements of ζ_n tends to 0 as $n \to \infty$ (Exercise 9.3.4).

PROPOSITION 9.3.4. *If (ζ_n) is a refining and generating sequence of finite partitions, then $h(T) = \lim_{n\to\infty} h(T, \zeta_n)$.*

Proof. Let ξ be a partition of X with m elements. Fix $\epsilon > 0$. Since (ζ_n) is a refining and generating sequence, for every $\delta > 0$ there is $n \in \mathbb{N}$ and a partition ξ_n with m elements such that $\xi_n \leqslant \bigvee_{i=1}^{n} \zeta_i$ and $d(\xi_n, \xi) < \delta$. By Proposition 9.2.2,

$$\rho(\xi, \zeta_n) = H(\xi|\zeta_n) + H(\zeta_n|\xi) < \epsilon.$$

By the Rokhlin inequality (Proposition 9.3.2(4)), $h(T, \xi) < h(T, \zeta_n) + \epsilon$. □

A (one-sided) *generator* for a non-invertible measure-preserving transformation T is a finite partition ξ such that the sequence $\xi^n = \bigvee_{k=0}^{n} T^{-k}(\xi)$ is generating. For an invertible T, a (two-sided) *generator* is a finite partition ξ such that the sequence $\bigvee_{k=-n}^{n} T^{k}(\xi)$ is generating. Equivalently, ξ is a generator if for any finite partition η there are partitions $\zeta_n \leqslant \bigvee_{k=0}^{n} T^{-k}(\xi)$ (or $\zeta_n \leqslant \bigvee_{k=-n}^{n} T^{k}(\xi)$) such that $d(\zeta_n, \eta) \to 0$.

The following corollary of Proposition 9.3.4 allows one to calculate the entropy of many measure-preserving transformations.

THEOREM 9.3.5 (Kolmogorov–Sinai). *Let ξ be a generator for T. Then $h(T) = h(T, \xi)$.*

Proof. We consider only the non-invertible case. Let η be a finite partition. Since ξ is a generator, there are partitions $\zeta_n \leqslant \bigvee_{i=0}^{n} T^{-i}(\xi)$ such that $d(\zeta_n, \eta) \to 0$. By Lemma 9.3.3 for any $\delta > 0$ there is $n \in \mathbb{N}$ and a partition $\xi_n \leqslant \zeta_n \leqslant \bigvee_{i=0}^{n} T^{-i}(\xi)$ with $H(\xi_n|\eta) < \delta$. By statements (3), (5) and (2) of Proposition 9.3.2,

$$h(T, \eta) \leqslant h(T, \xi_n) + H(\eta|\xi_n) \leqslant h\left(T, \bigvee_{i=0}^{n} T^{-i}(\xi)\right) + \delta = h(T, \xi) + \delta. \quad \square$$

PROPOSITION 9.3.6. *Let T and S be measure-preserving transformations of measure spaces $(X, \mathfrak{A}, \mu)$ and $(Y, \mathfrak{B}, \nu)$, respectively.*

1. *$h(T^k) = kh(T)$ for every $k \in \mathbb{N}$; if T is invertible, then $h(T^{-1}) = h(T)$ and $h(T^k) = |k|h(T)$ for every $k \in \mathbb{Z}$.*
2. *If T is a factor of S, then $h_\mu(T) \leqslant h_\nu(S)$.*
3. *$h_{\mu\times\nu}(T \times S) = h_\mu(T) + h_\nu(S)$.*

Proof. To prove statement (3), consider refining and generating sequences of partitions ξ_k and η_k in X and Y, respectively. Let $\boldsymbol{\nu}_X$ and $\boldsymbol{\nu}_Y$ be the trivial partitions of X and Y, respectively. Then

$$\zeta_k = (\xi_k \times \nu_Y) \vee (\nu_X \times \eta_k)$$

is a refining and generating sequence in $X \times Y$. Since

$$\zeta_k^n = \left(\xi_k^n \times \nu_Y\right) \vee \left(\nu_X \times \eta_k^n\right) \quad \text{and} \quad \left(\xi_k^n \times \nu_Y\right) \perp \left(\nu_X \times \eta_k^n\right),$$

we obtain, by Propositions 9.1.1 and 9.3.4, that

$$h(T \times S) = \lim_{k\to\infty} \lim_{n\to\infty} \frac{1}{n} H(\zeta_k^n) \lim_{k\to\infty} \lim_{n\to\infty} \frac{1}{n} \left(H(\xi_k^n) + H(\eta_k^n)\right) = h(T) + h(S).$$

The first two statements are left as exercises (Exercise 9.3.6). □

Let T be a measure-preserving transformation of a probability space $(X, \mathfrak{A}, \mu)$ and ζ a finite partition. As before let $m(x, \zeta^n)$ be the measure of the element of ζ^n containing $x \in X$. The amount of information conveyed by the fact that x lies in a particular element of ζ^n (or that the points x, $T(x), \ldots, T^{n-1}(x)$ lie in particular elements of ζ) is $I_{\zeta^n}(x) = -\log m(x, \zeta^n)$. A proof of the following theorem can be found in Petersen (1989) or Mañé (1988).

THEOREM 9.3.7 (Shannon–McMillan–Breiman). *Let T be an ergodic measure-preserving transformation of a probability space $(X, \mathfrak{A}, \mu)$ and ζ a finite partition. Then*

$$\lim_{n\to\infty} \frac{1}{n} I_{\zeta^n}(x) = h(T, \zeta) \quad \textit{for a.e. } x \in X \quad \textit{and in } L^1(X, \mathfrak{A}, \mu).$$

Theorem 9.3.7 implies that, for a typical point $x \in X$, the information $I_{\zeta^n}(x)$ grows asymptotically as $n \cdot h(T, \zeta)$ and the measure $m(x, \zeta^n)$ decays exponentially as $e^{-nh(T,\zeta)}$. The proof of the following corollary is left as an exercise (Exercise 9.3.8).

COROLLARY 9.3.8. *Let T be an ergodic measure-preserving transformation of a probability space $(X, \mathfrak{A}, \mu)$ and ζ a finite partition. Then for every $\epsilon > 0$ there is $n_0 \in \mathbb{N}$, and for every $n \geqslant n_0$ a subset S_n of the elements of ζ^n such that the total measure of the elements from S_n is $\geqslant 1 - \epsilon$ and for each element $C \in S_n$,*

$$-n(h(T, \zeta) + \epsilon) < \log \mu(C) < -n(h(T, \zeta) - \epsilon).$$

Exercise 9.3.1. Let T be a measure-preserving transformation of a non-atomic measure space $(X, \mathfrak{A}, \mu)$. For a finite partition ξ and $x \in X$, let $\xi_n(x)$ be the element of ξ^n containing x. Prove that $\mu(\xi^n(x)) \to 0$ as $n \to \infty$ for a.e.

x and every non-trivial finite partition ξ if and only if all powers $T^n, n \in \mathbb{N}$, are ergodic.

Exercise 9.3.2. Prove the remaining statements of Proposition 9.3.2.

Exercise 9.3.3. Prove that every Lebesgue space has a generating sequence of partitions.

Exercise 9.3.4. If ζ is a partition of a finite metric space, then we define the *diameter* of ζ to be $\mathrm{diam}(\zeta) = \sup_{C \in \zeta} \mathrm{diam}(C)$. Prove that a sequence (ζ_n) of finite partitions of a compact metric space X with a non-atomic Borel measure μ is generating if the diameter of ζ_n tends to 0 as $n \to \infty$.

Exercise 9.3.5. Suppose a measure-preserving transformation T has a generator with k elements. Prove that $h(T) \leqslant \log k$.

Exercise 9.3.6. Prove the first two statements of Proposition 9.3.6.

Exercise 9.3.7. Show that if an invertible transformation T has a one-sided generator, then $h(T) = 0$.

Exercise 9.3.8. Prove Corollary 9.3.8.

9.4 Examples of entropy calculation

Let (X, d) be a compact metric space and μ a non-atomic Borel measure on X. By Exercise 9.3.4, any sequence of finite partitions whose diameter tends to 0 is generating. We will use this fact repeatedly in computing the metric entropy of some topological maps.

Rotations of S^1. Let λ be the Lebesgue measure on S^1. If α is rational, then $R_\alpha^n = \mathrm{Id}$ for some n, so $h_\lambda(R_\alpha) = (1/n)h_\lambda(R_\alpha^n) = (1/n)h_\lambda(\mathrm{Id}) = 0$. If α is irrational, let ξ_N be a partition of S^1 into N intervals of equal length. Then ξ_N^n consists of nN intervals, so $H(\xi_N^n) \leqslant \log nN$. Thus $h(R_\alpha, \xi_N) \leqslant \lim_{n\to\infty}(\log nN)/n = 0$. The collection of partitions $\xi_N, N \in \mathbb{N}$ is clearly generating, so $h(R_\alpha) = 0$.

This result can also be deduced from Exercise 9.3.7 by noting that every forward semiorbit is dense, so any non-trivial partition is a one-sided generator for R_α.

Expanding maps. The partition

$$\xi = \{[0, 1/k), [1/k, 2/k), \ldots, [(k-1)/k, 1)\}$$

is a generator for the expanding map $E_k\colon S^1 \to S^1$ since the elements of ξ^n are of the form $[i/k^n, (i+1)/k^n)$. We have

$$H(\xi^n) = -\Sigma(1/|k|^n)\log(1/|k|^n) = n\log|k|,$$

so $h_\lambda(E_k) = \log|k|$.

Shifts. Let $\sigma\colon \Sigma_m \to \Sigma_m$ be the one or two-sided shift on m symbols, and let $p = (p_1, \dots, p_m)$ be a non-negative vector with $\sum_{i=1}^m p_i = 1$. The vector p defines a measure on the alphabet $\{1, 2, \dots, m\}$. The associated product measure μ_p on Σ_m is called a *Bernoulli measure*. For a cylinder set, we have

$$\mu_p\left(C^{n_1,\dots,n_k}_{j_1,\dots,j_k}\right) = \prod_{i=1}^{k} p_{j_i}.$$

Let $\xi = \{C^0_j\colon j = 1, \dots, m\}$. Then ξ is a (one- or two-sided) generator for σ since $\operatorname{diam}(\bigvee_{i=0}^m \sigma^i\xi) \to 0$ with respect to the metric $d(\omega, \omega') = 2^{-l}$, where $l = \min\{|i|\colon \omega_i \neq \omega'_i\}$. Thus

$$h_{\mu_p}(\sigma) = h_{\mu_p}(\sigma, \xi) = \lim_{n\to\infty} \frac{1}{n} H\left(\bigvee_{i=0}^{m} \sigma^{-i}\xi\right).$$

For $i \neq j, \sigma^i\xi$ and $\sigma^j\xi$ are independent, so

$$H\left(\bigvee_{i=0}^{m} \sigma^{-i}\xi\right) = nH(\xi).$$

Thus $h_{\mu_p}(\sigma) = H(\xi) = -\Sigma p_i \log p_i$.

Recall that the topological entropy of σ is $\log m$. Thus the metric entropy of σ with respect to any Bernoulli measure is less than or equal to the topological entropy, and equality holds if and only if $p = (1/n, \dots, 1/n)$.

We next calculate the metric entropy of σ with respect to the Markov measures defined in §4.4. Let A be an irreducible $m \times m$ stochastic matrix, and q the unique positive left eigenvector whose entries sum to 1. Recall that for the measure $P = P_{A,q}$, the measure of a cylinder set is

$$P\left(C^{n,n+1,\dots,n+k}_{j_0,j_1,\dots,j_k}\right) = q_{j_0} \prod_{i=0}^{k-1} A_{j_i j_{i+1}}.$$

By Proposition 9.3.1, we have $h_P(\sigma, \xi) = \lim_{n\to\infty} H(\xi|\sigma^{-1}(\xi^n))$. By definition,

$$H(\xi|\sigma^{-1}(\xi^n)) = -\sum_{C\in\xi, D\in\sigma^{-1}(\xi^n)} P(C \cap D)\log\frac{P(C\cap D)}{P(D)}.$$

For $C = C^0_{j_0} \in \xi$ and $D = C^{1,\dots,n}_{j_1,\dots,j_n} \in \sigma^{-1}(\xi^n)$, we have

$$P(C \cap D) = q_{j_0} \prod_{i=0}^{n-1} A_{j_i j_{i+1}} \quad \text{and} \quad P(D) = q_{j_1} \prod_{i=1}^{n-1} A_{j_i j_{i+1}}.$$

Thus

$$\begin{aligned} H(\xi|\sigma^{-1}(\xi^n)) &= -\sum_{j_0,j_1,\dots,j_n=1}^{m} q_{j_0} \prod_{i=1}^{n-1} A_{j_i j_{i+1}} \log\left(\frac{q_{j_0} A_{j_0 j_1}}{q_{j_1}}\right) \\ &= -\sum_{j_0,j_1,\dots,j_n=1}^{m} q_{j_0} \prod_{i=1}^{n-1} A_{j_i j_{i+1}} \left(\log A_{j_0 j_1} + \log q_{j_0} - \log q_{j_1}\right). \end{aligned} \tag{9.2}$$

Using the identities $\sum_{i=1}^n q_i A_{ik} = q_k$ and $\sum_{k=1}^n A_{ik} = 1$, we find that

$$\sum_{j_0,j_1,\dots,j_n} q_{j_0} \prod_{i=0}^{n-1} A_{j_i j_{i+1}} \log A_{j_0 j_1} = \sum_{j_0,j_1} q_{j_0} A_{j_0 j_1} \log A_{j_0 j_1}, \tag{9.3}$$

$$\sum_{j_0,j_1,\dots,j_n} q_{j_0} \prod_{i=0}^{n-1} A_{j_i j_{i+1}} \log q_{j_0} = \sum_{j_0} q_0 \log q_{j_0}, \tag{9.4}$$

$$\sum_{j_0,j_1,\dots,j_n=1}^{m} q_{j_0} \prod_{i=1}^{n-1} A_{j_i j_{i+1}} \log q_{j_1} = \sum_{j_1} q_{j_1} \log q_{j_1}. \tag{9.5}$$

It follows from (9.2)–(9.5) that:

$$h_P(\sigma) = -\sum_{j_0,j_1} q_{j_0} A_{j_0 j_1} \log A_{j_0 j_1}.$$

There are many Markov measures for a given subshift. We now construct a special Markov measure, called the *Shannon–Parry measure*, that maximizes the entropy. By the results of the next section, a Markov measure maximizes the entropy if and only if the metric entropy with respect to the measure is the same as the topological entropy of the underlying subshift.

Let B be a primitive matrix of 0's and 1's. Let λ be the largest positive eigenvalue of B and let q be a positive left eigenvector of B with eigenvalue λ. Let v be a positive right eigenvector of B with eigenvalue λ normalized so that $\langle q, v\rangle = 1$. Let V be the diagonal matrix whose diagonal entries are the coordinates of v, i.e. $V_{ij} = \delta_{ij} v_j$. Then $A = \lambda^{-1} V^{-1} B V$ is a stochastic matrix: all elements of A are positive and the rows sum to 1. The elements of A are $A_{ij} = \lambda^{-1} v_i^{-1} B_{ij} v_j$. Let $p = qV = (q_1 v_1, \dots, q_n v_n)$. Then p is a positive left eigenvector of A with eigenvalue 1, and $\sum_{i=1}^n p_i = \langle q, v\rangle = 1$.

The Markov measure $P = P_{A,p}$ is called the Shannon–Parry measure for the subshift σ_A. Recall that while P is defined on the full shift space Σ, its support is the subspace Σ_A. Thus $h_P(\sigma_A) = h_P(\sigma)$. Using the properties $qB = \lambda q$, $\langle q, v\rangle = 1$ and $B_{ij}\log B_{ij} = 0$, we have

$$
\begin{aligned}
h_P(\sigma) &= -\sum_{i,j} p_i A_{ij}\log A_{ij} \\
&= -\sum_{i,j} q_i v_i \lambda^{-1} v_i^{-1} B_{ij} v_j \log(\lambda^{-1} v_i^{-1} B_{ij} v_j) \\
&= -\sum_{i,j} \lambda^{-1} q_i v_j B_{ij} \log(\lambda^{-1} v_i^{-1} B_{ij} v_j) \\
&= \sum_{i,j} \lambda^{-1} q_i v_j B_{ij} \log\lambda + \sum_{i,j} \lambda^{-1} q_i v_j B_{ij} (\log v_i - \log B_{ij} v_j) \\
&= \log\lambda + \sum_j q_j v_j \log v_j - \sum_{i,j} \lambda^{-1} q_i v_j B_{ij} \log v_j - \sum_{i,j} \lambda^{-1} q_i v_j B_{ij} \log B_{ij} \\
&= \log\lambda + \sum_j v_j q_j \log v_j - \sum_i v_i q_i \log v_i = \log\lambda.
\end{aligned}
$$

Thus $h_P(\sigma_A) = \log\lambda$, which is the topological entropy of σ_A (Proposition 3.4.1).

Toral automorphisms. We consider only the two-dimensional case. Let $A\colon \mathbb{T}^2 \to \mathbb{T}^2$ be a hyperbolic toral automorphism. The Markov partition constructed in §5.12 gives a (measurable) semiconjugacy $\phi\colon \Sigma_A \to \mathbb{T}^2$ between a subshift of finite type and A. Since the image of the Lebesgue measure under ϕ^* is the Parry measure, the metric entropy of A (with respect to the Lebesgue measure) is the logarithm of the largest eigenvalue of A (Exercise 9.4.1).

Exercise 9.4.1. Let A be a hyperbolic toral automorphism. Prove that the image of the Lebesgue measure on $\mathbb{T}^2$ under the semiconjugacy ϕ is the Parry measure and calculate the metric entropy of A.

9.5 Variational principle[1]

In this section we establish the *variational principle for metric entropy* (Dinaburg, 1971; Goodwyn, 1969), which asserts that for a homeomorphism

[1] The proof of the variational principle below follows the argument of Misiurewicz (1976), see also Katok and Hasselblatt (1995) and Petersen (1989).

of a compact metric space, the topological entropy is the supremum of the metric entropies for all invariant probability measures.

Let f be a homeomorphism of a compact metric space X, and $\mathcal{M}$ the space of Borel probability measures on X.

LEMMA 9.5.1. *Let $\mu, \nu \in \mathcal{M}$ and $t \in (0,1)$. Then for any measurable partition of ξ of X,*

$$tH_\mu(\xi) + (1-t)H_\nu(\xi) \leqslant H_{t\mu+(1-t)\nu}(\xi).$$

Proof. The proof is a straightforward consequence of the convexity of $x \log x$ (Exercise 9.5.1). □

For a partition $\xi = \{A_1, \dots, A_k\}$, define the *boundary* of ξ to be the set $\partial\xi = \bigcup_{i=1}^k \partial A_i$, where $\partial A = \overline{A} \cap \overline{(X - A)}$.

LEMMA 9.5.2. *Let $\mu \in \mathcal{M}$. Then:*

1. *for any $x \in X$ and $\delta > 0$, there is $\delta' \in (0, \delta)$ such that $\mu(\partial B(x, \delta')) = 0$;*
2. *for any $\delta > 0$, there is a finite measurable partition $\xi = \{C_1, \dots, C_k\}$ with $\operatorname{diam}(C_i) < \delta$ for all i and $\mu(\partial\xi) = 0$;*
3. *if $\{\mu_n\} \subset \mathcal{M}$ is a sequence of probability measures that converges to μ in the weak* topology, and A is a measurable set with $\mu(\partial A) = 0$, then $\mu(A) = \lim_{n\to\infty} \mu_n(A)$.*

Proof. Let $S(x, \delta) = \{y \in X : d(x, y) = \delta\}$. Then $B(x, \delta) = \bigcup_{0 \leqslant \delta' < \delta} S(x, \delta')$. This is an uncountable union, so at least one of these must have measure 0. Since $\partial B(x, \delta) \subset S(x, \delta)$, statement (1) follows.

To prove statement (2), let $\{B_1, \dots, B_k\}$ be an open cover by balls of radius less than $\delta/2$ and $\mu(\partial B_i) = 0$. Let $C_1 = \overline{B_1}$, $C_2 = \overline{B_2} \backslash \overline{B_1}$, $C_i = \overline{B_i} \bigcup_{j=1}^{i-1} \backslash \overline{B_j}$. Then $\xi = \{C_1, \dots, C_k\}$ is a partition, and $\partial\xi = \cup \partial C_i \subset \bigcup_{i=1}^k \partial B_i$.

To prove statement (3), let A be a measurable set with $\mu(\partial A) = 0$. Since X is a normal topological space, there is a sequence $\{f_k\}$ of non-negative continuous functions on X such that $f_k \searrow \chi_{\overline{A}}$. Then, for fixed k,

$$\varlimsup_{n\to\infty} \mu_n(A) \leqslant \varlimsup_{n\to\infty} \mu_n(\overline{A}) \leqslant \lim_{n\to\infty} \mu_n(f_k) = \mu(f_k).$$

Taking the limit as $k \to \infty$, we obtain

$$\varlimsup_{n\to\infty} \mu_n(A) \leqslant \lim_{k\to\infty} \mu(f_k) = \mu(\overline{A}) = \mu(A).$$

Similarly,

$$\varlimsup_{n\to\infty} \mu_n(X \backslash A) \leqslant \mu(X \backslash A),$$

from which the result follows. □

Let $|E|$ denote the cardinality of a finite set E.

LEMMA 9.5.3. *Let E_n be an (n, ϵ)-separated set, $\nu_n = \frac{1}{|E_n|}\sum_{x\in E_n}\delta_x$, and $\mu_n = \frac{1}{n}\sum_{i=0}^{n-1} f_*^i \nu_n$. If μ is any weak* accumulation point of $\{\mu_n\}_{n\in\mathbb{N}}$, then μ is f-invariant and*

$$\overline{\lim_{n\to\infty}} \frac{1}{n}\log|E_n| \leqslant h_\mu(f).$$

Proof. Let μ be an accumulation point of $\{\mu_n\}_{n\in\mathbb{N}}$. Then μ is clearly f-invariant.

Let ξ be a measurable partition with elements of diameter less than ϵ and $\mu(\partial\xi) = 0$. If $C \in \xi^n$, then $\nu_n(C) = 0$ or $\frac{1}{|E_n|}$ since C contains at most one element of E_n. Thus $H_{\nu_n}(\xi^n) = \log|E_n|$.

Fix $0 < q < n$, and $0 \leqslant k < q$. Let $a(k) = [\frac{n-k}{q}]$.

Let $S = \{k + rq + i : 0 \leqslant r < a(k),\ 0 \leqslant i < q\}$, and let T be the complement of S in $\{0, 1, \ldots, n-1\}$. The cardinality of T is at most $k + q - 1 \leqslant 2q$. Since

$$\xi^n = \bigvee_{i=0}^{n-1} f^{-i}\xi = \left(\bigvee_{r=0}^{a(k)-1} f^{-rq-k}\xi^q\right) \vee \left(\bigvee_{i\in T} f^{-i}\xi\right),$$

it follows that

$$\begin{aligned}\log|E_n| = H_{\nu_n}(\xi^n) &\leqslant \sum_{r=0}^{a(k)-1} H_{\nu_n}(f^{-(rq+k)}\xi^q) + \sum_{i\in T} H_{\nu_n}(f^{-i}\xi)\\ &\leqslant \sum_{r=0}^{a(k)-1} H_{f_*^{rq+k}\nu_n}(\xi^q) + 2q\log|\xi|.\end{aligned}$$

Summing over k, and using Lemma 9.5.1, we get

$$\begin{aligned}\frac{q}{n}\log|E_n| = \frac{1}{n}\sum_{k=0}^{q-1} H_{\nu_n}(\xi^n) &\leqslant \sum_{k=0}^{q-1}\left(\sum_{r=0}^{a(k)-1}\frac{1}{n}H_{f_*^{rq+k}\nu_n}(\xi^q)\right) + \frac{2q^2}{n}\log|\xi|\\ &\leqslant H_{\mu_n}(\xi^q) + \frac{2q^2}{n}\log|\xi|.\end{aligned}$$

Thus, by Lemma 9.5.2(3), for fixed q,

$$\overline{\lim_{n\to\infty}}\frac{1}{n}\log|E_n| \leqslant \lim_{n\to\infty}\frac{1}{q}H_{\mu_n}(\xi^q) = \frac{1}{q}H_\mu(\xi^q).$$

Letting $q \to \infty$, we get $\overline{\lim}_{n\to\infty}\frac{1}{n}\log|E_n| \leqslant h_\mu(f, \xi)$. □

THEOREM 9.5.4 (Variational Principle). *Let f be a homeomorphism of a compact metric space (X, d). Then $h_{top}(f) = \sup\{h_\mu(f) \mid \mu \in \mathcal{M}_f\}$.*

Proof. Lemma 9.5.3 shows that $h_{\text{top}}(f) \leqslant \sup_{\mu\in\mathcal{M}_f} h_\mu(f)$, so we need only demonstrate the opposite inequality.

Let $\mu \in \mathcal{M}_f$ be an f-invariant Borel probability measure on X, and $\xi = \{C_1, \dots, C_k\}$ a measurable partition of X. By the regularity of μ and Lemma 9.3.3, we may choose compact sets $B_i \subset C_i$ so that the partition $\beta = \{B_0 = X \setminus \bigcup_{i=1}^k B_i, B_1, \dots, B_k\}$ satisfies $H(\xi|\beta) < 1$. Thus

$$h_\mu(f, \xi) \leqslant h_\mu(f, \beta) + H_\mu(\xi|\beta) \leqslant h_\mu(f, \beta) + 1.$$

The collection $\mathcal{B} = \{B_0 \cup B_1, \dots, B_0 \cup B_k\}$ is a covering of X by open sets. Moreover, $|\beta^n| \leqslant 2^n |\mathcal{B}^n|$ since each element of $\mathcal{B}^n$ intersects at most two elements of β. Thus

$$H_\mu(\beta^n) \leqslant \log|\beta^n| \leqslant n \log 2 + \log|\mathcal{B}^n|.$$

Let δ_0 be the Lebesgue number of $\mathcal{B}$, i.e. the supremum of all δ such that for all $x \in X, B(x, \delta)$ is contained in some $B_0 \cup B_i$. Then δ_0 is also the Lebesgue number of $\mathcal{B}^n$ with respect to the metric d_n.

No subcollection of $\mathcal{B}$ covers X, and the same is true of $\mathcal{B}^n$. Thus each element $C \in \mathcal{B}$ contains a point x_C that is not contained in any other element, so $B(x_C, \delta_0, n) \subset C$. If follows that the collection of all x_C is an (n, δ_0)-separated set. Thus $\text{sep}(n, \delta_0, f) \geqslant |\mathcal{B}^n|$, from which it follows that

$$\begin{aligned} h(f, \delta_0) &= \overline{\lim_{n\to\infty}} \frac{1}{n} \log(\text{sep}(n, \delta_0, f)) \geqslant \overline{\lim_{n\to\infty}} \frac{1}{n} \log|\mathcal{B}^n| \\ &\geqslant \overline{\lim_{n\to\infty}} \frac{1}{n} (\log|\beta^n| - n\log 2) \geqslant \overline{\lim_{n\to\infty}} \frac{1}{n} H_\mu(\beta^n) - \log 2 \\ &= h_\mu(f, \beta) - \log 2 \geqslant h_\mu(f, \xi) - \log 2 - 1. \end{aligned}$$

We conclude that $h_\mu(f) = h_\mu(f^n)/n \leqslant \frac{1}{n}(h_{\text{top}}(f^n) + \log 2 + 1)$ for all $n > 0$. Letting $n \to \infty$, we see that $h_\mu(f) \leqslant h_{\text{top}}(f)$ for all $\mu \in \mathcal{M}$, which proves the theorem. □

Exercise 9.5.1. Prove Lemma 9.5.1.

Exercise 9.5.2. Let f be an expansive map of a compact metric space with expansiveness constant δ_0. Show that f has a *measure of maximal entropy*, i.e. there is $\mu \in \mathcal{M}_f$ such that $h_\mu(f) = h_{\text{top}}(f)$. (Hint: Start with a measure supported on an (n, ϵ)-separated set, where $\epsilon \leqslant \delta_0$.)

Bibliography

Adler, R. L., and Weiss, B. 1967. Entropy, a complete metric invariant for automorphisms of the torus. *Proc. Nat. Acad. Sci. U.S.A.*, **57**, 1573–6.

Adler, Roy, and Flatto, Leopold. 1991. Geodesic flows, interval maps, and symbolic dynamics. *Bull. Amer. Math. Soc. (N.S.)*, **25**(2), 229–334.

Ahlfors, Lars V. 1973. *Conformal invariants: topics in geometric function theory*. New York: McGraw-Hill Book Co. McGraw-Hill Series in Higher Mathematics.

Anosov, D. V. 1967. Tangential fields of transversal foliations in Y-systems. *Math. Notes*, **2**, 818–23.

Anosov, D. V. 1969. *Geodesic flows on closed Riemann manifolds with negative curvature.* Providence, R.I.: American Mathematical Society.

Anosov, D. V., and Sinai, Ya. G. 1967. Some smooth ergodic systems. *Russian Math. Surveys*, **22**(5), 103–68.

Archibald, Ralph G. 1970. *An introduction to the theory of numbers*. Charles E. Merrill Publishing Co., Columbus, Ohio, 1970.

Beardon, Alan F. 1991. *Iteration of rational functions*. New York: Springer-Verlag.

Benedicks, M., and Carleson, L. 1991. The dynamics of the Hénon map. *Ann. of Math. (2)*, **133**, 73–169.

Berenstein, Carlos A., and Gay, Roger. 1991. *Complex variables*. New York: Springer-Verlag.

Bergelson, Vitaly. 1996. Ergodic Ramsey theory—an update. Pages 1–61 of: *Ergodic theory of* $\mathbf{Z}^d$ *actions (Warwick, 1993–1994)*. Cambridge: Cambridge University Press.

Bergelson, Vitaly. 2000. Ergodic theory and Diophantine problems. Pages 167–205 of: *Topics in symbolic dynamics and applications (Temuco, 1997)*. Cambridge: Cambridge University Press.

Berman, Abraham, and Plemmons, Robert J. 1994. *Nonnegative matrices in the mathematical sciences*. Philadelphia, PA: Society for Industrial and Applied Mathematics (SIAM).

Billingsley, Patrick. 1965. *Ergodic theory and information*. New York: John Wiley & Sons Inc.

Bowen, R., and Lanford, III., O. E. 1970. Zeta functions of restrictions of the shift transformation. Pages 43–9 of: *Global Analysis (Proc. Sympos. Pure Math., Vol. XIV, Berkeley, Calif., 1968)*. Providence, R.I.: Amer. Math. Soc.

Bowen, Rufus. 1970. Markov partitions for Axiom *A* diffeomorphisms. *Amer. J. Math.*, **92**, 725–47.

Boyle, Mike. 1993. Symbolic dynamics and matrices. Pages 1–38 of: *Combinatorial and graph-theoretical problems in linear algebra (Minneapolis, MN, 1991)*. IMA Vol. Math. Appl., vol. 50. New York: Springer.

Brin, Sergey, and Page, Lawrence. 1998. The Anatomy of a Large-Scale Hypertextual Web Search Engine. In: *Seventh International World Wide Web Conference (Brisbane, Australia, 1998)*. http://www7.scu.edu.au/1921/com1921.htm.

Carleson, Lennart, and Gamelin, Theodore W. 1993. *Complex dynamics*. Universitext: Tracts in Mathematics. New York: Springer-Verlag.

Chow, Shui Nee, and Hale, Jack K. 1982. *Methods of bifurcation theory*. New York: Springer-Verlag.

Collet, Pierre, and Eckmann, Jean-Pierre. 1980. *Iterated maps on the interval as dynamical systems*. Mass.: Birkhäuser Boston.

Conway, John B. 1995. *Functions of one complex variable. II*. New York: Springer-Verlag.

Cornfeld, I. P., Fomin, S. V., and Sinaĭ, Ya. G. 1982. *Ergodic theory*. Grundlehren der Mathematischen Wissenschaften, vol. 245. New York: Springer-Verlag.

de Melo, Welington, and van Strien, Sebastian. 1993. *One-dimensional dynamics*. Berlin: Springer-Verlag.

Denjoy, Arnaud. 1926. Sur l'itération des fonctions analytique. *C. R. Acad. Sci. Paris Sér. A-B*, **182**, 255–7.

Devaney, Robert L. 1989. *An introduction to chaotic dynamical systems*. Redwood City, CA: Addison-Wesley.

Dinaburg, Efim I. 1971. On the relation among various entropy characterizatistics of dynamical systems. *Mathematics of the USSR, Izvestia*, **5**, 337–78.

Douady, Adrien, and Hubbard, John Hamal. 1982. Itération des polynômes quadratiques complexes. *C. R. Acad. Sci. Paris Sér. I Math.*, **294**(3), 123–6.

Dunford, Nelson, and Schwartz, Jacob T. 1988. *Linear operators. Part II.* New York: John Wiley & Sons Inc.

Feigenbaum, Mitchell J. 1979. The universal metric properties of nonlinear transformations. *J. Statist. Phys.*, **21**(6), 669–706.

Folland, Gerald B. 1995. *A course in abstract harmonic analysis*. Boca Raton, FL: CRC Press.

Friedman, Nathaniel A. 1970. *Introduction to ergodic theory*. New York: Van Nostrand Reinhold Co. Van Nostrand Reinhold Mathematical Studies, No. 29.

Furstenberg, H. 1963. The structure of distal flows. *Amer. J. Math.*, **85**, 477–515.

Furstenberg, H. 1981a. *Recurrence in ergodic theory and combinatorial number theory*. Princeton, N.J.: Princeton University Press. M. B. Porter Lectures.

Furstenberg, H., and Weiss, B. 1978. Topological dynamics and combinatorial number theory. *J. Analyse Math.*, **34**, 61–85.

Furstenberg, Harry. 1977. Ergodic behavior of diagonal measures and a theorem of Szemerédi on arithmetic progressions. *J. Analyse Math.*, **31**, 204–56.

Furstenberg, Harry. 1981b. Poincaré recurrence and number theory. *Bull. Amer. Math. Soc. (N.S.)*, **5**(3), 211–34.

Gantmacher, F. R. 1959. *The theory of matrices. Vols. 1, 2.* New York: Chelsea Publishing Co. Translated by K. A. Hirsch.

Golubitsky, M., and Guillemin, V. 1973. *Stable mappings and their singularities*. New York: Springer-Verlag. Graduate Texts in Mathematics, vol. 14.

Goodwyn, L. Wayne. 1969. Topological entropy bounds measure-theoretic entropy. *Proc. Amer. Math. Soc.*, **23**, 679–88.

Gottschalk, W., and Hedlund, G. 1955. *Topological Dynamics*. A.M.S. Colloquim Publications, vol. XXXVI. Providence: Amer. Math. Soc.

Graham, R. L., Leeb, K., and Rothschild, B. L. 1972. Ramsey's theorem for a class of categories. *Advances in Math.*, **8**, 417–33.

Graham, R. L., Leeb, K., and Rothschild, B. L. 1973. Errata: "Ramsey's theorem for a class of categories". *Advances in Math.*, **10**, 326–7.

Hale, Jack K., and Koçak, Hüseyin. 1991. *Dynamics and bifurcations*. New York: Springer-Verlag.

Halmos, Paul R. 1944. In general a measure preserving transformation is mixing. *Ann. of Math. (2)*, **45**, 786–92.

Halmos, Paul R. 1950. *Measure Theory*. Princeton, N.J.: D. Van Nostrand Co.

Halmos, Paul R. 1960. *Lectures on ergodic theory*. New York: Chelsea Publishing Co.

Hardy, G. H., and Wright, E. M. 1979. *An introduction to the theory of numbers*. Fifth edn. New York: The Clarendon Press Oxford University Press.

Helson, Henry. 1995. *Harmonic analysis*. Second edn. Berkeley, CA: Henry Helson.

Hénon, M. 1976. A two-dimensional mapping with a strange attractor. *Comm. Math. Phys.*, **50**, 69–77.

Hirsch, Morris W. 1994. *Differential topology*. New York: Springer-Verlag. Corrected reprint of the 1976 original.

Katok, A. B. 1972. Dynamical systems with hyperbolic structure. 125–211. Three papers on smooth dynamical systems, Translations of the AMS (series 2), vol. 116, AMS, Providence, RI, 1981.

Katok, Anatole, and Hasselblatt, Boris. 1995. *Introduction to the modern theory of dynamical systems*. Encyclopedia of Mathematics and its Applications, vol. 54. Cambridge: Cambridge University Press. With a supplementary chapter by Katok and Leonardo Mendoza.

Kim, K. H., and Roush, F. W. 1999. The Williams conjecture is false for irreducible subshifts. *Ann. of Math. (2)*, **149**(2), 545–58.

King, J. L. 1990. A map with topological minimal self-joinings in the sense of del Junco. *Ergodic Theory Dynamical Systems*, **10**(4), 745–61.

Kolmogorov, A. N. 1958. A new metric invariant of transient dynamical systems and automorphisms in Lebesgue spaces. *Dokl. Akad. Nauk SSSR (N.S.)*, **119**, 861–4.

Kolmogorov, A. N. 1959. Entropy per unit time as a metric invariant of automorphisms. *Dokl. Akad. Nauk SSSR*, **124**, 754–5.

Koopman, B.O., and von Neumann, J. 1932. Dynamical systems of continuous spectra. *Proc. Nat. Acad. Sci. USA*, **18**, 255–63.

Krengel, Ulrich. 1985. *Ergodic theorems*. Berlin: Walter de Gruyter & Co. With a supplement by Antoine Brunel.

Lanford, III, Oscar E. 1984. A shorter proof of the existence of the Feigenbaum fixed point. *Comm. Math. Phys.*, **96**(4), 521–38.

Li, Tien Yien, and Yorke, James A. 1975. Period three implies chaos. *Amer. Math. Monthly*, **82**(10), 985–92.

Lind, Douglas, and Marcus, Brian. 1995. *An introduction to symbolic dynamics and coding*. Cambridge: Cambridge University Press.

Lorenz, E. N. 1963. Deterministic non-periodic flow. *J. of Atmos. Sci.*, **20**, 130–41.

Mañé, Ricardo. 1988. A proof of the C^1 stability conjecture. *Inst. Hautes Études Sci. Publ. Math.*, **66**, 161–210.

Marcus, Brian, Roth, Ron M., and Siegel, Paul H. 1995. Modulation codes for digital data storage. Pages 41–94 of: *Different aspects of coding theory (San Francisco, CA, 1995)*. Providence, RI: Amer. Math. Soc.

Mather, John N. 1968. Characterization of Anosov diffeomorphisms. *Nederl. Akad. Wetensch. Proc. Ser. A 71 = Indag. Math.*, **30**, 479–83.

Milnor, J. 1965. *Topology from the Differentiable Viewpoint*. Charlottesville: University Press of Virginia.

Milnor, John, and Thurston, William. 1988. On iterated maps of the interval. Pages 465–563 of: *Dynamical systems (College Park, MD, 1986–87)*. Lecture Notes in Mathematics, vol. 1342. Berlin: Springer.

Misiurewicz, Michał. 1976. A short proof of the variational principle for a $\mathbb{Z}_+^n$ action on a compact space. Pages 147–57. Astérisque, No. 40 of: *International Conference on Dynamical Systems in Mathematical Physics (Rennes, 1975)*. Paris: Soc. Math. France.

Moore, C. C. 1966. Ergodicity of flows on homogeneous spaces. *Amer. J. Math.*, **88**, 154–78.

Palis, Jacob, Jr., and de Melo, Welington. 1982. *Geometric theory of dynamical systems*. New York: Springer-Verlag. An introduction, Translated from the Portuguese by A. K. Manning.

Pesin, Ya. 1977. Characteristic Lyapunov exponents and smooth ergodic theory. *Russian Math. Surveys*, **32:4**, 55–114.

Petersen, Karl. 1989. *Ergodic theory*. Cambridge: Cambridge University Press.

Pontryagin, L. S. 1966. *Topological groups*. Inc., New York: Gordon and Breach Science Publishers. Translated from the second Russian edition by Arlen Brown.

Queffélec, Martine. 1987. *Substitution dynamical systems – spectral analysis*. Lecture Notes in Mathematics, vol. 1294. Berlin: Springer-Verlag.

Robbin, J. W. 1971. A structural stability theorem. *Ann. of Math. (2)*, **94**, 447–93.

Robinson, Clark. 1976. Structural stability of C^1 diffeomorphisms. *J. Differential Equations*, **22**(1), 28–73.

Robinson, Clark. 1995. *Dynamical systems*. Studies in Advanced Mathematics. Boca Raton, FL: CRC Press.

Rohlin, V. A. 1948. A "general" measure-preserving transformation is not mixing. *Doklady Akad. Nauk SSSR (N.S.)*, **60**, 349–51.

Rokhlin, V. 1967. Lectures on the entropy theory of measure preserving transformations. *Russian Math. Surveys*, **22**, 1–52.

Royden, H. L. 1988. *Real Analysis*. 3rd edn. Macmillian.

Rudin, Walter. 1987. *Real and complex analysis*. Third edn. New York: McGraw-Hill Book Co.

Rudin, Walter. 1991. *Functional analysis*. Second edn. New York: McGraw-Hill Inc.

Ruelle, David. 1989. *Elements of differentiable dynamics and bifurcation theory*. Boston, MA: Academic Press Inc.

Sárközy, A. 1978. On difference sets of sequences of integers. III. *Acta Math. Acad. Sci. Hungar.*, **31**(3-4), 355–86.

Shannon, Claude E., and Weaver, Warren. 1949. *The Mathematical Theory of Communication*. The University of Illinois Press, Urbana, Ill.

Sharkovsky, A. N. 1964. Co-existence of cycles of a continuous mapping of the line into itself. *Ukranian Math. J.*, **16**, 61–71.

Shishikura, Mitsuhiro. 1987. On the quasiconformal surgery of rational functions. *Ann. Sci. École Norm. Sup. (4)*, **20**(1), 1–29.

Sinai, Ya. G. 1959. On the concept of entropy of a dynamical system. *Dokl. Akad. Nauk SSSR (N.S.)*, **124**, 768–71.

Singer, D. 1978. Stable orbits and bifurcation of maps of the interval. *SIAM J. Appl. Math.*, **35**, 260–67.

Smale, S. 1967. Differentiable dynamical systems. *Bull. Amer. Math. Soc.*, **73**, 747–817.

Sullivan, Dennis. 1985. Quasiconformal homeomorphisms and dynamics. I. Solution of the Fatou–Julia problem on wandering domains. *Ann. of Math. (2)*, **122**(3), 401–18.

Szemerédi, E. 1969. On sets of integers containing no four elements in arithmetic progression. *Acta Math. Acad. Sci. Hungar.*, **20**, 89–104.

van Strien, Sebastian. 1988. Smooth dynamics on the interval (with an emphasis on quadratic-like maps). Pages 57–119 of: *New directions in dynamical systems*. Cambridge: Cambridge University Press.

Walters, Peter. 1975. *Ergodic theory – Introductory Lectures*. Berlin: Springer-Verlag. Lecture Notes in Mathematics, vol. 458.

Weiss, Benjamin. 1973. Subshifts of finite type and sofic systems. *Monatsh. Math.*, **77**, 462–74.

Williams, R. F. 1973. Classification of subshifts of finite type. *Ann. of Math. (2)*, **98**, 120–53. errata, ibid. (2) 99 (1974), 380–81.

Williams, R. F. 1984. Lorenz knots are prime. *Ergodic Theory Dynam. Systems*, **4**(1), 147–63.

Wolff, Julius. 1926. Sur l'itération des fonctions bornées. *C. R. Acad. Sci. Paris Sér. A-B*, **182**, 200–201.

Index

Printed in the United States
by Baker & Taylor Publisher Services